T0259968

Guide to Cloud Computing for Business and Technology Managers

From Distributed Computing
to Cloudware Applications

Guide to
Cloud Computing
for Business and
Technology Managers

From Distributed Computing
to Cloudware Applications

Vivek Kale

CRC Press
Taylor & Francis Group
Boca Raton London New York

CRC Press is an imprint of the
Taylor & Francis Group, an **informa** business

A CHAPMAN & HALL BOOK

CRC Press
Taylor & Francis Group
6000 Broken Sound Parkway NW, Suite 300
Boca Raton, FL 33487-2742

First issued in paperback 2019

ISBN-13: 978-1-4822-1922-7 (hbk)
ISBN-13: 978-0-367-37794-6 (pbk)

Library of Congress Cataloging-in-Publication Data

Kale, Vivek.
    Guide to cloud computing for business and technology managers : from distributed computing to cloudware applications / Vivek Kale.
       pages cm
    Includes bibliographical references and index.
    ISBN 978-1-4822-1922-7 (hardback)
    1. Cloud computing. I. Title.

QA76.585.K35 2014
004.67'82--dc23   2014027315

Visit the Taylor & Francis Web site at
http://www.taylorandfrancis.com

and the CRC Press Web site at
http://www.crcpress.com

To

My wife, Girija,

the keeper of my spirit,

my best friend,

a person of beauty and grace,

and above all,

courage and determination.

Contents

Section II   Road to Cloudware

Section III   Cloudware

Preface

Cloud computing is a major trend in information processing today. It solves real problems at compelling price levels. Consumers and enterprises alike are embracing the notion that what they need is *computing services*, something that happens, rather than *computing devices*, something that sits in the corner. Cloud computing's flexibility in service delivery makes it an ideal solution for companies faced with highly variable service demand situations or an uncertain financial environment.

Companies need IT infrastructures that enable them to operate more efficiently and accommodate continuous, incremental changes in business operations. To that end, many companies are already using server virtualization, and some are also using service-oriented architecture (SOA), to better leverage their existing IT investments and get additional flexibility and responsiveness from their existing systems infrastructure. Cloudware vendors are becoming more and more like utilities, offering reliable computing power and basic applications like e-mail, ERP, CRM, and a growing array of industry- and domain-specific applications.

Over the coming years, these vendors will develop economies of scale and expertise that would enable them to offer their services at a much lower cost than what most companies would spend to deliver those services internally. Consequently, companies will outsource more and more of their basic IT operations to manage their costs for basic IT services; this will enable companies to shift more of their time and attention to doing things with IT resources that add value to their products and provide meaningful differentiation in the eyes of their customers—IT will be used to deliver competitive advantage.

Cloud computing is a promising paradigm for delivering IT services as computing utilities. Cloud computing comes into focus only when you think about what IT always needs: a way to increase capacity or add capabilities on the fly without investing in new infrastructure, training new personnel, or licensing new software. Cloud computing refers to the hardware, systems software, and applications delivered as services over the Internet. Cloud computing is an information-processing model in which centrally administered computing capabilities are delivered as services, on an as-needed basis and charged on an as-utilized basis, across the network to a variety of user-facing devices.

While these clouds are the natural evolution of traditional data centers, they offer subscription-based access to infrastructure, platforms, and applications that are popularly referred to as IaaS (Infrastructure as a Service), PaaS (Platformas a Service), and SaaS (Software as a Service) and employ a utility pricing model where customers are charged based on their utilization of computational resources, storage, and transfer of data. While cloud computing has increased interoperability and usability and reduced the cost of computation, application hosting, and content storage and delivery by several orders of magnitude, there is still a significant amount of complexity involved in ensuring that applications and services can be scaled as needed to achieve consistent and reliable operation under peak loads.

Abstraction is a critical foundational concept for cloud computing because it allows us to think of a particular service—an application, a particular communication protocol, processing cycles within a CPU, or storage capacity on a hard disk—without thinking about the particular piece of hardware that will provide that service. Ultimately, this explains how migrating to cloud computing solutions engenders transition from managing technology to managing business processes, that is, from a fixed cost to a variable cost model. It is this variable cost operating model that allows companies to replace capital expenses with operating expenses, which is critical to any organization operating in high-change, unpredictable environments.

What Makes This Book Different?

This book interprets the cloud computing phenomenon of the 2000s from the point of view of business as well as technology. This book unravels the mystery of cloud computing environments and applications and their power and potential to transform the operating contexts of business enterprises. Customary discussions on cloud computing do not address the key differentiator of cloud computing environments, and applications from earlier enterprise applications like ERPs, CRMs, and SCMs: cloud computing, for the first time, is able to treat enterprise-level services not merely as discrete stand-alone services, but as Internet-locatable, composable, and repackageable building blocks for generating dynamically real-world (extended) enterprise business processes.

Reflecting the reality in the market, on balance, this book is focused more on the *what* rather than the *how* of cloud computing—even though it has multiple chapters related to the implementation of cloud computing applications like cloudware operations and management, cloudware security, and migrating to cloudware.

Here are the characteristic features of this book:

1. It enables IT managers and business decision-makers to get a clear understanding of what cloud really means, what it might do for them, and when it is practical to use cloud. It explains the context and demonstrates how the whole ecosystem works together to achieve the main objectives of cost reduction, business flexibility, and strategic focus.

2. It gives an introduction to the enterprise applications integration (EAI) solutions that were a first step toward enabling an integrated enterprise. It also gives a detailed description of service-oriented architecture (SOA) and related technology solutions that paved the road for cloud computing solutions, that is, cloudware applications.

3. It addresses the key differentiator of cloud computing environments, namely, that cloud computing, for the first time, is able to treat enterprise-level services not merely as discrete stand-alone services, but as Internet-locatable, composable, and repackageable building blocks for generating dynamically real-world enterprise business processes.

4. It provides a very wide treatment of cloud computing that covers delivery models like IaaS, PaaS, and SaaS, as well as deployment models like public, private, and hybrid clouds.

5. It also addresses some of the main concerns regarding the operations and management of cloud computing environments or cloudware, including performance, measurement, monitoring, and security.

6. It is not focused on any particular vendor or service offering. While there is good description of Amazon's, Google's, and Microsoft's cloud services, the text introduces solutions from several other players as well.

In the final analysis, cloud computing is an extension of the *network is computer* vision, namely, *network is service provider*. I wanted to write a book presenting cloud computing from this novel perspective; the outcome is the book that you are reading now. Thank you!

How Is This Book Organized?

This book traces the road to cloud computing, the detailed features and characteristics of cloud computing solutions and environments, and the last part of the book presents high-potential applications of cloud computing, that is, cloudware applications.

Right from its inception, the computer industry has been dominated by an unassailable trend toward increasing functional specificity over increasingly commoditized hardware. Chapter 1 presents an overview of this trend of progress toward massively parallel processing hardware accompanied by a parallel progression toward industry- and domain-specificity of the applications software. These two inexorable parallel drives have converged, resulting in the emergence of cloud computing environments, or cloudware.

Section I: Genesis of the Cloud

This section introduces the various milestones on the path to developing EAI solutions for enterprises. Maintaining competitiveness in the face of changing business environments necessitated changes in the enterprises' information systems. Evidently, few companies will attempt to replace their entire information systems. There have been many years of development invested into the existing applications, and reimplementing all of this functionality would require a lot of knowledge, time, and resources. Thus, replacing existing systems with new solutions will often not be a viable proposition. Therefore, standard ways to reuse existing systems and integrate them into the global, enterprise-wide information system must be defined. The resulting integrated information systems will improve the competitive advantage of the enterprises with a unified and efficient access to information.

This section starts with Chapter 2 tracing the genesis of the cloud to the networking and internetworking technologies of the 1980s. Chapter 3 describes the concepts of distributed systems. EAI methods and models are described in Chapter 4. Middleware and related integration technologies are described in Chapter 5. Finally, Chapter 6 introduces the game-changing J2EE environment, the model-view-controller (MVC) architecture and the corresponding reference architecture for enterprise applications.

Section II: Road to Cloud Computing

Enterprises require much more agility and flexibility than what could be provided by EAI solutions. IT can fulfill its role as a strategic differentiator only if it can enable enterprises to provide sustainable competitive advantage—the ability to change business processes in sync with changes in the business environment and that too at optimum costs! These will be built on a foundation of SOA that exposes the fundamental business capabilities as flexible, reusable services. Section II discusses SOA, which along with the constituting services, is the foundation of modern cloud computing solutions. The services support a layer of agile and flexible business processes that can be easily changed to provide new products and services to keep ahead of the competition. For a service, responsiveness trumps both efficiency and effectiveness. And, to realize this, people need to be able to conceptually describe what they want and not have to worry about how it is accomplished logically or physically. The most important value of SOA is that it provides an opportunity for IT and the business to communicate and interact with each other at a highly efficient and equally understandable level; that common, equally understood language is the language of services.

Chapter 7 presents the basic concepts and characteristics of SOA. The defining architecture of Web Services is described in Chapter 8. Chapter 9 explains the basic design of an enterprise service bus (ESB), and Chapter 10 introduces the principles of service composition and the related business process execution language (BPEL). This section wraps up with a description of two different service delivery models, namely, applications service providers (ASP) and grid computing in Chapters 11 and 12, respectively.

Section III: Cloudware

Section III presents a detailed discussion on various aspects of a cloud computing solution or cloudware. The approach adopted in this book will be useful to any professional who must present a case for realizing cloud computing solutions or to those who are involved in cloud computing projects. It provides a framework that will enable business and technical managers to make the optimal decisions necessary for successful migration to cloud computing environments and applications within their organizations. Chapter 13 introduces the basics of cloud computing, cloud delivery models, as well as deployment models such as public, private, and hybrid clouds. It details cloud delivery models such as IaaS, PaaS, and SaaS. IaaS involves services from enabling technologies such as virtual machines and virtualized storage, to sophisticated mechanisms for securely storing data in the cloud and managing virtual clusters. PaaS deals with the design and operation of sophisticated autoscaling environments. IaaS is related to the delivery of cloud-hosted software and applications.

Chapter 14 presents the advantages of cloud computing by detailing the economics of migrating to cloud computing solutions. Chapter 15 describes the various technologies employed for the realization of cloud computing solutions. Chapter 16 details with presently available cloud computing environments and the services they offer. Chapter 17 describes cloudware development paradigms and environments including Google MapReduce and the Hadoop ecosystem. Chapter 18 presents operations and management issues related to cloud computing that become critical as cloud computing environments become more complex and interoperable; these issues include risks, governance, resource management, and Service Level Agreements (SLAs). Chapter 19 provides an overview of governance, risks, and compliance issues confronted while availing cloud computing services. Chapter 20 presents details on how companies can successfully prepare and transition to cloud environments as well as achieve production readiness once such a transition is completed.

Section IV: Cloudware Applications

This section presents an overview of the areas of cloudware applications significant for the future, namely, big data, mobile (i.e., enterprise mobilization), and context-aware applications. Chapter 21 defines and identifies the common characteristics of big data applications along with corresponding tools, techniques, and technologies. Enterprise agility and its realization through mobilization applications are described in Chapter 22. Context-aware applications can significantly enhance the efficiency and effectiveness of even routinely occurring transactions. Chapter 23 introduces the concept of context as constituted by an ensemble of function-specific decision patterns. This chapter discusses location-based services applications as a particular example of context-aware applications. This chapter highlights the fact that any end-user application's effectiveness and performance can be enhanced by transforming it from a *bare* transaction to a transaction *clothed* by a surrounding context formed as an aggregate of all relevant decision patterns in the past. The generation of context itself is critically dependent on employing big data and mobilized applications, which in turn need cloud computing as a prerequisite.

Who Should Read This Book?

All stakeholders of a cloud computing project should read this book.

All those who are involved in any aspect of a cloud computing project will profit by using this book as a road map to make a more meaningful contribution to the success of their cloud computing initiative and the actual migration project(s).

The following is a list of recommended tracks of chapters that should be read by different categories of stakeholders:

- Executives and business managers should read Chapters 1 and 13 through 23.
- Operational managers should read Chapters 1, 4, and 11 through 23.
- Project managers and module leaders should read Chapters 1 through 4 and 11 through 23.
- Technical managers should read Chapters 1 through 23.
- Professionals interested in cloud computing should read Chapters 1 through 4, 11 through 16, and 18 through 23.
- Students of computer courses should read Chapters 1 through 13, 15, 17, and 21 through 23.

- Students of management courses should read Chapters 1, 4, 11 through 16, and 18 through 23.
- General readers interested in SOA and cloud computing phenomena should read Chapters 1, 2, 4, 11 through 14, 16, 18, and 20 through 23.

Acknowledgments

I would like to thank all those who have helped me with their clarifications, criticisms, and valuable information during the writing of this book, were patient enough to read the entire or parts of the manuscript, and made many valuable suggestions. They include Nitin Kadam, Mandar Barve and Shirish Panse.

I would like to thank my friend Nilesh Acharya, who has been the pillar of support for my various book projects in the last few years.

I would like to thank my editor, Aastha Sharma, for realizing this project and for her unfailing support in bringing this book project to completion. I would also like to thank the extended publishing team consisting of Richard Tressider, Marsha Pronin, S. Vinithan, A. Arunkumar, and a host of other people who were involved with the production of this book.

Most of all, I would like to thank my wife, Girija, and our beloved daughters, Tanaya and Atmaja, for all their love and support. Without them, I would not be where I am now in my life today.

Vivek Kale
Mumbai, India

Author

Vivek Kale has more than two decades of professional IT experience during which he has handled and consulted on various aspects of enterprise-wide information modeling, enterprise architectures, business process redesign, and, e-business architectures. He has been CIO of Essar Group, the steel/oil and gas major of India, as well as, Raymond Ltd., the textile and apparel major of India. He is a seasoned practitioner in transforming the business of IT, facilitating business agility and enabling the Process Oriented Enterprise. He is the author of *Implementing SAP R/3: The Guide for Business and Technology Managers*, Sams (2000), *A Guide to Implementing the Oracle Siebel CRM 8.x*, McGraw-Hill, India (2009), and *Implementing SAP CRM: The Guide for Business and Technology Managers*, Chapman & Hall (2014).

1

Increasing Functional Specificity over Increasingly Commoditized Hardware

A move toward clouds represents a fundamental change in how we handle information. It is almost the computing equivalent of the evolution in electricity supply from a hundred years ago, when farms and businesses closed down their own power generators and bought power instead from efficient industrial utilities. Until the end of the nineteenth century, businesses had to run their own power-generating facilities, producing all the energy to run their machinery. As industrial technology advanced, generators grew more sophisticated but were still located at the site of a business and maintained by its employees. Power generation was assumed to be an intrinsic part of running a business (much as data processing is now). The invention of the alternating-current electric grid at the turn of the century overturned that assumption. Supplying electricity to many users from central stations achieved huge economies of scale, and the price of electricity fell rapidly. The transformation in the supply of computing promises to be as dramatic as that in electricity supply (Figure 1.1).

1.1 Google's Vision of Utility Computing

The electricity utility comparison model fits neatly with Google's grand vision, established a decade ago, *to organize the world's information and make it universally accessible*. While its software and search strategy has already undermined Microsoft's dominance, Google has further plans to reshape the sector based on its search strengths. Google does not just use its computing grid to process Web searches but also to supply services such as word processing, spreadsheet, and e-mail programs—applications that have long been the mainstays of Microsoft's profitability. By supplying business computing as a set of simple services, Google (and other utility providers like Salesforce.com and Amazon Web Services) threatens to render large parts of the IT industry obsolete. Google operates a globe-spanning network of computers that answer search queries instantly by processing mountains

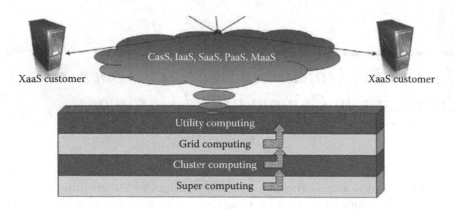

FIGURE 1.1
Evolution of computing services.

of data—whirring away in large, dark, refrigerated data centers. People at Google call this network *the cloud*.

One challenge of programming at Google is to leverage the cloud—to push it to do things that would overwhelm other machines and networks. For example, a partnership with IBM aims to plug universities around the world into Google-like computing clouds. Google engineers teach the cloud new tricks as it grows in size and sophistication—in 2007, they added four new data centers, at an average cost of $600 million a piece. Importantly, in building this cloud, Google is poised to take a new role in the computer industry. Google's cloud is a network of hundreds of thousands—by some estimates 1 million—of cheap servers, each storing huge amounts of data, to make searches faster. Unlike its predecessor, the supercomputer, Google's system never ages—as individual computers die, they are replaced individually with newer, faster boxes. This means the cloud regenerates as it grows, almost like a living thing.

As the concept of computing clouds spreads, it expands Google's footprint way beyond search, media, and advertising, and Google could become, in effect, the world's primary computer. No corporate computing system can match the efficiency, speed, and flexibility of resources of Google's cloud. It is estimated that Google can carry out a computing task for one-tenth of what it costs a typical company. Big data centers linked to form a cloud encapsulate the full disruptive potential of utility computing. If people and businesses can rely on central stations to fulfill all (or most) of their computing needs, they will be able to drastically reduce their expenditure on their own hardware and software—revenue that currently goes to Microsoft and the other tech giants.

Similarly, Amazon has opened up its own networks of computers to paying customers, initiating new users to cloud computing. Significantly, as the volumes of data from business and scientific research expand, computing power is turning into a strategic resource or a form of capital. For clouds to reach their potential, they need to be as easy to program and navigate as the Web. This suggests growing markets for cloud search and software tools—a natural business for Google and competitors like Amazon.com. As this strategy unfolds, people are starting to see Google as poised to become the dominant force in the next stage of computing. Companies and research organizations may eventually hand over most of their high-level computing tasks to a world-spanning network of computers forming a cloud. It is likely that all sorts of new business models will emerge.

The pioneers in a position to dominate this field are as follows:

1. *Google*—The only search company built from scratch around hardware, investing more than $2 billion a year in data centers, and the leader in cloud computing; Salesforce.com is partnering with Google in a joint venture—Google's cloud is best at sifting through data, but Salesforce.com has strengths in running business applications like accounting packages and lets companies write their own programs to run on its servers.

2. *Yahoo!*—Smaller and poorer than Google, with software not perfectly suited to cloud computing, but as the leading patron of Hadoop (a free software framework that supports distributed applications running on large clusters of commodity computers processing huge amounts of data), it could end up with a lead.

3. *IBM*—Dominant in business computing and traditional supercomputers, IBM is teaming up with Google to get a foothold in clouds. IBM is launching a pilot cloud system for the government of Vietnam and has built a showcase cloud center in Ireland.

4. *Microsoft*—Still currently dominated by its proprietary software, Microsoft is strong on the fundamentals of cloud science and is building massive data centers in Illinois and Siberia.

5. *Amazon*—The first to sell cloud computing as a service (Amazon Web Services); while smaller than rivals, its expertise in this area could provide a boost for the retailer in the next generation of Web Services from retail to media.

1.1.1 Drivers for Cloud Computing in Enterprises

1.1.1.1 Business Drivers

At present, companies have to survive and develop competitive advantage in a dynamic and turbulent environment of global competition and rapid business change. Companies are under constant pressure to simultaneously grow revenue and market share while reducing costs. To meet these requirements, companies have been changing, and three major trends can be observed that have impact on company requirements upon Information Technology (IT) support:

1. Striving toward high agility
2. Globalization of activities to be able to take advantage of opportunities provided by a global economy
3. Increased mobility

In dynamic business environments, agility is considered the key success factor for companies. Only companies with high agility can be successful in today's rapidly changing business environments. In literature, there are various definitions for the term *agility*: from general ones, for example, "the ability of firms to sense environmental change and respond readily," to more specific ones, for example, "…an innovative response to an unpredictable change." Business agility is the ability to sense highly uncertain external and internal changes and respond to them reactively or proactively, based on innovation of the internal operational processes, involving the customers in exploration and exploitation activities while leveraging the capabilities of partners in the business network. Business agility is therefore the ability to swiftly and easily change businesses and business processes beyond the normal level of flexibility to effectively manage unpredictable external and internal changes (see Section 22.1, "Agile Enterprises").

One basic obstacle for achieving agility is the prevailing IT infrastructure of enterprises. Despite efforts to increase flexibility of corporate IT, most prevailing corporate IT still involves hardwired processes and applications that cannot be changed quickly and easily. This results in long lead times before the IT infrastructure can follow and support new business process and product concepts. Thus, an agile company is only possible with an agile IT infrastructure that can quickly and efficiently be adjusted to new business ideas. Enterprises would like to have an IT infrastructure that can realign itself expeditiously to new business priorities. They require rapid and predictable turnaround times for provisioning computing power, storage, information, and application flows. Virtualization of resources (computers and data) and their flexible integration and combination to support changing business concepts has the potential to increase IT and business agility in companies (see Section 15.2, "Types of Virtualization").

Another development trend affecting the requirements upon the IT infrastructure in companies is the increasing globalization of companies. Companies are increasingly acting as global companies with activities spread over many locations worldwide. The globalization of companies resulted in globalization of their IT. To support the activities of remote company parts, IT resources and data as well as data centers are also scattered worldwide. Despite the global spread of activities, companies strive to use the competitive advantage of the involved regions in a synergetic way and to create a *Global One Company*. Thus, there is a growing need for IT support of global processes in an integrated and *follow-the-sun* principle (e.g., global supply chains) by relying on and integrating globally scattered IT resources. Virtualization and virtual centralization of available resources in cloud computing could provide the necessary integration of resources by keeping at the same time their physical distribution.

The third trend in companies that has impact on requirements upon their IT infrastructure is increasing mobility of employees and resources. Due to globalization, an increasing number of employees are mobile and require mobile support. At the same time, with the maturity of ubiquitous computing and the Internet of Things, an increasing number of external devices are expected to be involved as sensors in the IT infrastructure of companies (see Section 2.6 "Internet of Things"). Mobile computing resources and data as well as data sources as sensors need to be supported remotely in an efficient manner and at the same time need to be integrated into the existing infrastructure in a flexible way.

In order to support agility, flexible infrastructure is required that can be fast adapted to new processes. Virtualization and abstraction of the physical location of resources, support for services and their flexible bundling, as well as higher scalability and flexibility through inclusion of external resources based on Cloud Computing have the potential to provide an IT infrastructure that addresses the demands of business while utilizing the IT resources most efficiently and cost-effectively.

1.1.1.2 Technological Drivers

IT in companies has been constantly changing its shape in the last decades. This is driven by the changes in the way how companies conduct business described in the earlier section and by technological developments and innovation. At the beginning, there were centralized data centers with mainframes. More than a decade ago, a shift from large centralized mainframe computers toward more distributed systems started to transform corporate IT. First, PCs were added to support each single user in addition to mainframes that increasingly became distributed. Recently, mobile end devices have been added to support and enable greater mobility of employees. Initially, computing power and storage of mobile devices were limited, and mobile devices were mainly used for voice communication. Today, they have caught up and increasingly compete with PCs. A new trend is ubiquitous computing and the enhancement of the environment as well as products with sensors.

Overall, there is a trend toward distribution and decentralization of IT resources that at the same time is confronted with the need for consolidated and efficient use of IT resources. This results in several problems:

- Ever-increasing demand for storage and computing power at each data center

- Many and scattered data centers with underutilization of their resources
- Increasing maintenance costs of data centers

Business changes like globalization and mobility resulted in an increasing number of distributed data centers. At present, prevailing practice is to optimize each data center mostly independent of other data centers. This means that each data center is designed to accommodate high peak demand for computation power and data. As a result, there is an ever-increasing demand for storage and computing power. For example, the volume of digital content is constantly increasing. In 2007, the amount of information created exceeded available storage capacity for the first time ever. This implies technological challenges as well as challenges with regard to information governance for businesses.

The increasing number of data centers resulted in a disproportionate increase in their maintenance costs, in particular with respect to power and cooling costs. Energy efficiency of IT is a concern that becomes increasingly important. The continuously increasing amount of digital information requires increasing computing power, bigger storage capacities, and more powerful network infrastructure to transmit information. This ultimately results in increasing carbon footprint of IT. By 2020, ICTs are estimated to become among the biggest greenhouse gas emitters, accounting for around 3% of all emissions. Growth in the number and size of data centers is estimated to be the fastest increasing contributor to greenhouse emissions.

Cloud computing has been among the first attempts to manage the high number of computing nodes in distributed data centers and to achieve better utilization of distributed and heterogeneous computing resources in companies. Advances in virtualization technology enable greater decoupling between physical computing resources and software applications and promise higher industry adoption of distributed computing concepts such as the Cloud. The continuous increase of maintenance costs and demand for additional resources as well as for scalability and flexibility of resources is leading many companies to consider outsourcing their data centers to external providers. Cloud computing has emerged as one of the enabling technologies that allow such external hosting efficiently.

It is important to consider that Cloud Computing is not only changing the IT infrastructure in a company but has the potential to provide significant business value. As mentioned earlier, increased agility, that is, an enterprise's increased ability to respond and adjust quickly and efficiently to external market stimuli, is considered a key success factor for companies today. Existing IT infrastructure is considered to be a major obstacle to a company's agility. Prevailing IT infrastructure reflects the inflexible built-to-order structure: thousands of application silos, each with its own custom-configured hardware, and diverse and often incompatible assets that greatly limit a company's flexibility and thus reduce time to market;

what is therefore needed is an architecture that, in a similar way as the electricity grid, decouples the means of supporting the day-to-day operations of users from the underlying functional infrastructure that underpins them. This would also allow the business to reconfigure its operational strategy without necessarily amending its underlying IT systems. With the functionality described earlier, Cloud Computing has the potential to provide the decoupling layer in companies. In conclusion, the biggest benefit of cloud computing is the increased potential for companies to achieve new levels of innovation capabilities that can differentiate their business from competitors. Cloud Computing enables implementing of new business processes and applications that companies would not be able to implement by using conventional information technology. Cloud computing provides a virtual, resilient, responsive, flexible, and cost-effective infrastructure that fosters innovation and collaboration.

1.2 Modern On-Demand Computing

On-demand computing is an increasingly popular enterprise model in which computing resources are made available to the user as needed. Computing resources that are maintained on a user's site are becoming fewer and fewer, while those made available by a service provider are on the rise. The on-demand model evolved to overcome the challenge of being able to meet fluctuating resource demands efficiently. Because demand for computing resources can vary drastically from one time to another, maintaining sufficient resources to meet peak requirements can be costly. Over-engineering a solution can be just as adverse as a situation where the enterprise cuts costs by maintaining only minimal computing resources, resulting in insufficient resources to meet peak load requirements. Concepts such as clustered computing, grid computing, and parallel computing may all seem very similar to the concept of on-demand computing, but they can be better understood if one thinks of them as building blocks that evolved over time and with techno-evolution to achieve the modern cloud computing model we think of and use today.

1.2.1 Grid Computing

In the late 1980s, computers were clustered together to form a single larger computer in order to simulate a supercomputer and harness greater processing power. This technique was common and was used by many IT departments. Clustering, as it was called, allowed one to configure computers using special protocols so they could *talk* to each other. The purpose was to balance the computational load across several machines, divvying up units of work and spreading it across multiple processors. To the user, it made little difference which CPU executed an application. Cluster management software ensured that the CPU with the most available processing capability at that time was used to run the code. A key to efficient cluster management was engineering where the data were to be held. This process became known as data residency. Computers in the cluster were usually physically connected to magnetic disks that stored and retrieved data while the CPUs performed input/output (I/O) processes quickly and efficiently.

In the early 1990s, Ian Foster and Carl Kesselman presented their concept of *The Grid*. They used an analogy to the electricity grid, where users could plug in and use a (metered) utility service. They reasoned that if companies cannot generate their own power, it would be reasonable to assume they would purchase that service from a third party capable of providing a steady electricity supply. The same should apply to computing resources: if one node could plug itself into a grid of computers and pay only for the resources it used, it would be a more cost-effective solution for companies than buying and managing their own infrastructure. Grid computing expands on the techniques used in clustered computing models, where multiple independent clusters appear to act like a grid simply because they are not all located within the same domain. A major obstacle to overcome in the migration from a clustering model to grid computing was data residency. Because of the distributed nature of a grid, computational nodes could be anywhere in the world.

The issues of storage management, migration of data, and security provisioning were key to any proposed solution in order for a grid model to succeed. A toolkit called Globus was created to solve these issues, but the infrastructure hardware available still has not progressed to a level where true grid computing can be wholly achieved. The Globus Toolkit is an open-source software toolkit used for building grid systems and applications. It is being developed and maintained by the Globus Alliance and many others all over the world. The Globus Alliance has grown into a community of organizations and individuals developing fundamental technologies to support the grid model. The toolkit provided by Globus allows people to share computing power, databases (DBs), instruments, and other online tools securely across corporate, institutional, and geographic boundaries without sacrificing local autonomy.

In 2002, EMC offered a Content-Addressable Storage (CAS) solution called Centera as yet another cloud-based data storage service that competes with Amazon's offering. EMC's product creates a global network of data centers, each with massive storage capabilities. When a user creates a document, the application server sends it to the Centera storage system. The storage system then returns a unique content address to the server. The unique address allows the system to verify the integrity of the documents whenever a user moves or copies them. From that point, the application can request the document by submitting the address. Duplicates of documents are saved only once under the same address, leading to reduced storage requirements. Centera then retrieves the document regardless of where it may be physically located. EMC's Centera product takes the sensible approach that no one can afford the risk of placing all of their data in one place, so the data are distributed around the globe. Their cloud will monitor data usage and automatically move data around in order to load-balance data requests and better manage the flow of Internet traffic. Centera is constantly self-tuning to react automatically to surges in demand. The Centera architecture functions as a cluster that automatically configures itself upon installation. The system also handles failover, load balancing, and failure notification.

1.2.2 Server Virtualization

Virtualization is a method of running multiple independent virtual operating systems on a single physical computer. This approach maximizes the return on investment for the computer. The term was coined in the 1960s in reference to a virtual machine (sometimes called a pseudo-machine). The creation and management of virtual machines has often been called platform virtualization. Platform virtualization is performed on a given computer (hardware platform) by software called a control program. The control program creates a simulated environment, a virtual computer, which enables the device to use hosted software specific to the virtual environment, sometimes called guest software.

The guest software, which is often itself a complete operating system, runs just as if it were installed on a stand-alone computer. Frequently, more than one virtual machine is able to be simulated on a single physical computer, their number being limited only by the host device's physical hardware resources. Because the guest software often requires access to specific peripheral devices in order to function, the virtualized platform must support guest interfaces to those devices. Examples of such devices are the hard disk drive, CD-ROM, DVD, and network interface card. Virtualization technology is a way of reducing the majority of hardware acquisition and maintenance costs, which can result in significant savings for any company.

1.3 Computer Hardware

The mechanical loom invented by a Frenchman named Joseph Jacquard was an invention that made a profound impact on the history of industrialization, as well as in the history of computing. With the use of cards punched with holes, it was possible for the Jacquard loom to weave fabrics in a variety of patterns. Jacquard's loom was controlled by a program encoded into the punched cards. The operator created the program once and was able to duplicate it many times with consistency and accuracy. Herman Hollerith eventually adapted Jacquard's concept of the punched card to record census data in the late 1880s. Hollerith's machine was highly successful; it cut the time it took to tabulate the result of the census by two-thirds, and it made money for the company that manufactured it. In 1911, this company merged with its competitor to form International Business Machines (IBM).

First Generation: Vacuum Tube Technology, 1946–1956

The first generation of computers relied on vacuum tubes to store and process information. These tubes consumed huge amount of power, were short-lived, and generated a great deal of heat. First-generation computers were colossal in size, had extremely limited memory and processing capability, and their usage was restricted to limited areas in science and engineering. The maximum main memory size was approximately 5000 bytes (5 kilobytes), with a processing speed of 10 kilo instructions per second. This generation employed rotating magnetic drums for internal storage and punched cards for external storage. Jobs such as running programs or printing reports had to be coordinated manually.

Second Generation: Transistors, 1957–1963

In the second generation, the vacuum tubes were replaced by transistors for storing and processing information. Transistors were much more stable and reliable than vacuum tubes; they generated less heat and consumed less power. However, each transistor had to be custom made and wired into a printed circuit board—a slow and tedious process. Magnetic core memory was the primary storage technology of this period. It was composed of small magnetic doughnuts with 1 mm diameter, which could be polarized in either of two possible directions to represent a bit (*binary digit*) of data. This whole system had to be assembled by hand and, hence, was time consuming and very expensive. Second-generation computers had a random access memory (RAM) of up to 32 kilobytes and processing speeds of about 200 kilo instructions per sec to 300 kilo instructions per second. The enhanced processing power and memory of the second-generation computers enabled them to be used most widely not only for scientific and engineering work, but also for business work (like payroll and billing).

Third Generation: Integrated Circuits, 1964–1979

Third-generation computers relied on integrated circuits, which were made by printing hundreds and later thousands of tiny transistors on small silicon chips. These devices came to be called *semiconductors*. Computer memories expanded to 2 megabytes of RAM memory, and processing speeds accelerated to 5 million instructions per second (MIPS). The third-generation computers introduced software that could be used by people without extensive technical training, making it possible for a much larger section of people to use them in their respective areas in business.

Fourth Generation: Very-Large-Scale Integrated Circuits, 1980–1990

Fourth-generation computers relied on very large-scale integrated circuits (VLSIC), which were packed with hundreds of thousands, and later millions, of circuits per chip. These devices came to be called *microprocessors*. Computer memory sizes ballooned to over 2 gigabytes (GB) or more, while processing speeds exceeded 200 MIPS or more. Correspondingly, costs fell precipitously making possible inexpensive desktop computers that were widely used in business and everyday life. The fourth generation of computers was characterized by further miniaturization of circuits, increased multiprogramming, and virtual storage memory.

VLSIC technology has fuelled a growing movement toward microminiaturization, entailing the proliferation of computers so small, fast, and cheap that they have become ubiquitous and almost *invisible*. For example, many of the intelligent features that have made automobiles, stereos, toys, watches, cameras, mobiles, and other equipment easier to use are enabled by microprocessors.

Fifth Generation: Non–von Neumann Architectures, 1990–Present

Fifth-generation computers are based on non-von Neumann architectures entailing massively parallel processing for handling multimedia data (voice, graphics, images, and so on). Processing speeds exceed 500 MIPS or more. Conventional computers are based on the von Neumann architecture, which processes information serially, one instruction at a time. Massively parallel computers have a huge network of processor chips interwoven in a complex and flexible manner. As opposed to parallel processing, where a small number of powerful but expensive specialized chips are linked together, massively parallel processing (MPP) machines chain hundreds or even thousands of inexpensive, commonly used chips to attack large computing problems, attaining supercomputer speeds. MPP have cost and speed advantages over conventional computers because they can take advantage of off-the-shelf chips to accomplish processing at one-tenth to one-twentieth the cost of conventional supercomputers.

The Mauchly–Eckert–von Neumann concept of the stored program computer used the basic technical idea that a binary number system could be directly mapped to the two physical states of a flip-flop electronic circuit. In this circuit, the logical concept of the binary unit "1" could be interpreted as the on (or conducting state) and the binary unit "0" could be interpreted as the off (or not conducting state) of the electric circuit. In this way, the functional concept of numbers (written on the binary base) could be directly mapped into the physical states (physical morphology) of a set of electronic flip-flop circuits. The number of these circuits together would express how large a number could be represented. This is what is meant by word length in the digital computer. Binary numbers must not only encode data but also the instructions that perform the computational operations on the data. One of the points of progress in computer technology has been how long a word length could be built into a computer.

The design of the early computer used a hierarchy of logical operations. The lowest level of logic was the mapping of a set of bistable flip-flop circuits to a binary number system. A next step-up had circuits mapped to a Boolean logic (AND, OR, NOT circuits). A next step-up had these Boolean logic circuits connected together for arithmetic operations (such as add and subtract, multiply and divide). Computational instructions were then encoded as sequences of Boolean logic operations and/or arithmetic operations. Finally, at the highest logic level, von Neumann's stored program concept was expressed as a clocked cycle of fetching and performing computational instructions on data. This is now known as a von Neumann computer architecture—sequential instruction operated as a calculation cycle, timed to an internal clock.

The modern computer has four hierarchical levels of schematic logics mapped to physical morphologies (forms and processes) of transistor circuits:

1. Binary numbers mapped to bistable electronic circuits
2. Boolean logic operations mapped to electronic circuits of bistable circuits
3. Mathematical basic operations mapped (through Boolean constructions) to electronic circuits
4. Program instructions mapped sequentially into temporary electronic circuits (of Boolean and/or arithmetic instructions)

1.3.1 Types of Computer Systems

Today's computer systems come in a variety of sizes, shapes, and computing capabilities. The *Apollo 11* spacecraft that enabled landing men on the moon and returning them safely to earth was equipped with a computer that assisted them in everything from navigation to systems monitoring, and it had a 2.048 MHz CPU built by MIT. Today's standards can be measured in the 4 GHz in many home PCs (megahertz [MHz] is 1 million computing cycles per second and gigahertz [GHz] is 1 billion computing cycles per second). Further, the *Apollo 11* computer weighed 70 pounds versus today's powerful laptops weighing as little as 1 pound—we have come a long way. Rapid hardware and software developments and changing end user needs continue to drive the emergence of new models of computers, from the smallest handheld personal digital assistant/cell phone combinations to the largest multiple-CPU mainframes for enterprises. Categories such as *microcomputer, midrange, mainframe*, and supercomputer systems are still used to help us express the relative processing power and number of end users that can be supported by different types of computers. These are not precise classifications, and they do overlap each other.

Microcomputers

Microcomputers are the most important category of computer systems for both business and household consumers. Although usually called a *personal computer*, or PC, a microcomputer is much more than a small computer for use by an individual as a communication device. The computing power of microcomputers now exceeds that of the mainframes of previous computer generations, at a fraction of their cost. Thus, they have become powerful networked *professional workstations* for business professionals.

Midrange Computers

Midrange computers are primarily high-end network servers and other types of servers that can handle the large-scale processing of many business applications. Although not as powerful as mainframe computers, they are less costly to buy, operate, and maintain than mainframe systems and thus meet the computing needs of many organizations. Midrange systems first became popular as minicomputers in scientific research, instrumentation systems, engineering analysis, and industrial process monitoring and control. Minicomputers were able to easily handle such functions because these applications are narrow in scope and do not demand the processing versatility of mainframe systems. Today, midrange systems include servers used in industrial process-control and manufacturing plants and play major roles in computer-aided manufacturing (CAM). They can also take the form of powerful technical workstations for computer-aided design (CAD) and other computation and graphics-intensive applications. Midrange systems are also used as *front-end servers* to assist mainframe computers in telecommunications processing and network management.

Midrange systems have become popular as powerful network servers (computers used to coordinate communications and manage resource sharing in network settings) to help manage large Internet websites, corporate intranets and extranets, and other networks. Internet functions and other applications are popular high-end server applications, as are integrated enterprise-wide manufacturing, distribution, and financial applications. Other applications, like data warehouse management, data mining, and online analytical processing are contributing to the demand for high-end server systems.

Mainframe Computers

Mainframe computers are large, fast, and powerful computer systems; they can process thousands of million instructions per second (MIPS). They can also have large primary storage capacities with main memory capacity ranging from hundreds of gigabytes to many terabytes. Mainframes have downsized drastically in the last few years, dramatically reducing their air-conditioning needs, electrical power consumption, and floor space requirements—and thus their acquisition, operating, and ownership costs. Most of these improvements are the result of a move from the cumbersome water-cooled mainframes to a newer air-cooled technology for mainframe systems.

Mainframe computers continue to handle the information processing needs of major corporations and government agencies with high transaction processing volumes or complex computational problems. For example, major international banks, airlines, oil companies, and other large corporations process millions of sales transactions and customer inquiries every day with the help of large mainframe systems. Mainframes are still used for computation-intensive applications, such as analyzing seismic data from oil field explorations or simulating flight conditions in designing aircraft.

Mainframes are also widely used as *superservers* for the large client/server networks and high-volume Internet websites of large companies. Mainframes are becoming a popular business computing platform for data mining and warehousing, as well as electronic commerce applications.

Supercomputers

Supercomputers are a category of extremely powerful computer systems specifically designed for scientific, engineering, and business applications requiring extremely high speeds for massive numeric computations. Supercomputers use *parallel processing* architectures of interconnected microprocessors (which can execute many parallel instructions). They can easily perform arithmetic calculations at speeds of billions of floating-point operations per second (*gigaflops*)—a floating point operation is a basic computer arithmetic operation, such as addition, on numbers that include a decimal point. Supercomputers that can calculate in trillions of floating-point operations per second (*teraflops*), which use massive parallel processing (MPP) designs of thousands of microprocessors, are now in use (see Section 1.4.4 below).

The market for supercomputers includes government research agencies, large universities, and major corporations. They use supercomputers for applications such as global weather forecasting, military defence systems, computational cosmology and astronomy, microprocessor research and design, and large-scale data mining.

Experts continue to predict the merging or disappearance of several computer categories. They think, for example, that many midrange and mainframe systems have been made obsolete by the power and versatility of networks composed of microcomputers and servers. Other industry experts have predicted that the emergence of network computers and *information appliances* for applications on the Internet and corporate intranets will replace many personal computers, especially in large organizations and in the home computer market.

Interconnecting microprocessors to create *minisupercomputers* is a reality. The next wave was looking at harnessing the virtually infinite amount of unused computing power that exists in the myriad of desktops and laptops within the boundaries or outside of a modern organization. *Distributed* or *grid* or *cloud computing* in general is a special type of parallel computing that relies on complete or virtual computers (with onboard CPU, storage, power supply, network interface, and so forth) connected to a network (private, public, or the Internet) by a conventional or virtual network interface. This is in contrast to the traditional notion of a supercomputer, which has many processors connected together in a single machine. While the grid could be formed by harnessing the unused CPU power in all of the desktops and laptops in a single division of a company (or in the entire company, for that matter), the cloud could be formed by harnessing CPU, storage, and other resources in a company or external service providers. Chapter 12 discusses Grid Computing, while Section III, Chapters 13 to 20, discusses the nature and characteristics of Cloud Computing.

The primary advantage of distributed computing is that each node can be purchased as commodity hardware; when combined, it can produce computing resources similar to a multiprocessor supercomputer, but at a significantly lower cost. This is due to the economies of scale of producing desktops and laptops, compared with the lower efficiency of designing and constructing a small number of custom supercomputers.

1.4 Parallel Processing

Parallel processing is performed by the simultaneous execution of program instructions that have been allocated across multiple processors with the objective of running a program in less time. On the earliest computers, a user could run only one program at a time. This being the case, a computation-intensive program that took X minutes to run, using a tape system for data I/O that took X minutes to run, would take a total of X + X minutes to execute. To improve performance, early forms of parallel processing were developed to allow interleaved execution of both programs simultaneously. The computer would start an I/O operation (which is typically measured in milliseconds), and while it was waiting for the I/O operation to complete, it would execute the processor-intensive program (measured in nanoseconds). The total execution time for the two jobs combined became only slightly longer than the X minutes required for the I/O operations to complete.

1.4.1 Multiprogramming

The next advancement in parallel processing was multiprogramming. In a multiprogramming system, multiple programs submitted by users are each allowed to use the processor for a short time, each taking turns and having exclusive time with the processor in order to execute instructions. This approach is known as *round-robin scheduling* (RR scheduling). It is one of the oldest, simplest, fairest, and most widely used scheduling algorithms, designed especially for time-sharing systems. In RR scheduling, a small unit of time called a time slice is defined. All executable processes are held in a circular queue. The time slice is defined based on the number of executable processes that are in the queue. For example, if there are five user processes held in the queue and the time slice allocated for the queue to execute in total is 1 s, each user process is allocated 200 ms of process execution time on the CPU before the scheduler begins moving to the next process in the queue. The CPU scheduler manages this queue, allocating the CPU to each process for a time interval of one time slice. New processes are always added to the end of the queue. The CPU scheduler picks the first process from the queue, sets its timer to interrupt the process after the expiration of the timer, and then dispatches the next process in the queue. The process whose time has expired is placed at the end of the queue. If a process is still running at the end of a time slice, the CPU is interrupted and the process goes to the end of the queue. If the process finishes before the end of the time slice, it releases the CPU voluntarily. In either case, the CPU scheduler assigns the CPU to the next process in the queue. Every time a process is granted the CPU, a context switch occurs, which adds overhead to the process execution time. To users, it appears that all of the programs are executing at the same time.

Resource contention problems often arose in these early systems. Explicit requests for resources led to a condition known as deadlock. Competition for resources on machines with no tie-breaking instructions led to the critical section routine. Contention occurs when several processes request access to the same resource. In order to detect deadlock situations, a counter for each processor keeps track of the number of consecutive requests from a process that have been rejected. Once that number reaches a predetermined threshold, a state machine that inhibits other processes from making requests to the main store is initiated until the deadlocked process is successful in gaining access to the resource.

1.4.2 Vector Processing

The next step in the evolution of parallel processing was the introduction of multiprocessing. Here, two or more processors share a common workload. The earliest versions of multiprocessing were designed as a master/slave model, where one processor (the master) was responsible for all the tasks to be performed and it only off-loaded tasks to the other processor (the slave) when the master processor determined, based on a predetermined threshold, that work could be shifted to increase performance. This arrangement was necessary because it was not then understood how to program the machines, so they could cooperate in managing the resources of the system. Vector processing was developed to increase processing performance by operating in a multitasking manner. Matrix operations were added to computers to allow a single instruction to manipulate two arrays of numbers performing arithmetic operations. This was valuable in certain types of applications in which data occurred in the form of vectors or matrices. In applications with less well-formed data, vector processing was less valuable.

1.4.3 Symmetric Multiprocessing Systems

The next advancement was the development of symmetric multiprocessing (SMP) systems to address the problem of resource management in master/slave models. In SMP systems, each processor is equally capable and responsible for managing the workflow as it passes through the system. The primary goal is to achieve sequential consistency, in other words, to make SMP systems appear to be exactly the same as a single-processor, multiprogramming platform. Engineers discovered that system performance could be increased nearly 10%–20% by executing some instructions out of order.

However, programmers had to deal with the increased complexity and cope with a situation where two or more programs might read and write the same operands simultaneously. This difficulty, however, is limited to a very few programs, because it only occurs in rare circumstances. To this day, the question of how SMP machines should behave when accessing shared data remains unresolved.

Data propagation time increases in proportion to the number of processors added to SMP systems. After a certain number (usually somewhere around 40–50 processors), performance benefits gained by using even more processors do not justify the additional expense of adding such processors. To solve the problem of long data propagation times, message passing systems were created. In these systems, programs that share data send messages to each other to announce that particular operands have been assigned a new value. Instead of a global message announcing an operand's new value, the message is communicated only to those areas that need to know the change. There is a network designed to support the transfer of messages between applications. This allows a great number of processors (as many as several thousand) to work in tandem in a system. These systems are highly scalable and are called massively parallel processing (MPP) systems.

1.4.4 Massively Parallel Processing

MPP is used in computer architecture circles to refer to a computer system with many independent arithmetic units or entire microprocessors, which run in parallel. *Massive* connotes hundreds if not thousands of such units. In this form of computing, all the processing elements are interconnected to act as one very large computer. This approach is in contrast to a distributed computing model, where massive numbers of separate computers are used to solve a single problem such as in the SETI (see Chapter 12, "Grid Computing"). Early examples of MPP systems were the Distributed Array Processor, the Connection Machine, and the Ultracomputer. In data mining, there is a need to perform multiple searches of a static database. The earliest massively parallel processing systems all used serial computers as individual processing units in order to maximize the number of units available for a given size and cost. Single-chip implementations of massively parallel processing arrays are becoming ever more cost-effective due to the advancements in integrated circuit technology.

MPP machines are not easy to program, but for certain applications, such as data mining, they are the best solution.

1.5 Enterprise Systems

The Enterprise System (ES) is an information system that integrates business processes with the aim of creating value and reducing costs by making the right information available to the right people at the right time to help them make good decisions in managing resources proactively and productively. An ERP is comprised of multimodule application software packages that serve and support multiple business functions. These large automated cross-functional systems were designed to bring about improved operational efficiency and effectiveness through integrating, streamlining, and improving fundamental back-office business processes.

Traditional ESs (like ERP systems) were called back-office systems because they involved activities and processes in which the customer and general public were not typically involved, at least not directly. Functions supported by ES typically included accounting, manufacturing, human resource management, purchasing, inventory management, inbound and outbound logistics, marketing, finance, and to some extent engineering. The objective of traditional ESs in general was greater efficiency and to a lesser extent effectiveness. Contemporary ESs have been designed to streamline and integrate operation processes and information flows within a company to promote synergy and greater organizational effectiveness and innovation. These newer ESs have moved beyond the back-office to support front-office processes and activities like those fundamental to customer relationship management.

1.5.1 Evolution of ES

ESs have evolved from simple Materials Requirement Planning (MRP) to ERP, Extended Enterprise Systems (EES), and beyond. Table 1.1 gives a snapshot of the various stages of Enterprise Systems (ES).

1.5.1.1 Materials Requirement Planning (MRP)

The first practical efforts in the ES field occurred at the beginning of the 1970s, when computerized applications based on MRP methods were developed to support purchasing and production scheduling activities. MRP is a heuristic based on three main inputs: the *master production schedule*, which specifies how many products are going to be produced during a period of time; the *bill of materials*, which describes how those products are going to be built and what materials are going to be required; and the *inventory record file*, which reports how many products, components, and materials are held in-house. The method can easily be programmed in any basic computerized application, as it follows deterministic assumptions and a well-defined algorithm.

MRP employed a type of backward scheduling wherein lead times were used to work backward from a due date to an order release date. While the primary objective of MRP was to compute material requirements, the MRP system proved also to be a useful scheduling tool. Order placement and order delivery were planned by the MRP system. Not only were orders for materials and components generated by an MRP system but also production orders for manufacturing operations that used those materials and components to make higher-level items like subassemblies and finished products.

TABLE 1.1

Evolution of Enterprise Systems

System	Primary Business Need(s)	Scope	Enabling Technology
MRP	Efficiency	Inventory management and production planning and control	Mainframe computers, batch processing, traditional file systems
MRP II	Efficiency, effectiveness, and integration of manufacturing systems	Extending to the entire manufacturing firm (becoming cross-functional)	Mainframes and minicomputers, real-time (time-sharing) processing, database management systems (relational)
ERP	Efficiency (primarily back-office), effectiveness, and integration of all organizational systems	Entire organization (increasingly cross-functional), both manufacturing and nonmanufacturing operations	Mainframes, mini- and microcomputers, client/server networks with distributed processing and distributed databases, data warehousing, mining, knowledge management
ERP II	Efficiency, effectiveness, and integration within and among enterprises	Entire organization extending to other organizations (cross-functional and cross-enterprise partners, suppliers, customers, etc.)	Mainframes, client/server systems, distributed computing, knowledge management, Internet technology (includes intranets, extranets, portals)
Inter-Enterprise Resource Planning, Enterprise Systems, Supply Chain Management, or whatever label gains common acceptance	Efficiency, effectiveness, coordination, and integration within and among all relevant supply chain members as well as other partners or stakeholders on a global scale	Entire organization and its constituents (increasingly global and cross-cultural) comprising global supply chain from beginning to end as well as other industry and government constituents	Internet, Service Oriented Architecture, Application Service Providers, wireless networking, mobile wireless, knowledge management, grid computing, artificial intelligence

As MRP systems became popular and more and more companies started using them, practitioners, vendors, and researchers started to realize that the data and information produced by the MRP system in the course of material requirements planning and production scheduling could be augmented with additional data and used for other purposes. One of the earliest add-ons was the

Capacity Requirement Planning module, which could be used in developing capacity plans to produce the master production schedule. Manpower planning and support for human resources management were incorporated into MRP. Distribution management capabilities were added. The enhanced MRP and its many modules provided data useful in the financial planning of manufacturing operations; thus, financial planning capabilities were added. Business needs, primarily for operational efficiency and to a lesser extent for greater effectiveness, and advancements in computer processing and storage technology brought about MRP and influenced its evolution. What started as an efficiency-oriented tool for production and inventory management was becoming increasingly a cross-functional system.

1.5.1.2 Closed-Loop Materials Requirement Planning

A very important capability to evolve in MRP systems was the ability to close the loop (control loop). This was largely because of the development of real-time (closed-loop) MRP systems to replace regenerative MRP systems in response to changing business needs and improved computer technology—time sharing was replacing batch processing as the dominant computer processing mode. With time-sharing mainframe systems, the MRP system could run 24/7 and update continuously. Use of the corporate mainframe that performed other important computing task for the organization was not practical for some companies, because MRP consumed too many system resources; subsequently, some companies opted to use mainframes (now growing smaller and cheaper but increasing in processing speed and storage capability) or minicomputers (could do more, faster than old mainframes) that could be dedicated to MRP. MRP could now respond (update relevant records) to timely data fed into the system and produced by the system. This closed the control loop with timely feedback for decision making by incorporating current data from the factory floor, warehouse, vendors, transportation companies, and other internal and external sources, thus giving the MRP system the capability to provide current (almost real-time) information for better planning and control. These closed-loop systems better reflected the realities of the production floor, logistics, inventory, and more. It was this transformation of MRP into a planning and control tool for manufacturing by closing the loop, along with all the additional modules, that did more than plan materials— they planned and controlled various manufacturer resources—that led to MRP II. Here too, improved computer technology and evolving business needs for more accurate and timely information to support decision making and greater organizational effectiveness contributed to the evolution from MRP to MRP II.

1.5.1.3 Manufacturing Resource Planning II (MRP II)

The MRP in MRP II stands for Manufacturing Resource Planning rather than materials requirements planning. The MRP system had evolved from a material requirements planning system into a planning and control system for resources in manufacturing operations—an enterprise information system for manufacturing. As time passed, MRP II systems became more widespread, and more sophisticated, particularly when used in manufacturing to support and complement computer-integrated manufacturing (CIM). Databases started replacing traditional file systems, allowing for better systems integration and greater query capabilities to support decision makers, and the telecommunications network became an integral part of these systems in order to support communications between and coordination among system components that were sometimes geographically distributed but still within the company.

1.5.1.4 Enterprise Resource Planning (ERP)

During the late 1970s and early 1980s, new advances in IT, such as local area networks, personal computers, object-orientated programming, and more accurate operations management heuristics, allowed some of MRP's deterministic assumptions to be relaxed, particularly the assumption of infinite capacity. MRP II was developed based on MRP principles but incorporated some important operational restrictions, such as available capacity, maintenance turnaround time, and financial considerations. MRP II also introduced simulation options to enable the exploration and evaluation of different scenarios. MRP II is defined as business planning, sales and operations planning, production scheduling, MRP, capacity requirement planning, and the execution support systems for capacity and material. Output from these systems is integrated with financial reports such as the business plan, purchase commitment report, shipping budget, and inventory projections in dollars. An important contribution of the MRP II approach was the integration of financial considerations, improving management control and performance of operations and making different manufacturing approaches comparable. However, while MRP II allowed the integration of sales, engineering, manufacturing, storage, and finance, these areas continued to be managed as isolated systems. In other words, there was no real online integration, and the system did not provide integration with other critical support areas, such as accounting, human resource management, quality control, and distribution.

The need for greater efficiency and effectiveness in back-office operations was not unique to manufacturing but was also common to nonmanufacturing operations. Companies in nonmanufacturing sectors such as health care, financial services, air transportation, and consumer goods started to use MRP II–like systems to manage critical resources. Early ERP systems typically ran on mainframes like their predecessors, MRP, and MRP II, but many migrated to client/server systems where networks were central and distributed databases more common. The growth of ERP and the migration to client/server systems really got a boost from the Y2K scare. Many companies were convinced of the need to replace older mainframe-based systems, some ERP and some not, with the newer client/server architecture.

An analysis of the performance of ES shows that a key indicator is the level of enterprise integration. First-generation MRP systems only provided limited integration for sales, engineering, operations, and storage. Second-generation MRP II solutions enhanced that integration and included financial capabilities. ERP systems enabled the jump to full enterprise integration. Finally, CRM and SCM systems are expanding that integration to include customers and suppliers. In this history, there is a clear positive trend of performance improvement, coinciding with the diffusion of ES functional innovations. If we assume that ERP, CRM, and SCM systems achieve real integration, the next stage is likely to be an ES that allows for the integration of a group of businesses.

1.5.2 Extended Enterprise Systems (EES)

The most salient trend in the continuing evolution of ES is the focus on front-office applications and interorganizational business processes, particularly in support of supply chain management (SCM). At present, greater organizational effectiveness in managing the entire supply chain all the way to the end customer is a priority in business. The greater emphasis on front-office functions and cross-enterprise communications and collaboration via the Internet simply reflects changing business needs and priorities. The demand for specific modules/capabilities in particular shows that businesses are looking beyond the enterprise. This external focus is encouraging vendors to seize the moment by responding with the modules/systems that meet evolving business needs. In this renewed context, ESs enable organizations to integrate and coordinate their business processes. They provide a single system that is central to the organization and ensure that information can be shared across all functional levels and management hierarchies.

ES is creeping out of the back-office into the front and beyond the enterprise to customers, suppliers, and more, in order to meet changing business needs. Key players like Baal, Oracle, PeopleSoft, and SAP have incorporated Advanced Planning and Scheduling (APS), Sales Force Automation (SFA), Customer Relationship Management (CRM), SCM, Business Intelligence, and E-commerce modules/capabilities into their systems or repositioned their ESs as part of broader Enterprise Systems suites incorporating these and other modules/capabilities. ES products reflect the evolving business needs of clients and the capabilities of IT, perhaps most notably those related to the Web. Traditional ES (i.e., ERP) has not lost its significance because back-office efficiency, effectiveness, and flexibility will continue to be important. However, current focus seems more external as organizations look for ways to support and improve relationships and interactions with customers, suppliers, partners, and other stakeholders. While the integration of internal functions is still important and in many enterprises still has not been achieved to a great extent, external integration is now receiving much attention.

1.5.2.1 Extended Enterprise Systems (EES) Framework

The conceptual framework of EES consists of four distinct layers:

1. Foundation layer
2. Process layer
3. Analytical layer
4. E-business layer

Each layer consists of collaborative components described in Table 1.2.

1.5.2.1.1 Foundation Layer

The foundation layer consists of the core components of EES, which shape the underlying architecture and also provide a platform for EES systems. EES does not need to be centralized or monolithic. One of the core components is the integrated database, which may be a distributed database. Another core

TABLE 1.2

Four Layers of EES

Layer		Components
Foundation	Core	Integrated Database (DB)
		Application Framework (AF)
Process	Central	Enterprise Resource Planning (ERP)
		Business Process Management (BPM)
Analytical	Corporate	Supply Chain Management (SCM)
		Customer Relationship Management (CRM)
		Supplier Relationship Management (SRM)
		Product Lifecycle Management (PLM)
		Employee Lifecycle Management (ELM)
		Corporate Performance Management (CPM)
Portal	Collaborative	Business-to-consumer (B2C)
		Business-to-business (B2B)
		Business-to-employee (B2E)
		Enterprise Application Integration (EAI)

component is the application framework (AF), which can also be distributed. The integrated database and the application framework provide an open and distributed platform for EES.

1.5.2.1.2 Process Layer

The process layer of the concept is the central component of EES, which is Web based, open, and componentized (this is different from being Web enabled) and may be implemented as a set of distributed Web Services. This layer corresponds to the traditional transaction-based systems. ERP still makes up the backbone of EES along with the additional integrated modules aimed at new business sectors outside the manufacturing industries. The backbone of ERP is the traditional ERP modules like financials, sales and distribution, logistics, manufacturing, or HR.

The EES concept is based on Business Process Management (BPM). ERP has been based on *best practice* process reference models, but EES systems primarily build on the notion of the process as the central entity. EES includes tools to manage processes: design (or orchestrate) processes, to execute and to evaluate processes (Business Activity Monitoring), and redesigning processes get effect in real time. The BPM component allows for EES to be accommodated to suit different business practices for specific business segments that otherwise would require effort-intensive customization. EES further includes vertical solutions for specific segments like apparel and footwear or the public sector. Vertical solutions are sets of standardized preconfigured systems and processes with *add-on* to match the specific requirements of a specific sector.

1.5.2.1.3 Analytical Layer

The analytical layer consists of the corporate components that extend and enhance the central ERP functions by providing decision support to manage relations and corporate issues. Corporate components are not necessarily synchronized with the integrated database, and the components may easily be *add-ons* instituted by acquiring third-party products/vendors. In the future, the list of components for this layer can get augmented by newer additions like Product Lifecycle Management (ERP for R&D function) and Employee Lifecycle Management (ERP for human resources).

1.5.2.1.4 E-Business Layer

The e-business layer is the portal of EESs, and this layer consists of a set of collaborative components. The collaborative components deal with the communication and the integration between the corporate ERP II system and actors like customers, business partners, employees, and even external systems.

1.5.2.2 Extended Functionality

E-Commerce is arguably one of the most important developments in business in the last 50 years, and M-Commerce is poised to take its place alongside or within the rapidly growing area of E-Commerce. Internet technology has made E-Commerce in its many forms (B2B, B2C, C2C, etc.) possible. Mobile and wireless technologies are expected to make *always on* Internet and *anytime/anywhere* location-based services (also requiring global positioning systems) a reality, as well as a host of other capability characteristics of M-Business. One can expect to see ES geared more to the support of both E-Commerce and M-Commerce. Internet, mobile, and wireless technologies should figure prominently in new and improved system modules and capabilities.

The current business emphasis on intra- and interorganizational process integration and external collaboration should remain a driving force in the evolution of ES in the foreseeable future. Some businesses are attempting to transform themselves from traditional, vertically integrated organizations into multienterprise, *recombinant entities* reliant on core-competency-based strategies. Integrated SCM and business networks will receive great emphasis, reinforcing the importance of IT support for cross-enterprise collaboration and interenterprise processes. ESs will have to support the required interactions and processes among and within business entities and work with other systems/modules that do the same. There will be a great need for business processes to span organizational boundaries (some do at present), possibly requiring a single shared inter-enterprise system that will do it, or at least communicate with and coprocess (share/divide processing tasks) with other ES systems.

Middleware, ASPs, and enterprise portal technologies may play an important role in the integration of such modules and systems. The widespread adoption of a single ASP solution among supply chain partners may facilitate interoperability as all supply chain partners essentially use the same system. Alternatively, a supply chain portal (vertical portal), jointly owned by supply chain partners or a value-added service provider that coordinates the entire supply chain and powered by a single system serving all participants, could be the model for the future. ASP solutions are moving the ES within the reach of SMEs, as it costs much less to *rent* than to *buy*.

The capability of Web Services to allow businesses to share data, applications, and processes across the Internet may result in ES systems of the future relying heavily on the Service-Oriented Architecture (SOA), within which Web Services are created and stored, providing the building blocks for programs and systems. The use of *best-in-breed* Web Service–based solutions might be more palatable to businesses, since it might be easier and less risky to plug in a new Web Service–based solution than replace or add on a new product module. While the *one source* alternative seems most popular at present, the *best-in-breed* approach will be good if greater interoperability/ integration among vendor products is achieved. There is a need for greater *out-of-the-box* interoperability, thus a need for standards.

Data warehouses and Knowledge Management System (KMS) should enable future ERP systems to support more automated business decision making, and they should be helpful in the complex decision making needed in the context of fully integrated supply chain management. More automated decision making in both front-office and back-office systems should eliminate/minimize human variability and error, greatly increase decision speed, and hopefully improve decision quality. Business Intelligence (BI) tools, which are experiencing a significant growth in popularity, take internal and external data and transform them into information used in building knowledge that helps decision makers to make more *informed* decisions.

> Greater interoperability of diverse systems and more thorough integration within and between enterprise systems is likely to remain the most priority for all enterprises. An environment for business applications much like the *plug-and-play* environment for hardware environment would make it easier for organizations to integrate their own systems and have their systems integrated with other organizations' systems. Such an environment necessitates greater standardization. This ideal *plug-and-play* environment would make it easier for firms to opt for a *best-in-breed* strategy for application/module acquisition as opposed to reliance on a single vendor for a complete package of front-office, back-office, and strategic systems.

1.6 Autonomic Computing

The software systems of today are constantly faced with new and demanding requirements in terms of their availability, robustness, dynamism, and pervasiveness. Pressures on the maintenance of systems, services, and software are increasing by the day—maintenance tasks are becoming increasingly difficult and correspondingly more time consuming to carry out. It is believed that we may be reaching a barrier in terms of complexity and that innovative practices are needed to ensure the continual delivery of software-based services.

IBM introduced the term *autonomic computing* to characterize the notion of a computer system that is able to adapt to internal and external changes with or without minimal human intervention. As the development, maintenance and operation of computing systems are becoming more complex, we are witnessing development of more and more automated deployment and maintenance strategies. Based on dedicated tools, such approaches aim to automate a number of administrative tasks such as installing packages and modules, defining authorizations and updating configurations.

IBM's mission was to enhance and impart systems with self-management capabilities. Thus, systems are able to evolve in an autonomous manner, fixing undesirable behaviors and adapting to their changing requirements and environment. The level of autonomy given to a system is a product of the ability to map the administration function to a machine-executable process, and the ease of implementing such a process into the system. Accordingly, the system administrators can delegate a part of their workload to the system itself; consequently, they can focus on the system's fundamentals and off-load more mundane tasks to the automatic administration software. Moreover, administration tasks performed by the systems are of higher quality, minimizing the possibility of introducing bugs during such maintenance activities.

The benefits of autonomic computing are

- Decrease in maintenance risk and expenditure: The goal is to obtain systems that are able to configure themselves automatically, and with a potential to achieve zero configuration for the administrator, hence reducing costs. Thus, it enables a revaluation of the human system administrator's tasks, allowing them to focus on more strategic or value-adding aspects of the system support function.
- Increase in service availability: Anticipation of potential problems and automatic system diagnosis can provide increased application dependability and other nonfunctional benefits. For instance, increased security can enable the system to be better prepared to counter malicious acts.

However, implementing autonomic computing systems is challenging because it implies the following dimensions:

a. A computer system must be aware, that is, be able to keep some knowledge about its envisaged goals, its past and current situation. Accordingly, it has to assimilate some type of reasoning capabilities to decide on corrective actions *en route* whenever needed.

b. A computer system should provide a high-level interface, allowing human administrators to specify or modify system goals, tune reasoning processes, and oversee the system's ability to manage and attain its objectives.

c. A computer system must be able to monitor itself at runtime in order to know its internal situation. It must also be able to monitor part of its execution environment to enable relevant evolutions.

d. A computer system must be able to adapt itself at runtime in order to implement the requisite corrective administrative actions without disrupting or engendering ongoing operations.

1.7 Summary

This chapter provides a bird's eye view of the dominating trend of the last 50 years in the computer industry, namely, increasing functional specificity over increasingly commoditized hardware. The first half of the chapter presents an overview of the evolution from Grid Computing through Parallel Processing to Symmetric Multiprocessing Systems toward Massive Parallel Processing using commoditized hardware. The latter part of the chapter traces the evolution from batch Materials Requirement Planning to the current Extended Enterprise Systems incorporating Web Services and SOA. In the following parts, the book presents the genesis, roadmap, and characteristics of Cloudware, that is, the cloud computing area. The last part of the book describes Cloudware Applications in the areas of big data, mobile, and context-aware systems.

Section I

Genesis of Cloudware

This section introduces the various milestones on the path to developing EAI solutions for enterprises. Maintaining competitiveness in the face of changing business environments necessitated ways to reuse existing systems and integrate them into the global, enterprise-wide information system must be defined. The resulting integrated information systems will improve the competitive advantage of the enterprises with a unified and efficient access to information. Chapter 2 traces the genesis of the cloud to the networking and Internetworking technologies of the 1980s. Chapter 3 describes the concepts of distributed systems. Chapter 4 describes EAI methods and models. Chapter 5 deals with middleware and related integration technologies. Finally, Chapter 6 introduces the reference architecture for enterprise J2EE applications.

2

Networking and Internetworking

The origin of what has developed to become the Internet and World Wide Web (WWW) goes back to work done in the early 1940s by Vannevar Bush. He was the originator of the concept of an associative system for the organization of data within networks (the other being the concept of breaking messages into discrete smaller packages to maximize efficiency by utilizing the networks effectively). Bush was an American scientist who had done work on submarine detection for the US Navy. Bush, who was C. Shannon's professor at MIT, had built the differential analyzer that sat in the basement of the Moore School and had facilitated further computing developments during the Second World War from his position in the US Office of Scientific Research and Planning.

Bush's influence on the development of the Internet is due to his visionary description of an information management system that he called the *memex*. The memex (memory extender) is described in his famous essay "As We May Think," which was published in the *Atlantic Monthly* in 1945. This article represents the first published articulation of the idea of a web. Bush conceived of what he called the memex, essentially a microfilm/audio recording device electronically linked to a library and able to display books and films that would allow *selection by association rather than by indexing*. The memex was to be a sort of multiscreened microfilm reader operated by a keyboard into which a user could scan an entire personal library as well as all notes, letters, and communications. When data of any sort are placed in storage, they are filed alphabetically or numerically, and information is found (when it is) by tracing it down from subclass to subclass. It can only be in one place. The human mind does not work that way. It operates by association. With one item in its grasp, it snaps instantly to the next that is suggested by the association of thoughts, in accordance with some intricate web of trails carried by the cells of the brain.

This description motivated Ted Nelson and Douglas Engelbart to independently formulate the various ideas that would become hypertext. Doug Engelbart, had demonstrated a prototype information retrieval system at the 1968 Fall Joint Computer Conference, and Ted Nelson was developer of a similar system called Xanadu. In his self-published manifesto, Nelson defined *hypertext* as "forms of writing which branch or perform on request; they are best presented on computer display screens." Nelson praised Engelbart's On-Line System (NLS) but noted that Engelbart believed in tightly structuring information in outline formats. Nelson wanted something closer to Vannevar Bush's earlier concept, which Bush hoped would replicate the mind's ability to make associations across subject boundaries. Nelson worked tirelessly through the 1970s and 1980s to bring Xanadu to life. He remained close to but always outside of the academic and research community, and his ideas inspired the work at Brown University, led by Andries van Dam. Independently of these researchers, Apple introduced a program called HyperCard for the Macintosh in 1987. HyperCard implemented only a fraction of the concepts of hypertext as understood by van Dam or Nelson, but it was simple, easy to use, and even easy for a novice to program. For all its limits, HyperCard brought the notion of nonlinear text and graphics out of the laboratory setting.

Tim Berners-Lee would later use hypertext as part of the development of the WWW.

2.1 ARPANET

In the 1960s, there were approximately 10,000 computers in the world. These computers were very expensive ($100K–$200K) and had very primitive processing power. They contained only a few thousand words of magnetic memory, and programming and debugging of these computers was difficult. Further, communication between computers was virtually nonexistent. However, several computer scientists had dreams of worldwide networks of computers, where every computer around the globe is interconnected to all of the other computers in the world. For example, Licklider wrote memos in the early 1960s on his concept of an intergalactic network. This concept envisaged that everyone around the globe would be interconnected and able to access programs and data at any site from anywhere.

The US Department of Defense founded the Advance Research Projects Agency (ARPA) in the late 1950s. ARPA embraced high-risk, high-return research and laid the foundation for what became ARPANET and later the Internet. J.C.R. Licklider was an early pioneer of AI, and he also formulated the idea of a global computer network. He wrote his influential paper, "Man–Computer Symbiosis" in 1960, and this paper outlined the need for simple interaction between users and computers. Licklider became the first head of the computer research program at ARPA, which was called the Information Processing Techniques Office (IPTO). He developed close links with MIT, UCLA, and BBN Technologies and started work on his vision. BBN Technologies (originally Bolt, Beranek, and Newman) was a high-technology research and development company. It was especially famous for its work in the development of packet switching for the ARPANET and the Internet. It also did defense work for Defense Advanced Research

Projects Agency DARPA. BBN played an important part in the implementation and operation of ARPANET.

 The "@" sign used in an e-mail address was a BBN innovation. Ray Tomlinson of BBN Technologies developed a program that allowed electronic mail to be sent over the ARPANET. Tomlinson developed the "user@host" convention, and this was eventually to become the standard for electronic mail in the late 1980s.

Various groups, including National Physical Laboratory (NPL), the RAND Corporation, and MIT, commenced work on packet-switching networks. The concept of packet switching was invented by Donald Davies at the NPL in 1965. Packet switching is a fast message communication system between computers. Long messages are split into packets that are then sent separately so as to minimize the risk of congestion.

At that time, a basic problem for Licklider was the lack of language and machine standardization. The early computers had different standards for representing data, and this meant that the data standard of each computer would need to be known for effective communication to take place. There was a need to establish a standard for data representation, and a US government committee developed the ANSII (American Standard Code for Information Interchange) in 1963. This became the first universal standard for data for computers, and it allowed machines from different manufacturers to exchange data. The standard allowed 7-bit binary strings to stand for a letter in the English alphabet, an Arabic numeral, or a punctuation symbols; the use of 7 bits allowed 128 distinct characters to be represented. The development of the IBM-360 mainframe standardized the use of 8 bits for a word, and 12-bit or 36-bit words became obsolete.

Davies also worked on the ACE computer (one of the earliest stored program computers) that was developed at the NPL in the United Kingdom in the late 1940s. The first wide area network connection was created in 1965. It involved the connection of a computer at MIT to a computer in Santa Monica via a dedicated telephone line. This result showed that telephone lines could be used for the transfer of data although they were expensive in their use of bandwidth. The need to build a network of computers became apparent to ARPA in the mid-1960s, and this led to work commencing on the ARPANET project in 1966. The plan was to implement a packet-switched network based on the theoretical work done on packet switching at NPL and MIT. The goal was to have a network speed of 56 Kbps. ARPANET was to become the world's first packet-switched network.

BBN Technologies was awarded the contract to implement the network. Two nodes were planned for the network initially, and the goal was to eventually have 19 nodes. The first two nodes were based at UCLA and SRI. The network management was performed by interconnected *Interface Message Processors* (IMPs) in front of the major computers. Each site had a team to produce the software to allow its computers and the IMP to communicate. The IMPs eventually evolved to become the network routers that are used today. The team at UCLA called itself the Network Working Group and saw its role as developing the Network Control Protocol (NCP). This was essentially a rule book that specified how computers on the network should communicate.

The first host-to-host connection was made between a computer at UCLA and a computer at SRI in late 1969. Several other nodes were added to the network until it reached its target of 19 nodes in 1971. The Network Working Group developed the Telnet Protocol and file transfer protocol (FTP) in 1971. The Telnet program allowed the user of one computer to remotely log in to another computer. The file transfer protocol allows the user of one computer to send or receive files from another computer. A public demonstration of ARPANET was made in 1972 and it was a huge success. One of the earliest demos was that of Weizenbaum's ELIZA program. This is a famous AI program that allows a user to conduct a typed conversation with an artificially intelligent machine (psychiatrist) at MIT. The viability of packet switching as a standard for network communication had been clearly demonstrated.

By the early 1970s, over 30 institutions were connected to the ARPANET. These included consulting organizations such as BBN, Xerox, and the MITRE Corporation and government organizations such as NASA. Bob Metacalfe developed a wire-based local area network (LAN) at Xerox that would eventually become Ethernet in the mid-1970s.

2.1.1 Ethernet

If the Internet of the 1990s became the *Information Superhighway*, then Ethernet became the equally important network of local roads to feed it. As a descendent of ARPA research, the global networks we now call the Internet came into existence before the local Ethernet was invented at Xerox. But Ethernet transformed the nature of office and personal computing even before the Internet had a significant effect.

Ethernet was invented at Xerox PARC in 1973 by Robert Metcalfe and David Boggs. He moved to Xerox PARC in 1972; one of his first tasks there was to hook up PARC's PDP-10 clone, the MAXC, to ARPANET. Metcalfe connected Xerox's MAXC to ARPANET, but the focus at Xerox was on local networking: to connect a single-user computer (later to become the Alto) to others like it and to a shared high-quality printer, all within the same building. The ARPANET model, with its expensive, dedicated Interface Message Processors (IMPs), was not appropriate.

When Metcalfe arrived at PARC, there was already a local network established, using Data General minicomputers linked in a star-shaped topology.

Metcalfe and his colleagues felt that even the Data General network was too expensive and not flexible enough to work in an office setting, where one may want to connect or disconnect machines frequently. He also felt it was not robust enough—the network's operation depended on a few critical pieces not failing. He recalled a network he saw in Hawaii that used radio signals to link computers among the Hawaiian islands, called ALOHAnet. With this system, files were broken up into *packets*, no longer than 1000 bits long, with an address of the intended recipient attached to the head of each. Other computers on the net were tuned to the UHF frequency and listened for the packets, accepting the ones that were addressed to it and ignoring all the others.

What made this system attractive for Metcalfe was that the medium—in this case radio—was passive. It simply carried the signals, with the computers at each node doing the processing, queuing, and routing work. The offices at Xerox PARC were not separated by water, but the concept was perfectly suited for a suite of offices in a single building. Metcalfe proposed substituting a cheap coaxial cable for the *ether* that carried ALOHAnet's signals. A new computer could be added to the *Ethernet* simply by tapping into the cable. To send data, a computer first listened to make sure there were no packets already on the line; if not, it sent out its own. If two computers happened to transmit at the same time, each would back off for a random interval and try again. If such collisions started to occur frequently, the computers themselves would back off and not transmit so often. By careful mathematical analysis, Metcalfe showed that such a system could handle a lot of traffic without becoming overloaded. He wrote a description of it in May 1973 and recruited David Boggs to help build it. They had a fairly large network running by the following year. Metcalfe recalled that its speed, around three million bits per second, was unheard of at the time, when "the 50-kilobit-per-second (Kbps) telephone circuits of the ARPANET were considered fast."

Those speeds fundamentally altered the relationship between small and large computers. Clusters of small computers now, finally, provided an alternative to the classic model of a large central system that was timeshared and accessed through dumb terminals. Ethernet would have its biggest impact on the workstation, and later PC, market, but its first success came in 1979, when Digital Equipment Corporation, Intel, and Xerox joined to establish it as a standard, with DEC using it for the VAX. UNIX-based workstations nearly all adopted Ethernet, although token ring and a few alternate schemes are also used.

DOS-based personal computers were late in getting networked. Neither the Intel processors they used nor DOS was well suited for it.

Purchasers of PCs and PC software were driven by personal, not corporate, needs. The PC and DOS standards led to commercial software that was not only inexpensive but also better than what came with centralized systems. By the mid-1980s, it was clear that no amount of corporate policy directives could keep the PC out of the office, especially among those employees who already had a PC at home. The solution was a technical fix: network the PCs to one another, in a local-area network (LAN). The company that emerged with over half the business by 1989 was Novell, located in the Salt Lake City area. Novell's Netware was a complex—and expensive—operating system that overlaid DOS, seizing the machine and directing control to a *file server*, typically a PC with generous mass storage and I/O capability (the term *server* originated in Metcalfe and Boggs's 1976 paper on the Ethernet). By locating data and office automation software on this server rather than on individual machines, some measure of central control could be reestablished. Networking of PCs lagged behind the networking that UNIX workstations enjoyed from the start, but the personal computer's lower cost and better office software drove this market.

> Local networking took the *personal* out of personal computing, at least in the office environment. Still, the networked office computers of the 1990s gave their users a lot more autonomy and independence than the timeshared mainframes accessed through *dumb terminals* or *glass Teletypes* in the 1970s.

2.1.2 TCP/IP Protocol

ARPA became DARPA (Defence Advanced Research Projects Agency) in 1973, and it commenced work on a project connecting seven computers on four islands using a radio-based network, as well as a project to establish a satellite connection between sites in Norway and in the United Kingdom. This led to a need for the interconnection of the ARPANET with other networks. The key challenge was to investigate ways of achieving convergence between ARPANET, radio-based networks, and satellite networks, as these all had different interfaces, packet sizes, and transmission rates. Therefore, there was a need for a network-to-network connection protocol, and its development would prove to be an important step toward developing the Internet.

An international network working group (INWG) was formed in 1973. The concept of the Transmission Control Protocol (TCP) was developed at DARPA by Bob Kahn and Vint Cerf, and they presented their ideas at an INWG meeting at the University of Sussex in England in 1974. TCP allowed cross network connections, and it began to replace the original NCP protocol used in ARPANET. However, it would take some time for the existing and new networks to adopt the TCP protocol. TCP is a set of network standards that specify the details of how computers communicate, as well as the standards for interconnecting networks and computers. It was designed to be flexible and provides a transmission standard that deals with physical differences in host computers, routers, and networks. TCP is designed to transfer data over networks that support different packet sizes and that may sometimes lose packets. It allows the Internetworking of very different networks that then act as one network.

The new protocol standards were known as the Transmission Control Protocol (TCP) and the Internet protocol (IP). TCP details how information is broken into packets and reassembled on delivery, whereas IP is focused on sending the packet across the network. The protocol that resides at the network layer in the TCP/IP protocol suite is called Internet Protocol (IP). IP's primary function is to perform the routing necessary to move data packets across the Internet. IP is a connectionless protocol that does not concern itself with keeping track of lost, duplicated, or delayed packets or packets delivered out of order. Furthermore, the sender and receiver of these packets may not be informed that these problems have occurred. Thus, IP is also referred to as an unreliable service. If an application requires a reliable service, then the application needs to include a reliable transport service *above* the connectionless, unreliable packet delivery service. The reliable transport service is provided by software called Transmission Control Protocol, which turns an unreliable network into a reliable one, free from lost and duplicate packets. This combined service is known as TCP/IP. These standards allow users to send electronic mail or to transfer files electronically, without needing to concern themselves with the physical differences in the networks.

TCP/IP is a family or suite of protocols consisting of four layers.

a. Network interface layer: This layer is responsible for formatting packets and placing them on to the underlying network. It is equivalent to the physical and data link layers on OSI Model.

b. Internet layer: This layer is responsible for network addressing. It includes the Internet protocol and the address resolution protocol. It is equivalent to the network layer on the OSI Model.

c. Transport layer: This layer is concerned with data transport and is implemented by TCP and the user datagram protocol (UDP). It is equivalent to the transport and session layers in the OSI Model.

d. Application layer: This layer is responsible for liaising between user applications and the transport layer. The applications include the file transfer protocol (FTP), Telnet, domain name system (DNS), and simple mail transfer protocol (SMTP). It is equivalent to the application and presentation layers on the OSI Model.

The Internet protocol (IP) is a connectionless protocol that is responsible for addressing and routing packets. It is responsible for breaking up and assembling packets, with large packets broken down into smaller packets when they are traveling through a network that supports smaller packets. A connectionless protocol means that a session is not established before data are exchanged. Packet delivery with IP is not guaranteed as packets may be lost or delivered out of sequence. An acknowledgment is not sent when data are received, and the sender or receiver is not informed when a packet is lost or delivered out of sequence. A packet is forwarded by the router only if the router knows a route to the destination. Otherwise, it is dropped. Packets are dropped if their checksum is invalid or if their time to live is zero. The acknowledgment of packets is the responsibility of the TCP protocol.

The ARPANET employed the TCP/IP protocols as a standard from 1983.

The TCP/IP application layer includes several frequently used applications:

1. Hypertext Transfer Protocol (HTTP) to allow Web browsers and servers to send and receive WWW pages
2. Simple Mail Transfer Protocol (SMTP) to allow users to send and receive electronic mail
3. File Transfer Protocol (FTP) to transfer files from one computer system to another
4. Telnet to allow a remote user to log in to another computer system
5. Simple Network Management Protocol (SNMP) to allow the numerous elements within a computer network to be managed from a single point

> The late 1970s saw the development of newsgroups that aimed to provide information about a particular subject. A newsgroup started with a name that is appropriate with respect to the content that it is providing. Newsgroups were implemented via USENET and were an early example of client–server architecture. A user dials in to a server with a request to forward a certain newsgroup posting; the server then *serves* the request.

2.2 Computer Networks

The merging of computers and communications has had a profound influence on the way computer systems are organized. Although data centers holding thousands of Internet servers are becoming common, the once-dominant concept of the *computer center* as a room with a large computer to which users bring their work for processing is now totally obsolete. The old model of a single computer serving all of the organization's computational needs has been replaced by the one in which a large number of separate but interconnected computers do the job. These systems are called computer networks.

Two computers are said to be networked if they are able to exchange information. The connection need not be via a copper wire; fiber optics, microwaves, infrared, and communication satellites can also be used. Networks come in many sizes, shapes, and forms, as we will see later. They are usually connected together to make larger networks, with the Internet being the most well-known example of a network of networks.

Computer Network and a Distributed System: The key distinction between them is that in a distributed system, a collection of independent computers appears to its users as a single coherent system. Usually, it has a single model or paradigm that it presents to the users. Often, a layer of software on top of the operating system, called middleware, is responsible for implementing this model. A well-known example of a distributed system is the WWW. It runs on top of the Internet and presents a model in which everything looks like a document (Web page). On the other hand, in a computer network, coherence, model, and software are absent. Users are exposed to the actual machines, without any attempt by the system to make the machines look and act in a coherent way. If the machines have different hardware and different operating systems, that is fully visible to the users. If a user wants to run a program on a remote machine, it entails logging onto that machine and run it there.

In effect, a distributed system is a software system built on top of a network. The software gives it a high degree of cohesiveness and transparency. Thus, the distinction between a network and a distributed system lies with the software (especially the operating system), rather than with the hardware. Nevertheless, there is considerable overlap between the two subjects. For example, both distributed systems and computer networks need to move files around. The difference lies in who invokes the movement, the system, or the user.

2.2.1 Network Principles

2.2.1.1 Protocol

The term *protocol* is used to refer to a well-known set of rules and formats to be used for communication between processes in order to perform a given task. The definition of a protocol has two important parts to it:

1. A specification of the sequence of messages that must be exchanged
2. A specification of the format of the data in the messages

The existence of well-known protocols enables the separate software components of distributed systems to be developed independently and implemented in different programming languages on computers that may have different order codes and data representations. A protocol is implemented by a pair of software modules located in the sending and receiving computers. For example, a transport protocol transmits messages of any length from a sending process to a receiving process. A process wishing to transmit a message to another process, issues a call to a transport protocol module, passing it a message in the specified format. The transport software then concerns itself with the transmission of the message to its destination, subdividing it into packets of some specified size and format that can be transmitted to the destination via the network Protocol—another, lower-level protocol. The corresponding transport protocol module in the receiving computer receives the packet via the network-level protocol module and performs inverse transformations to regenerate the message before passing it to a receiving process.

2.2.1.2 Protocol Layers

Network software is arranged in a hierarchy of layers. Each layer presents an interface to the layers above it that extends the properties of the underlying communication system. A layer is represented by a module in every computer connected to the network. Each module appears to communicate directly with a module at the same level in another computer in the network, but in reality, data are not transmitted directly between the protocol modules at each level. Instead, each layer of network software communicates by local procedure calls with the layers above and below it. On the sending side, each layer (except the topmost, or application layer) accepts items of data in a specified format from the layer above it and applies transformations to encapsulate the data in the format specified for that layer before passing it to the layer below for further processing. On the receiving side, the converse transformations are applied to data items received from the layer below before they are passed to the layer above. The protocol type of the layer above is included in the header of each layer, to enable the protocol stack at the receiver to select the correct software components to unpack the packets. The data items are received and passed upward through the hierarchy of software modules, being transformed at each stage until they are in a form that can be passed to the intended recipient process.

2.2.1.3 Protocol Suite

A complete set of protocol layers is referred to as a protocol suite or a protocol stack, reflecting the layered structure. Figure 2.1 shows a protocol stack that conforms to the seven-layer Reference Model for Open Systems Interconnection (OSI) adopted by the International Organization for Standardization (ISO). The OSI Reference Model was adopted in order to encourage the development of protocol standards that would meet the requirements of open systems. The purpose of each level in the OSI Reference Model is summarized in Table 2.1. As its name implies, it is a framework for the definition of protocols and not a definition for a specific suite of protocols.

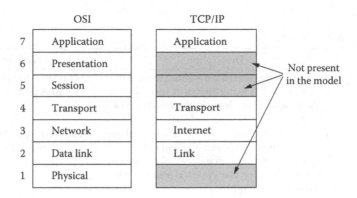

FIGURE 2.1
OSI versus TCP/IP network reference model.

Protocol suites that conform to the OSI model must include at least one specific protocol at each of the seven levels that the model defines (Table 2.1).

2.2.1.4 Datagram

Datagram is essentially another name for data packet. The term *datagram* refers to the similarity of this delivery mode to the way in which letters and telegrams are delivered. The essential feature of datagram networks is that the delivery of each packet is a *one-shot* process; no setup is required, and once the packet is delivered, the network retains no information about it.

2.2.2 Types of Network

There is no generally accepted taxonomy into which all computer networks fit, but two dimensions stand out as important: transmission technology and scale. We will now examine each of these in turn. Broadly speaking, there are two types of transmission technology that are in widespread use: broadcast links and point-to-point links.

Point-to-point links connect individual pairs of machines. To go from the source to the destination on a network made up of point-to-point links, short messages, called packets in certain contexts, may have to first visit one or more intermediate machines. Often, multiple routes, of different lengths, are possible, so finding good ones is important in point-to-point networks. Point-to-point transmission with exactly one sender and exactly one receiver is sometimes called unicasting. In contrast, on a broadcast network, the communication channel is shared by all the machines on the network; packets sent by any machine are received by all the others. An address field within each packet specifies the intended recipient. Upon receiving a packet, a machine checks the address field. If the packet is intended for the receiving

TABLE 2.1

OSI Layers or Stack

Layer	Description	Examples
Application	Protocols at this level are designed to meet the communication requirements of specific applications, often defining the interface to a service.	HTTP, FTP, SMTP, CORBA HOP
Presentation	Protocols at this level transmit data in a network representation that is independent of the representations used in individual computers, which may differ. Encryption is also performed in this layer, if required.	TLS security, CORBA data representation
Session	At this level, reliability and adaptation measures are performed, such as detection of failures and automatic recovery.	SIP
Transport	This is the lowest level at which messages (rather than packets) are handled. Messages are addressed to communication ports attached to processes. Protocols in this layer may be connection oriented or connectionless.	TCP, LDP
Network	Transfers data packets between computers in a specific network. In a WAN or an Internetwork, this involves the generation of a route passing through routers. In a single LAN, no routing is required.	IP, ATM virtual circuits
Data link	Responsible for transmission of packets between nodes that are directly connected by a physical link. In a WAN, transmission is between pairs of routers or between routers and hosts. In a LAN, it is between any pair of hosts.	Ethernet MAC, ATM cell transfer, PPP
Physical	The circuits and hardware that drive the network. It transmits sequences of binary data by analogue signaling, using amplitude or frequency modulation of electrical signals (on cable circuits), light signals (on fiber optic circuits), or other electromagnetic signals (on radio and microwave circuits).	Ethernet baseband signaling, ISDN

machine, that machine processes the packet; if the packet is intended for some other machine, the packet is just ignored.

Broadcast systems usually also allow the possibility of addressing a packet to all destinations by using a special code in the address field. When a packet with this code is transmitted, it is received and processed by every machine on the network. This mode of operation is called broadcasting. Some broadcast systems also support transmission to a subset of the machines, which is known as multicasting. An alternative criterion for classifying networks is by scale. Distance is important as a classification metric because different technologies are used at different scales. At the top are personal area networks, networks that are meant for one person. Beyond these come longer-range networks. These can be divided into local, metropolitan, and wide area networks, each with increasing scale. Finally, the connection of two or more networks is called an Internetwork. The worldwide Internet is certainly the best-known (but not the only) example of an Internetwork.

2.2.2.1 Personal Area Networks

PANs (Personal Area Networks) let devices communicate over the range of a person. A common example is a wireless network that connects a computer with its peripherals. Almost every computer has an attached monitor, keyboard, mouse, and printer. Without using wireless, this connection must be done with cables. So many new users have a hard time finding the right cables and plugging them into the right little holes (even though they are usually color coded) that most computer vendors offer the option of sending a technician to the user's home to do it. To help these users, some companies got together to design a short-range wireless network called Bluetooth to connect these components without wires. The idea is that if your devices have Bluetooth, then you need no cables. You just put them down, turn them on, and they work together. For many people, this ease of operation is a big plus. PANs can also be built with other technologies that communicate over short ranges, such as RFID on smartcards and library books.

2.2.2.2 Local Area Networks

Wireless LANs are very popular these days, especially in homes, older office buildings, cafeterias, and other places where it is too much trouble to install cables. A LAN is a privately owned network that operates within and nearby a single building like a home, office, or factory. LANs are widely used to connect personal computers and consumer electronics to let them share resources (e.g., printers) and exchange information. When LANs are used by companies, they are called enterprise networks.

In these systems, every computer has a radio modem and an antenna that it uses to communicate with other computers. In most cases, each computer talks to a device in the ceiling called AP (access point), wireless router, or base station, relays packets between wireless computers and also between them and the Internet. However, if other computers are close enough, they can communicate directly with one another in a peer-to-peer configuration.

There is a standard for wireless LANs called IEEE 802.11, popularly known as Wi-Fi, which has become very widespread. It runs at speeds anywhere from eleven to hundreds of megabytes per second. Wired LANs use a range of different transmission technologies. Most of them use copper wires, but some use optical fiber. LANs are restricted in size, which means that the worst-case transmission time is bounded and known in advance. Knowing these bounds helps with the task of designing network protocols. Typically, wired LANs run at speeds of 100 Mbps to 1 Gbps, have low delay (microseconds or nanoseconds), and make very few errors. Newer LANs can operate at up to 10 Gbps. Compared to wireless networks, wired LANs exceed them in all dimensions of performance. It is just easier to send signals over a wire or a fiber than through the air.

The topology of many wired LANs is built from point-to-point links. IEEE 802.3, popularly called Ethernet, is, by far, the most common type of wired LAN. In switched Ethernet, each computer speaks the Ethernet protocol and connects to a box called a switch with a point-to-point link, hence the name. A switch has multiple ports, each of which can connect to one computer. The job of the switch is to relay packets between computers that are attached to it, using the address in each packet to determine which computer to send it to. Switched Ethernet is a modern version of the original classic Ethernet design that broadcasts all the packets over a single linear cable. At most one machine could successfully transmit at a time, and a distributed arbitration mechanism was used to resolve conflicts. It used a simple algorithm: computers could transmit whenever the cable was idle. If two or more packets collided, each computer just waited a random time and tried later. To build larger LANs, switches can be plugged into each other using their ports.

Both wireless and wired broadcast networks can be divided into static and dynamic designs, depending on how the channel is allocated. A typical static allocation would be to divide time into discrete intervals and use a round-robin algorithm, allowing each machine to broadcast only when its time slot comes up. Static allocation wastes channel capacity when a machine has nothing to say during its allocated slot, so most systems attempt to allocate the channel dynamically (i.e., on demand). Dynamic allocation methods for a common channel are either centralized or decentralized. In the centralized channel allocation method, there is a single entity, for example, the base station in cellular networks, which determines who goes next. It might do this by accepting multiple packets and prioritizing them according to some internal algorithm. In the decentralized channel allocation method, there is no central entity; each machine must decide for itself whether to transmit. You might think that this approach would lead to chaos, but it does not. Later, we will study many algorithms designed to bring order out of the potential chaos.

2.2.2.3 Metropolitan Area Networks

A MAN (Metropolitan Area Network) covers a city. The best-known examples of MANs are the cable television networks available in many cities. These systems grew from earlier community antenna systems used in areas with poor over-the-air television reception. In those early systems, a large antenna was placed on top of a nearby hill and a signal was then piped to the subscribers' houses. Cable television is not the only MAN, though. Recent developments in high-speed wireless Internet access have resulted in another MAN, which has been standardized as IEEE 802.16 and is popularly known as WiMAX.

2.2.2.4 Wide Area Networks

A WAN (Wide Area Network) spans a large geographical area, often a country or continent. Each of these offices contains computers intended for running user (i.e., application) programs. The rest of the network that connects these hosts is then called the communication subnet, or just subnet for short. The job of the subnet is to carry messages from host to host. In most WANs, the subnet consists of two distinct components: transmission lines and switching elements. Transmission lines move bits between machines. They can be made of copper wire, optical fiber, or even radio links. Most companies do not have transmission lines lying about, so instead, they lease the lines from a telecommunications company. Switching elements, or just switches or routers, are specialized computers that connect two or more transmission lines. When data arrive on an incoming line, the switching element must choose an outgoing line on which to forward them.

The WAN, as we have described it, looks similar to a large wired LAN, but there are some important differences that go beyond long wires. Usually in a WAN, the hosts and subnet are owned and operated by different people. Second, the routers will usually connect different kinds of networking technology. Lastly, subnet(s) can be entire LANs themselves. This means that many WANs will in fact be Internetworks or composite networks that are made up of more than one network.

Rather than lease dedicated transmission lines, a company might connect its offices to the Internet. Virtual Private Network (VPN) allows connections to be made between the offices as virtual links that use the underlying capacity of the Internet. Compared to the dedicated arrangement, a VPN has the usual advantage of virtualization, which is that it provides flexible reuse of a resource (Internet connectivity), and the usual disadvantage of virtualization, which is a lack of control over the underlying resources.

2.2.3 Network Models

Now that we have discussed layered networks in the abstract, it is time to look at some examples. We will discuss two important network architectures: the OSI reference model and the TCP/IP reference model. Although the protocols associated with the OSI model are not used any more, the model itself is actually quite general and still valid, and the features discussed at each layer are still very important. The TCP/IP model has the opposite properties: the model itself is not of much use but the protocols are widely used.

2.2.3.1 OSI Reference Model

The OSI model (minus the physical medium) is shown in Figure 2.1. This model is based on a proposal developed by the International Standards Organization (ISO) as a first step toward international standardization of the protocols used in the various layers. The model is called the ISO OSI (Open Systems Interconnection) Reference Model because it deals with connecting open systems—that is, systems that are open for communication with other systems. We will just call it the OSI model for short.

The OSI model has seven layers. The principles that were applied to arrive at the seven layers can be briefly summarized as follows:

1. A layer should be created where a different abstraction is needed.
2. Each layer should perform a well-defined function.
3. The function of each layer should be chosen with an eye toward defining internationally standardized protocols.
4. The layer boundaries should be chosen to minimize the information flow across the interfaces.
5. The number of layers should be large enough that distinct functions need not be thrown together in the same layer out of necessity and small enough that the architecture does not become unwieldy.

We describe each layer of the model in turn, starting at the bottom layer. Note that the OSI model itself is not a network architecture because it does not specify the exact services and protocols to be used in each layer. It just tells what each layer should do. However, ISO has also produced standards for all the layers, although these are not part of the reference model itself. Each one has been published as a separate international standard. The model (in part) is widely used although the associated protocols have been long forgotten.

2.2.3.1.1 Physical Layer

The physical layer is concerned with transmitting raw bits over a communication channel. The design issues have to do with making sure that when one side sends a 1 bit, it is received by the other side as a 1 bit, not as a 0 bit. Typical questions here are what electrical signals should be used to represent a 1 and a 0, how many nanoseconds a bit lasts, whether transmission may proceed simultaneously in both directions, how the initial connection is established, how it is torn down when both sides are finished, how many pins the network connector has, and what each pin is used for. These design issues largely deal with mechanical, electrical, and timing interfaces, as well as the physical transmission medium, which lies below the physical layer.

2.2.3.1.2 Data Link Layer

The main task of the data link layer is to transform a raw transmission facility into a line that appears free of undetected transmission errors. It does so by masking the real errors so the network layer does not see them. It accomplishes this task by having the sender break up the input data into data frames (typically a few hundred or a few thousand bytes) and transmit the frames sequentially. If the service is reliable, the receiver confirms correct receipt of each frame by sending back an acknowledgement frame.

Another issue that arises in the data link layer (and most of the higher layers as well) is how to keep a fast transmitter from drowning a slow receiver in data. Some traffic regulation mechanism may be needed to let the transmitter know when the receiver can accept more data.

Broadcast networks have an additional issue in the data link layer: how to control access to the shared channel. A special sublayer of the data link layer, the medium access control sublayer, deals with this problem.

2.2.3.1.3 Network Layer

The network layer controls the operation of the subnet. A key design issue is determining how packets are routed from source to destination. Routes can be based on static tables that are *wired into* the network and rarely changed, or more often, they can be updated automatically to avoid failed components. They can also be determined at the start of each conversation, for example, a terminal session, such as a log-in to a remote machine. Finally, they can be highly dynamic, being determined anew for each packet to reflect the current network load. If too many packets are present in the subnet at the same time, they will get in one another's way, forming bottlenecks. Handling congestion is also a responsibility of the network layer, in conjunction with higher layers that adapt the load they place on the network. More generally, the quality of service provided (delay, transit time, jitter, etc.) is also a network layer issue.

When a packet has to travel from one network to another to get to its destination, many problems can arise. The addressing used by the second network may be different from that used by the first one. The second one may not accept the packet at all because it is too large. The protocols may differ, and so on. It is up to the network layer to overcome all these problems to allow heterogeneous networks to be interconnected. In broadcast networks, the routing problem is simple, so the network layer is often thin or even nonexistent.

2.2.3.1.4 Transport Layer

The basic function of the transport layer is to accept data from above it, split them up into smaller units if need be, pass these to the network layer, and ensure that the pieces all arrive correctly at the other end. Furthermore, all this must be done efficiently and in a way that isolates the upper layers from the inevitable changes in the hardware technology over the course of time. The transport layer also determines what type of service to provide to the session layer and, ultimately, to the users of the network. The most popular type of transport connection is an error-free point-to-point channel that delivers messages or bytes in the order in which they were sent. However, other possible kinds of transport service exist, such as the transporting of isolated messages with no guarantee about the order of delivery and the broadcasting of messages to multiple destinations. The type of service is determined when the connection is established.

 As an aside, an error-free channel is completely impossible to achieve; what people really mean by this term is that the error rate is low enough to ignore in practice.

The transport layer is a true end-to-end layer; it carries data all the way from the source to the destination. In other words, a program on the source machine carries on a conversation with a similar program on the destination machine, using the message headers and control messages. In the lower layers, each protocols is between a machine and its immediate neighbors, and not between the ultimate source and destination machines, which may be separated by many routers. The difference between layers 1–3, which are chained, and layers 4–7, which are end-to-end, is illustrated in Figure 2.1.

2.2.3.1.5 Session Layer

The session layer allows users on different machines to establish sessions between them. Sessions offer various services, including dialog control (keeping track of whose turn it is to transmit), token management (preventing two parties from attempting the same critical operation simultaneously), and synchronization (checkpointing long transmissions to allow them to pick up from where they left off in the event of a crash and subsequent recovery).

2.2.3.1.6 Presentation Layer

Unlike the lower layers, which are mostly concerned with moving bits around, the presentation layer is concerned with the syntax and semantics of the information transmitted. In order to make it possible for computers with different internal data representations to communicate, the data structures to be exchanged can be defined in an abstract way, along with a standard encoding to be used *on the wire*. The presentation layer manages these abstract data structures and allows higher-level data structures (e.g., banking records) to be defined and exchanged.

2.2.3.1.7 *Application Layer*

The application layer contains a variety of protocols that are commonly needed by users. One widely used application protocol is (HyperText Transfer Protocol), which is the basis for the WWW. When a browser wants a Web page, it sends the name of the page it wants to the server hosting the page using HTTP. The server then sends the page back. Other application protocols are used for file transfer, electronic mail, and network news.

2.2.3.2 TCP/IP Reference Model

Let us now turn from the OSI reference model to the reference model used in the grandparent of all wide area computer networks, the ARPANET, and its successor, the worldwide Internet. Although we will give a brief history of the ARPANET later, it is useful to mention a few key aspects of it now. The ARPANET was a research network sponsored by the DoD (U.S. Department of Defense). It eventually connected hundreds of universities and government installations, using leased telephone lines. When satellite and radio networks were added later, the existing protocols had trouble interworking with them, so a new reference architecture was needed. Thus, from nearly the beginning, the ability to connect multiple networks in a seamless way was one of the major design goals. This architecture later became known as the TCP/IP Reference Model, after its two primary protocols. Given the DoD's worry that some of its precious hosts, routers, and Internetwork gateways might get blown to pieces at a moment's notice by an attack from the Soviet Union, another major goal was that the network be able to survive loss of subnet hardware, without existing conversations being broken off. In other words, the DoD wanted connections to remain intact as long as the source and destination machines were functioning, even if some of the machines or transmission lines in between were suddenly put out of operation. Furthermore, since applications with divergent requirements were envisioned, ranging from transferring files to real-time speech transmission, a flexible architecture was needed.

2.2.3.2.1 *Link Layer*

All these requirements led to the choice of a packet-switching network based on a connectionless layer that runs across different networks. The lowest layer in the model, the link layer, describes what links such as serial lines and classic Ethernet must do to meet the needs of this connectionless Internet layer. It is not really a layer at all, in the normal sense of the term, but rather an interface between hosts and transmission links. Early material on the TCP/IP model has little to say about it.

2.2.3.2.2 *Internet Layer*

The Internet layer is the linchpin that holds the whole architecture together. Its job is to permit hosts to inject packets into any network and have them travel independently to the destination (potentially on a different network). They may even arrive in a completely different order than they were sent, in which case, it is the job of higher layers to rearrange them, if in-order delivery is desired. Note that *Internet* is used here in a generic sense, even though this layer is present in the Internet.

The analogy here is with the (snail) mail system. A person can drop a sequence of international letters into a mailbox in one country, and with a little luck, most of them will be delivered to the correct address in the destination country. The letters will probably travel through one or more international mail gateways along the way, but this is transparent to the users. Furthermore, that each country (i.e., each network) has its own stamps, preferred envelope sizes, and delivery rules is hidden from the users. The Internet layer defines an official packet format and protocol called IP, plus a companion protocol called ICMP (Internet Control Message Protocol), that helps it function. The job of the Internet layer is to deliver IP packets where they are supposed to go. Packet routing is clearly a major issue here, as is congestion (though IP has not proven effective at avoiding congestion).

2.2.3.2.3 Transport Layer

The layer above the Internet layer in the TCP/IP model is now usually called the transport layer. It is designed to allow peer entities on the source and destination hosts to carry on a conversation, just as in the OSI transport layer. Two end-to-end transport protocols have been defined here. The first one, TCP (Transmission Control Protocol), is a reliable connection-oriented protocol that allows a byte stream originating on one machine to be delivered without error on any other machine in the Internet. It segments the incoming byte stream into discrete messages and passes each one on to the Internet layer. At the destination, the receiving TCP process reassembles the received messages into the output stream. TCP also handles flow control to make sure a fast sender cannot swamp a slow receiver with more messages than it can handle.

The second protocol in this layer, UDP (User Datagram Protocol), is an unreliable, connectionless protocol for applications that do not want TCP's sequencing or flow control and wish to provide their own. It is also widely used for one-shot, client–server-type request–reply queries and applications in which prompt delivery is more important than accurate delivery, such as transmitting speech or video. Since the model was developed, IP has been implemented on many other networks.

2.2.3.2.4 Application Layer

The TCP/IP model does not have session or presentation layers. No need for them was perceived. Instead, applications simply include any session and presentation functions that they require. Experience with the OSI model has proven this view correct: these layers are of little use to most applications.

On top of the transport layer is the application layer. It contains all the higher-level protocols. The early ones included virtual terminal (TELNET), file transfer (FTP), and electronic mail (SMTP). Many other protocols have been added to these over the years. Some important ones include the Domain Name System (DNS), for mapping host names onto their network addresses; HTTP, the protocol for fetching pages on the WWW; and RTP, the protocol for delivering real-time media such as voice or movies.

OSI VERSUS TCP/IP REFERENCE MODEL

The OSI reference model was devised before the corresponding protocols were invented. This ordering meant that the model was not biased toward one particular set of protocols, a fact that made it quite general. The downside of this ordering was that the designers did not have much experience with the subject and did not have a good idea of which functionality to put in which layer. For example, the data link layer originally dealt only with point-to-point networks. When broadcast networks came around, a new sublayer had to be hacked into the model. Furthermore, no thought was given to Internetworking. With TCP/IP, the reverse was true: the protocols came first, and the model was really just a description of the existing protocols. There was no problem with the protocols fitting the model. They fit perfectly. The only trouble was that the model did not fit any other protocol stacks. Consequently, it was not especially useful for describing other non-TCP/IP networks.

2.3 Internet

The origins of the Internet can be traced to the U.S. government support of the ARPANET project. Computers in several U.S. universities were linked via packet switching, and this allowed messages to be sent between the universities that were part of the network. The use of ARPANET was limited initially to academia and to the U.S. military, and in the early years, there was little interest from industrial companies.

However, by the mid-1980s, there were over 2000 hosts on the TCP/IP-enabled network, and the ARPANET was becoming more heavily used and congested. It was decided to shut down the network by the late 1980s, and the National Science Foundation in the United States commenced work on the NSFNET. This work commenced in the mid-1980s, and the network consisted of multiple regional networks connected to a major backbone. The original links in NSFNET were 56 Kbps, but these were later updated to the faster T1 (1.544 Mbps) links. The NSFNET T1 backbone initially connected 13 sites, but this increased due to a growing interest from academic and industrial sites in the United States and from around the world. The NSF began to realize from the mid-1980s onward that the Internet had significant commercial potential.

The NSFNET backbone was upgraded with T1 links in 1988, and the Internet began to become more international. Sites in Canada and several European countries were connected to the Internet. DARPA formed the Computer Emergency Response Team (CERT) to deal with any emergency incidents arising during the operation of the network. Advanced Network Services (ANS) was founded in 1991. This was an independent not-for-profit company, and it installed a new network that replaced the NSFNET T1 network. The ANSNET backbone operated over T3 (45 Mbps) links, and it was different from previous networks such as ARPANET and NSFNET in that it was owned and operated by a private company rather than the U.S. government. The NSF decided to focus on research aspects of networks rather than the operational side. The ANSNET network was a move away from a core network such as NSFET to a distributive network architecture operated by commercial providers such as Sprint, MCI, and BBN. The network was connected by major network exchange points, termed Network Access Points (NAPs). There were over 160,000 hosts connected to the Internet by the late 1980s.

2.3.1 Internet Services

2.3.1.1 Electronic Mail (E-Mail)

Electronic mail, or e-mail, is the computerized version of writing a letter and mailing it at the local post office. Many people are so committed to using e-mail that if it were taken away tomorrow, some serious social and economic repercussions would be felt throughout the United States and the rest of the world. Many commercial e-mail programs are in existence, as well as a number of free ones that can be downloaded from the Internet. Although each e-mail program has its own unique feel and options, most offer the following services:

- Creating an e-mail message
- Sending an e-mail message to one recipient, multiple recipients, or a mailing list
- Receiving, storing, replying to, and forwarding e-mail messages
- Attaching a file, such as a word-processing document, a spreadsheet, an image, or a program, to an outgoing e-mail message

2.3.1.2 File Transfer Protocol (FTP)

The File Transfer Protocol, or FTP, was one of the first services offered on the Internet. Its primary functions are to allow a user to download a file from a remote site to the user's computer and to upload a file from the user's computer to a remote site. These files could contain data, such as numbers, text, or images, or executable computer programs. Although the WWW has become the major vehicle for retrieving text- and image-based documents, many organizations still find it useful to create an FTP repository of data and program files.

2.3.1.3 Remote Log-In (Telnet)

Remote log-in, or Telnet, is a terminal emulation program for TCP/IP networks, such as the Internet, that allows users to log in to a remote computer. The Telnet program runs on your computer and connects your workstation to a remote server on the Internet. Once you are connected to the remote server or host, you can enter commands through the Telnet program, and those commands will be executed as if you were entering them directly at the terminal of the remote computer.

2.3.1.4 Voice-Over-IP

One of the newer services that is attracting the interest of companies and home users alike is the sending of voice signals over an IP-based network such as the Internet. The practice of making telephone calls over the Internet has had a number of different names, including packet voice, voice-over packet, voice over the Internet, Internet telephony, and Voice-over-IP (VoIP). But it appears the industry has settled on the term *Voice over IP*, in reference to the Internet Protocol (IP), which controls the transfer of data over the Internet. Whatever its title, Voice over IP has emerged as one of the hottest Internet services and has certainly drawn the attention of many companies.

2.3.1.5 Listservs

A listserv is a popular software program used to create and manage Internet mailing lists. Listserv software maintains a table of e-mail addresses that reflects the current members of the listserv. When an individual sends an e-mail to the listserv address, the listserv sends a copy of this e-mail message to every e-mail address stored in the listserv table. Thus, every member of the listserv receives the e-mail message.

2.3.1.6 Streaming Audio and Video

Streaming audio and video involves the continuous download of a compressed audio or video file, which can then be heard or viewed on the user's workstation. Typical examples of streaming audio are popular and classical music, live radio broadcasts, and historical or archived lectures, music, and radio broadcasts. Typical examples of streaming video include prerecorded television shows and other video productions, lectures, and live video productions. Businesses can use streaming audio and video to provide training videos, product samples, and live feeds from corporate offices, to name a few examples.

2.3.1.7 Instant Messages, Tweets, and Blogs

Instant messaging (IM) allows a user to see if people are currently logged in on the network and, if they are, to send them short messages in real time. Many users, especially those in the corporate environment, are turning away from e-mail and using instant messaging as a means of communicating. The advantages of instant messaging include real-time conversations, server storage savings (because you are not storing and forwarding instant messages, as you would e-mails), and the capability to carry on a *silent* conversation between multiple parties. Service providers such as AOL, Microsoft's MSN, and Yahoo!, as well as a number of other software companies, incorporate instant messaging into their products.

2.4 World Wide Web

The WWW was invented by Tim Berners-Lee in 1990 at CERN in Geneva, Switzerland. CERN is a key European and international center for research in the nuclear field, and several thousand physicists and scientists work there. Berners-Lee first came to CERN in 1980 for a short contract programming assignment. He came from a strong scientific background as both his parents had been involved in the programming of the Mark I computer at Manchester University in the 1950s. He graduated in physics in the mid-1970s at Oxford University in England. Berners-Lee's invention of the WWW was a revolutionary milestone in computing. It has transformed the way that businesses operate as well as transforming the use of the Internet from mainly academic (with some commercial use) to an integral part of peoples' lives.

One of the problems that scientists at CERN faced was that of keeping track of people, computers, documents, databases, etc. This problem was more acute due to the international nature of CERN, as the center had many visiting scientists from overseas who spent several months there. It also had a large pool of permanent staff. Visiting scientists used to come and go, and in the late 1980s, there was no efficient and effective way to share information among scientists. It was often desirable for a visiting scientist to obtain information or data from the CERN computers. In other cases, the scientist wished to make results of their research available to CERN in an easy manner. Berners-Lee developed a program called *Enquire* to assist with information sharing, and the program also assisted in keeping track of the work of visiting scientists. He returned to CERN in the mid-1980s to work on other projects, and he devoted part of his free time to consider solutions to the information sharing problem. This was eventually to lead to his breakthrough and his invention of the WWW in 1990.

His vision and its subsequent realization was beneficial to both CERN and the wider world. He envisioned that all information stored on computers everywhere was linked and computers were programmed to create a space where everything could be linked to everything. Berners-Lee essentially created a system to give every *page* on a computer a standard address. This standard address is called the universal resource locator and is better known by its acronym URL. Each page is accessible via the hypertext transfer protocol (HTTP), and the page is formatted with the hypertext markup language (HTML). Each page is visible using a Web browser.

The characteristic features of the WWW are as follows:

1. *Universal Resource Identifier (later renamed to Universal Resource Locator [URL])*. This provides a unique address code for each Web page. Browsers decode the URL location to access the Web page. For example, www.amazon.com uniquely identifies the Amazon.com host website in the United States.

2. *Hyper Text Markup Language (HTML)* is used for designing the layout of Web pages. It allows the formatting of pages containing hypertext links. HTML is standardized and controlled by the World Wide Web Consortium (http://www.w3.org).

3. The *Hypertext Transport Protocol (HTTP)* allows a new Web page to be accessed from the current page.

4. A *browser* is a client program that allows a user to interact with the pages and information on the WWW. It uses the HTTP protocol to make requests of Web servers throughout the Internet on behalf of the browser user.

Inventors tend to be influenced by existing inventions and especially inventions that are relevant to their areas of expertise. The Internet was a key existing invention, and it allowed worldwide communication via electronic e-mail, the transfer of files electronically via FTP, and newsgroups that allowed users to make postings on various topics. Another key invention that was relevant to Berners-Lee was that of hypertext. This was invented by Ted Nelson in the 1960s, and it allows links to be present in text. For example, a document such as a book contains a table of contents, an index, and a bibliography. These are all links to material that is either within the book itself or external to the book. The reader of a book is able to follow the link to obtain the internal or external information. The other key invention that was relevant to Berners-Lee was that of the mouse. This was invented by Doug Engelbart in the 1960s, and it allowed the cursor to be steered around the screen.

The major leap that Berners-Lee made was essentially a marriage of the Internet, hypertext, and the mouse into what has become the World Wide Web. He was especially concerned with allowing communication across computers and software of different types. He also wanted to avoid the structure of most databases, which forced people to put information into categories before they knew if such classifications were appropriate or not. To these ends, he devised a Universal Resource Identifier (later called the Uniform Resource Locator or URL) that could point to any document (or any other type of resource) in the universe of information. In place of the File Transfer Protocol then in use, he created a more sophisticated Hypertext Transfer Protocol (HTTP), which was faster and had more features. Finally, he defined an Hypertext Markup Language (HTML) for the movement of hypertext across the network. Within a few years, these abbreviations, along with WWW for the World Wide Web itself, would be as common as RAM, K, or any other jargon in the computer field.

In order to create and display Web pages, some type of markup language is necessary. While there are many types of markup languages, we will briefly introduce three common types here: Hypertext Markup Language (HTML), dynamic Hypertext Markup Language (dynamic HTML), and eXtensible Markup Language (XML). HTML, D-HTML, and XML are members of a family of markup languages called Standard Generalized Markup Language (SGML). Despite the name, SGML itself is not a markup language, but a description of how to create a markup language. To put it another way, SGML is a metalanguage. Hypertext Markup Language (HTML) is a set of codes inserted into a document that is intended for display on a Web browser. The codes, or markup symbols, instruct the browser how to display a Web page's text, images, and other elements. The individual markup codes are often referred to as tags and are surrounded by brackets (< >). Most HTML tags consist of an opening tag, followed by one or more attributes, and a closing tag. Closing tags are preceded by a forward slash (/). Attributes are parameters that specify various qualities that an HTML tag can take on. For example, a common attribute is HREF, which specifies the URL of a file in an anchor tag (<A>).

2.4.1　Origin of the World Wide Web Browser

The invention of the WWW by Berners-Lee was a revolution in the use of the Internet. Users could now surf the Web: that is, hyperlink among the millions of computers in the world and obtain information easily. The WWW creates a space in which users can access information easily in any part of the world. This is done using only a Web browser and simple Web addresses. Browsers are used to connect to remote computers over the Internet and to request, retrieve, and display the Web pages on the local machine. The user can then click on hyperlinks on Web pages to access further relevant information that may be on an entirely different continent. Berners-Lee developed the first Web browser called the World-Wide Web browser. He also wrote the first browser program, and this allowed users to access Web pages throughout the world.

The early browsers included Gopher developed at the University of Minnesota and Mosaic developed at the University of Illinois. These were replaced in later years by Netscape, and the objective of its design was to create a graphical-user interface browser that would be easy to use and would gain widespread acceptance in the Internet community. Initially, the Netscape browser dominated the browser market, and this remained so until Microsoft developed its own browser called Internet Explorer. Microsoft's browser would eventually come to dominate the browser market, after what became known as the browser wars. The eventual dominance of Microsoft Internet explorer was controversial, and it was subject to legal investigations in the United States. The development of the graphical browsers led to the commercialization of the WWW.

The WWW got off to a slow start. Its distinctive feature, the ability to jump to different resources through hyperlinks, was of little use until there were at least a few other places besides CERN that supported it. Until editing software was written, users had to construct the links in a document by hand, a very tedious process. To view Web materials, one used a program called a *browser* (the term may have originated with Apple's HyperCard). Early Web browsers (including two called Lynx and Viola) presented screens that were similar to Gopher's, with a list of menu selections.

Around the fall of 1992, Marc Andreessen and Eric Bina began discussing ways of making it easier to navigate the Web. While still a student at the University of Illinois, Andreessen took a job programming for the National Center for Supercomputing Applications, a center set up with NSF that was also the impetus for the original ARPANET. By January 1993, Andreessen and Bina had written an early version of a browser they would later call Mosaic, and they released a version of it over the Internet. Mosaic married the ease of use of Hypercard with the full hypertext capabilities of the WWW. To select items, one used a mouse (thus circling back to Doug Engelbart, who invented it for that purpose). One knew an item had a hyperlink by its different color. A second feature of Mosaic, the one that most impressed the people who first used it, was its seamless integration of text and images. With the help of others at NCSA, Mosaic was rewritten to run on Windows-based machines and Macintoshes as well as workstations. As a product of a government-funded laboratory, Mosaic was made available free or for a nominal charge.

Andreessen managed to commercialize his invention quickly. In early 1994, he was approached by Jim Clark, the founder of Silicon Graphics, who suggested that they commercialize the invention. Andreessen agreed, but apparently the University of Illinois objected to this idea. Like the University of Pennsylvania a half-century before it, Illinois saw the value of the work done on its campus, but it failed to see the much greater value of the people who did that work. Clark left Silicon Graphics and with Andreessen founded Mosaic Communications that spring. The University of Illinois asserted its claim to the name Mosaic, so the company changed its name to Netscape Communications Corporation. Clark and Andreessen visited Champaign-Urbana and quickly hired many of the programmers who had worked on the software. Netscape introduced its version of the browser in September 1994. The University of Illinois continued to offer Mosaic, in a licensing agreement with another company, but Netscape's software quickly supplanted Mosaic as the most popular version of the program.

2.4.2 Applications of the World Wide Web

Berners-Lee used the analogy of the market economy to describe the commercial potential of the WWW. He realized that the WWW offered the potential to conduct business in cyberspace without human interaction, rather than the traditional way of buyers and sellers coming together to do business in the market place. Anyone can trade with anyone else except that they do not have to go to the market square to do so. The invention of the WWW was announced in August 1991, and the growth of the Web has been phenomenal since then. The growth has often been exponential, and exponential growth rate curves became a feature of newly formed Internet companies and their business plans.

The WWW is revolutionary in that

- No single organization is controlling the Web
- No single computer is controlling the Web
- Millions of computers are interconnected
- It is an enormous market place of millions (billions) of users
- The Web is not located in one physical location
- The Web is a space and not a physical thing

The WWW has been applied to many areas including

- Travel industry (booking flights, train tickets, and hotels)
- e-marketing
- Portal sites (such as Yahoo! and Hotmail)

- Ordering books and CDs over the Web (such as www.amazon.com)
- Recruitment services (such as www.jobserve.com)
- Internet banking
- Online casinos (for gambling)
- Newspapers and news channels
- Online shopping and shopping malls

Berners-Lee invented the well-known terms such as URL, HTML, and WWW, and these terms are ubiquitous today. Berners-Lee is now the director of the World Wide Web Consortium, and this MIT-based organization sets the software standards for the Web.

2.5 Semantic Web

While the Web keeps growing at an astounding pace, most Web pages are still designed for human consumption and cannot be processed by machines. Similarly, while Web search engines help retrieve Web pages, they do not offer support to interpret the results—for that, human intervention is still required. As the size of search results is often just too big for humans to interpret, finding relevant information on the Web is not as easy as we would desire. The existing Web has evolved as a medium for information exchange among people, rather than machines. As a consequence, the semantic content, that is, the meaning of the information on a Web page, is coded in a way that is accessible to human beings only. Today's Web may be defined as the *Syntactic Web*, where information presentation is carried out by computers, and the interpretation and identification of relevant information is delegated to human beings. With the volume of available digital data growing at an exponential rate, it is becoming virtually impossible for human beings to manage the complexity and volume of the available information. This phenomenon, often referred to as *information overload*, poses a serious threat to the continued usefulness of today's Web.

As the volume of Web resources grows exponentially, researchers from industry, government, and academia are now exploring the possibility of creating a Semantic Web in which meaning is made explicit, allowing machines to process and integrate Web resources intelligently. Biologists use a well-defined taxonomy, the Linnaean taxonomy, adopted and shared by most of the scientific community worldwide. Likewise, computer scientists are looking for a similar model to help structure Web content. In 2001, T. Berners-Lee, J. Hendler, and O. Lassila published a revolutionary article in *Scientific American* titled "The Semantic Web: A New Form of Web Content That Is Meaningful to Computers Will Unleash a Revolution of New Possibilities"*.

The semantic web is an extension of the current Web in which information is given well-defined meaning, enabling computers and people to work in cooperation. In the lower part of the architecture, we find three building blocks that can be used to encode text (Unicode), to identify resources on the Web (URIs) and to structure and exchange information (XML). Resource Description Framework (RDF) is a simple, yet powerful data model and language for describing Web resources. The SPARQL Protocol and RDF Query Language (SPARQL) is the de facto standard used to query RDF data. While RDF and the RDF Schema provide a model for representing Semantic Web data and for structuring semantic data using simple hierarchies of classes and properties, respectively, the SPARQL language and protocol provide the means to express queries and retrieve information from across diverse Semantic Web data sources. The need for a new language is motivated by the different data models and semantics at the level of XML and RDF, respectively.

Ontology is a formal, explicit specification of a shared conceptualization of a particular domain—concepts are the core elements of the conceptualization corresponding to entities of the domain being described, and properties and relations are used to describe interconnections between such concepts. Web Ontology Language (OWL) is the standard language for representing knowledge on the Web. This language was designed to be used by applications that need to process the content of information on the Web instead of just presenting information to human users. Using OWL, one can explicitly represent the meaning of terms in vocabularies and the relationships between those terms. The Rule Interchange Format (RIF) is the W3C Recommendation that defines a framework to exchange rule-based languages on the Web. Like OWL, RIF defines a set of languages covering various aspects of the rule layer of the Semantic Web.

2.6 Internet of Things

Just like the Internet and Web connecting humans, the Internet of Things (IoT) is a revolutionary way of architecting and implementing systems and services based on evolutionary changes. The Internet as we know it is transforming radically, from an academic network in the 1980s and early 1990s to

---

* Berners-Lee, T.; Lassila, O.; Hendler, J. (2001) The semantic web: A new form of Web content that is meaningful to computers will unleash a revolution of new possibilities. *Scientific American*, 284(5), pp. 34–43.

a mass-market, consumer-oriented network. Now, it is set to become fully pervasive, connected, interactive, and intelligent. Real-time communication is possible not only by humans but also by *things* at any time and from anywhere.

It is quite likely that sooner or later the majority of items connected to the Internet will not be humans, but things. IoT will primarily expand communication from the 7 billion people around the world to the estimated 50 to 70 billion machines. This would result in a world where everything is connected and can be accessed from anywhere—this has a potential of connecting 100 trillion things that are deemed to exist on Earth. With the advent of IoT, the physical world itself will become a connected information system. In the world of the IoT, sensors and actuators embedded in physical objects are linked through wired and wireless networks that connect the Internet. These information systems churn out huge volumes of data that flow to computers for analysis. When objects can both sense the environment and communicate, they become tools for understanding the complexity of the real world and responding to it swiftly.

 This would also mean significant opportunities for the telecom industry to develop new IoT subscribers that would easily overtake the number of current subscribers based on population.

Internet of Things (IoT) can be defined as the network formed by things or objects having identities and virtual personalities that interact using intelligent interfaces to connect and communicate with the users, social and environmental contexts. IoT is also referred as pervasive or ubiquitous computing systems. The goal of IoT is to achieve pervasive IoT connectivity and grand integration and to provide secure, fast, and personalized functionalities and services such as monitoring, sensing, tracking, locating, alerting, scheduling, controlling, protecting, logging, auditing, planning, maintenance, upgrading, data mining, trending, reporting, decision support, dashboard, back-office applications, and so on.

IoT would be closely associated with environmental, societal, and economic issues such as climate change, environment protection, energy saving, and globalization. For these reasons the IoT would be increasingly used in a large number of sectors like health care, energy and environment, safety and security, transportation, logistics, and manufacturing.

Major IoT applications in various sectors are as follows:

- Energy and Power: Supply/Alternatives/Demand. Turbines, generators, meters, substations, switches
- Healthcare: Care/Personal/Research. Medical devices, imaging, diagnostics, monitor, surgical equipment

- Buildings: Institutional/Commercial/Industrial/Home. HVAC, fire and safety, security, elevators, access control systems, lighting
- Industrial: Process Industries/Forming/Converting/Discrete Assembly/Distribution/Supply Chain. Pumps, valves, vessels, tanks, automation and control equipment, capital equipment, pipelines
- Retail: Stores/Hospitality/Services. Point-of-sale terminals, vending machines, RFID tags, scanners and registers, lighting and refrigeration systems
- Security and Infrastructure: Homeland Security/Emergency Services/National and Regional Defense. GPS systems, radar systems, environmental sensors, vehicles, weaponry, fencing
- Transportation: On-Road Vehicles/Off-Road Vehicles/ Nonvehicular/Transport Infrastructure. Commercial vehicles, airplanes, trains, ships, signage, tolls, RF tags, parking meters, surveillance cameras, tracking systems
- Information Technology and Network Infrastructure: Enterprise/ Data Centers. Switches, servers, storage
- Resources: Agriculture/Mining/Oil/Gas/Water. Mining equipment, drilling equipment, pipelines, agricultural equipment
- Consumer/Professional: Appliances/White Goods/Office Equipment/Home Electronics. M2M devices, gadgets, smartphones, tablet PCs, home gateways

In terms of the type of technological artifacts involved, the IoT applications can be subdivided into four categories:

a. The Internet of Devices: Machine-to-Machine (M2M)

M2M refers to technologies that allow both wireless and wired devices to communicate with each other or, in most cases, a centralized server. An M2M system uses devices (such as sensors or meters) to capture events (such as temperature or inventory level), which are relayed through a network (wireless, wired, or hybrid) to an application (software program) that translates the captured events into meaningful information. M2M communication is a relatively new business concept, born from the original telemetry technology, utilizing similar technologies but modern versions.

b. The Internet of Objects: Radio-frequency Identification (RFID)

RFID uses radio waves to transfer data from an electronic tag attached to an object to a central system through a reader for the purpose of identifying and tracking the object.

c. The Internet of Transducers: Wireless Sensor Networks (SNS)

SNS consists of spatially distributed autonomous sensors to monitor physical or environmental conditions, such as temperature, sound, vibration, pressure, motion, or pollutants, and to cooperatively pass their data through the network to a main location. The more modern networks are bidirectional, becoming wireless sensor and actuator networks (WSANs) enabling the control of sensor activities.

d. The Internet of Controllers: Supervisory Control and Data Acquisition (SCADA)

SCADA is an autonomous system based on closed-loop control theory or a smart system or a cyber physical system (CPS) that connects, monitors, and controls equipment via the network (mostly wired short-range networks, sometimes wireless or hybrid) in a facility such as a plant or a building.

2.7 Summary

This chapter describes the genesis of computer networks at ARPANET, the nature and type of networks, and the standard network models. It recounts briefly the history of invention of the Internet and the WWW, ending with some practical applications of the WWW.

3

Distributed Systems

3.1 Distributed Applications

Distributed applications consist of a collection of heterogeneous but fully autonomous components that can execute on different computers. While each of these components has full control over its constituent subparts, there is no master component that possesses control over all the components of a distributed system. Thus, for the system to appear as a single and integrated whole, the various components need to be able to interact with each other via predefined interfaces through a computer network.

The characteristic global features of a successful distributed application are as follows:

- Distributed systems are heterogeneous, arising from the need to, say, integrate components on a legacy IBM mainframe with the components newly created to operate on a UNIX workstation or Windows NT machine.

- Distributed systems are scalable in that when a component becomes overloaded with too many requests or users, another replica of the same component can be instantiated and added to the distributed system to share the load among them. Moreover, these instantiated components can be located closer to the local users and other interacting components to improve the performance of the overall distributed system.

- Distributed systems execute components concurrently in a multi-threaded mode via multiply invoked components corresponding to the number of simultaneously invoked processes.

- Distributed systems are fault tolerant in that they duplicate components on different computers so that if one computer fails, another can take over without affecting the availability of the overall system.

- Distributed systems are more resilient in that whereas distributed systems have multiple points of failure, the unaffected components are fully operational even though some of the components are not functional or are malfunctioning. Moreover, the distributed system could invoke another instance of the failed components along with the corresponding state of the process (characterized by the program counter, the register variable contents, and the state of the virtual memory used by the process) to continue with the process.

- Distributed systems demonstrate invariance or transparency with reference to characteristics like
  - Accessibility, either locally or across networks to the components
  - Physical location of the components
  - Migration of components from one host to another
  - Replication of components, including their states
  - Concurrency of components requesting services from shared components
  - Scalability in terms of the actual number of requests or users at any instance
  - Performance in terms of the number and type of available resources
  - Points of failure, be it failure of the component, network, or response

Performance transparency is achievable through the technique of load balancing that is based on the replication transparency. The middleware layer transparently decides on the balancing decision to select the replica with the least load to provide the requested service. Furthermore, the performance can also be prevented from degrading by continuously monitoring the patterns of access to components and migrating the components appropriately to minimize remote access.

3.1.1 N-Tier Application Architecture

In the 1980s, the prior monolithic architecture began to be replaced by the client/server architecture, which split applications into two pieces in an attempt to leverage new inexpensive desktop machines. Distributing the processing loads across many inexpensive clients allowed client/server applications to scale more linearly than single host/single process applications could, and the use of off-the-shelf software like relational database management systems (RDBMS) greatly reduced application development time. While the client could handle the user interface and data display tasks of the application, and the server could handle all the data management tasks, there was no clear solution for storing the logic corresponding to the business processes being automated. Consequently, the business logic tended to split between the client and the server; typically, the rules for displaying data became embedded inside the user interface, and the rules for integrating several different data sources became stored procedures inside the database. Whereas this division of logic made it difficult to reuse the user interface code with a different data source, it also made it equally difficult to use the logic stored in the database with a different front-end user interface (like ATM and mobile) without being required to redevelop the logic implemented in the earlier interface. Thus, a customer service system developed for a particular client system (like a 3270 terminal, a PC, or a workstation) would have great difficulty in providing telephony and Internet interfaces with the same business functionality.

The client/server architecture failed to recognize the importance of managing the business rules applicable to an enterprise independent of both the user interface and the storage and management of enterprise data. The three-tiered application architecture of the 1990s resolved this problem by subdividing the application into three distinct layers:

1. Data management, which stores and manages data independent of how they are processed and displayed by the other layers
2. Business logic, which implements the business logic to process data independent of how they are stored or displayed by the other two layers
3. Presentation, which formats and displays the data independent of the way they are interpreted/processed and stored by the other two layers

With the advent of the Internet in the past few years, the three tiers were split even further to accommodate the heterogeneity in terms of the user interfaces, processing systems, or databases existing in various parts of an enterprise.

 The power of the n-tier architecture derives from the fact that instead of treating components as integral parts of applications, components are treated as stand-alone entities, which can provide services for applications. Applications exist only as cooperating constellation of components, and each component in turn can simultaneously be part of many different applications.

3.1.1.1 N-Tier Architecture Advantage

The n-tier architecture has many advantages over the more traditional client/server architecture:

- *Agile software*: The n-tier architecture is useful in creating more flexible and easily modifiable software. By treating software components as stand-alone data providers, middleware service providers, business service providers, and service consumers, the n-tier architecture creates software infrastructure of reusable parts.

- *Maintainable software*: The n-tier architecture is useful in creating more maintainable and easily upgradeable software. Because software components are stand-alone reusable parts of business logic, they are used from the same place without the need for multiplication or replication and are therefore easier to change and upgrade, rendering the application as a whole more easily maintainable.

- *Reliable software*: The n-tier architecture is useful in creating more testable, more easily debuggable, and thus more reliable software. Flexible and maintainable software does not automatically imply reliable software, but because software components are stand-alone packets of business logic, bugs can be localized more easily and their functionality can be calibrated more accurately, rendering the application as a whole more reliable.

- *Reduced complexity*: The n-tier architecture is useful in creating more streamlined, simplified, and standardized software because the software component paradigm eliminates the need for custom interconnections between disparate constituents of a composite application (which includes existing and legacy systems) that increase in complexity rapidly with the increase in the number of disparate constituents. For instance, for a composite application constituted of n applications and m data sources, the problem of corresponding $n \times m$ interconnections is barely manageable even for small values of n and m. However, in the n-tier architecture, this problem is resolved to a great extent by interfacing all components to (say) a single standardized data bus—this reduces the problem of $m \times n$ interconnections to that of only $n + m$ interconnections! All components can connect with each other via connections to this singular data bus without the need for multiple customized single-purpose interconnections between each pair of components.

The interfacing approach of point-to-point interfaces between two applications would be prohibitively expensive for EAIs that may involve tens and hundreds of such interfaces. EAIs also adopt the alternate approach of instituting an information broker whereby all systems communicate with the information broker by uploading data into the same while simultaneously translating them into a single format and protocols native to this central broker. Because information is routed through the information broker, rather than going directly among different systems, this simplifies the problem considerably and it becomes easy to connect disparate systems via their respective adapters for this broker that uses the singular format and protocol of this central broker. Any future systems have to devise only one *adapter* to integrate with the central broker to start communicating transparently with all other systems. The exchange of data between the various systems interconnected by EAI is governed by the business rules determined by the user; and the message broker routes the messages according to these rules. However, the data in the messages are translated en route into whatever format is required by the concerned application.

- *Simplified systems management*: The n-tier architecture is useful in reducing the effort of systems management especially of the software on client machines, particularly for large enterprises that may have tens of thousands of client machines or even for enterprises that have multilocated and decentralized IT/IS operations. For instance, for conventional client/server systems, the plan for deploying a new version of any application would immediately run into a difficulty of choosing between
  - Changing the entire installed base of clients in a single massive effort during which the normal operations come to a complete standstill
  - Undergoing a long and expensive but more regulated phase-in of the new software, during which IT/IS is required to support multiple and mutually incompatible versions of the server and client codes, and above all, the new application is available to only a part of the target user base

N-tier architecture by reason of its software-on-demand paradigm reduces the need for physical updates of the client machines considerably; in many of these cases, the updates can be distributed through HTTP and a Web browser or through separately available automatic application distribution systems. But in the case of applications with zero footprint clients like SAP CRM, this need is completely eliminated. Moreover, as the majority of the code resides in the business logic, middleware, and data layers that are typically deployed on centralized, back-end servers supported by professional staff, the updates and enhancements to these layers are relatively painless.

3.1.1.2 Limits of the N-Tier Architecture

While n-tier architectures deliver all the advantages associated with distributed systems, they also have a downside. N-tier systems are workable only because of a network-based data bus for communications between the various tiers.

Such a communications layer will have the following adverse effects:

- Add to the latency of the system and degrade overall performance.
- Libraries of software components (and classes) required for interfacing with the data bus will typically increase the size of the application.

However, the system architects usually take these problems into account at the time of planning and designing the overall enterprise infrastructure and architecture. And, in the event that these problems become noticeable (because of dramatic increase in business and transaction volumes) or foreseeable (because of envisaged M&A activities, divestures, etc.), the enterprise architecture is revisited in its entirety.

3.1.2 Enterprise Component Architecture

Large corporations that have already invested vast sums of money in existing enterprise applications, infrastructure, hardware, and employees cannot change overnight. Moreover, the software skills and techniques that are essential for Web-based application development are different from those prevalent in most of the companies. Consequently, enterprises need to learn new development techniques, new methodologies, and new ways of fitting all these solutions together if their transition to e-commerce is to be successful. In this section, we will briefly examine why a successful enterprise should be based on enterprise component architecture as embodied in SAP CRM applications.

Flexible applications need to be constructed by assembling a variety of components (functional modules) to create newer products and services. Although the initial components will have to be developed from scratch, over time and with accumulation of basic components, applications will be assembled more and more from increasing reuse of the existing components. However, to accomplish this, the software needs to be designed as component-based applications, and there must be an infrastructure to support the component-oriented development of applications. An organizational structure that enables change is based on reusable elements that are reconfigurable in a scalable framework. In the componentized architecture, a system is considered as a group of components sharing a common interaction framework and serving a common purpose. A framework is a set of standards constraining and enabling the interactions of compatible system components, where each component is itself an autonomous system subunit with a self-contained identity, purpose, and capability and is capable of interacting with other components.

Component architecture then is a high-level description of the major components of a system and the relationship between them. Enterprise component architecture defines the enterprise's infrastructure for components as well as defines

- How compliant components are built
- How components are stored and cataloged
- How components are located and reused

3.1.3 Enterprise Component Model

A component model can be described as a set of application programming interfaces (APIs) and an architecture, which enables developers to define and create software components, which also have an ability to be connected dynamically together and interoperate with other components. Most components operate within a possible nested hierarchy of containers, that is, components.

 Typically, an architecture is a more abstract description, while a model is a more concrete description—similar to the distinction between classes and instances in object-oriented development environments (see Section 3.2.1.1, "Difference between Objects and Components").

Component models provide an array of mechanisms and services for interoperation of components:

- Self-discovery of components enables rendering of information on supported interfaces and methods to other components. This enables components to publish its capabilities as well as interact with other components dynamically.
- Properties of components determine the state, behavior, or appearance of the component.
- Customization of components enables the properties of a component to be set or modified from within a component or externally by another component/container.
- Persistence of components involves the process of saving and restoring a component's current state.
- Event control of components enables a component to create (or generate) an event for another component or to respond to an event. Predefining various kinds of events can also differentiate the occurrence of different kinds of events.

3.1.4 Distributed Application Requirements

As described earlier, an application is made up of three main constituents, namely, data, servers, and clients. A large-scale system is one that is capable of supporting simultaneously hundreds of thousands of components that run in a server, dozens or hundreds of servers, and tens of thousands of clients. The key to building large-scale applications is optimum resource utilization involving two aspects: .

1. Determining the critical resources in a system
2. Devising a mechanism not only to use the resources efficiently but also to enable selective addition of more resources as required

3.1.4.1 Resource Management

Enterprise Server manages a pool of application servers, mediates the coordination between them, and balances the workload among them. The modularity of the enterprise component architecture enhances scalability by adding, as required, more resources to the system configuration.

The application server itself uses several different resource pools to manage critical resources:

- Data connection links the server to the database; each server has one or more connections to the database. For applications to scale well, a server employs a database connection pool to reuse connections among the component/client rather than using one connection for each component/client.
- Components' process threads are effectively the flow of control corresponding to the one that the component is using while processing resources on the server. A component has two aspects— behavior as demonstrated via executing threads and the state of the business entity being processed. Each component uses several additional resources such as memory, database connections, and locks. The server apparently supports many components concurrently by maintaining a pool of threads that are shared among the many components that need to be executed. When needed, the server creates the instance of (or restores the state of) the component for use in any application; subsequently, the server saves the component's state, flushes the component from the server, and redeploys those resources for other components. This process of saving and restoring a component's state is termed *state management*.
- Client connection links the client to the application via a server. For each client, there is an associated connection on the server and the corresponding information regarding the client and connection is also stored on the server. For the application to support tens of thousands of clients, the number of such client connections in the pool should be minimized.

3.1.4.2 Application Management

Application management refers to the tasks necessary to set up the application and to keep the application running successfully and includes the following:

- *Configuration*: The ability to set up the system and assign resources, including specifications of the number of servers, the number of clients supported concurrently, assignment of servers to particular machines, and assignment of components to servers. For high availability, the configuration is changeable dynamically while the system is in operation.

- *Monitoring*: The ability to get statistics about the system's configuration and operation, including the number of connected users, the status of servers, resources utilization of servers, and throughput and response times. For high availability, tracing, and troubleshooting, systems automatically monitor information such as server status and also respond to detected errors or predefined events.

- *Controlling*: The ability to effect system configuration and operations, including the ability to start and stop servers or components and modify parameters.

3.1.4.3 Application Deployment

Application deployment relates to the issues of initial deployment, securing the application and upgrading it. The ideal deployment scenario involves no special software on the client at all, as is the case with browser-based thin clients.

The application architecture must support application versions and the ability for multiple versions of the same application to operate simultaneously. Security is also a critical requirement for the deployment of an application (see Sections 19.1.2 through 19.1.4).

3.2 Component-Based Technologies

In traditional procedural programming, the attributes (data) and the behavior (operations on data) are usually separated. Since there is no higher-level grouping (e.g., object) that binds the attributes and the behavior, a developer can never really know the entire result of a procedure call and therefore never knows exactly all the data that are affected by that procedure call. The detachment of the procedure from the data means that several unrelated procedures could change the same piece of data. This complicates the tasks of managing changes to data by procedures and testing to ensure that a new procedure does not adversely impact existing procedures. In the procedural programming world, developers have to understand the procedural code to know whether it is affecting the target data. If the developer does not perform detailed due diligence, it is likely the developer might develop a procedure that is very similar to one that already exists.

The advent of object-oriented programming represented the generational change in the IT arena. In the object-oriented world, data and the code that supports that data are encapsulated in an object. The use of an object as a higher-level grouping of data and behavior implies that there is less risk of similar methods or procedures that perform the same operations. It is easier to manage the data and access to the data because all operations on the same data are encapsulated in an object. The enhanced management of data and operations on data promotes higher code reuse. In the object-oriented world, the methods are contained in the object and they are easy for the developer to find and use. Furthermore, if an object does not satisfy the needs of the developer, he or she can create a new object from an existing object. The new object would inherit all of the attributes and methods of the existing object. The ability to inherit attributes and methods tremendously enhances code reuse. It makes maintenance easier because changes in the parent object propagate to the children.

In terms of design and modeling, object orientation is a more realistic representation of the real world than procedural programming. In real life, physical (e.g., car) and nonphysical (e.g., sales order) objects have attributes and behaviors. Attributes are changed by behavior based on some trigger (i.e., event). Objects can interact with each other such that one object's attribute might be changed by its behavior to another object's behavior. An example would be the customer placing a sales order. In this example, the customer object performs an action that results in a sales order object. The sales order object has attributes (i.e., customer name, address, credit worthiness) that come from the customer object through the interaction. Object-oriented analysis and design allow dynamic and interactive modeling of objects that is hard to achieve with structured design methodology commonly associated with procedural programming.

3.2.1 Advent of Component-Based Technologies

Components take code reuse to a higher level than objects. The concept behind components is to create pluggable objects to ease software development efforts. In component-based programming, there is a library of components the developer can use. In general terms, a component is usually a set of objects (though it could be procedural programs) that are self-contained and perform functions that are not specific to any context. The nonspecificity property of a component means it can be used by any application, even a future application, which needs the functions it contains. It is meant to be used in a plug and play fashion, where the developer could use a component to perform desired functions without having to worry about how the component works. To the developer, the component is a black box. The developer only needs to know what a component does and not the implementation details.

3.2.1.1 Difference between Objects and Components

Object and component are two distinct concepts. Objects are typically finer grained; an object usually involves one data entity, for instance, the purchase order item entity, and it makes visible all the implementation details for the behavior of that entity. In contrast to objects, a component is typically more coarse grained; a component is at a higher level of abstraction than an object. It is accessible to the outside only through well-defined interfaces. It is well encapsulated and cannot be used partially, that is, when an application uses a component, it consumes the entire component. The granularity of a component depends on the number of tasks it performs. A coarse-grained component performs multiple tasks, while a fine-grained component typically performs one task. An example of a coarse-grained component would be the purchasing component. The purchasing component could contain a purchase order, a purchase order item, and vendor objects. In this case, a call to a component that creates a purchase order would perform all the steps and updates to complete the purchase order without implementing and knowing the updates to the individual objects.

 Objects and components are two distinct concepts unrelated to each other; while objects do not have to be grouped into components to be useful, components can also be developed using a procedural programming language.

Objects use transparent or white box reuse. The source code of the objects is available to the developer. The developer can modify the source code of an object to achieve the desired effects. This implies there is weak control over the services an object might perform. Depending on the way an object is used, the internals of an object might affect its services. Thus, the services an object offers might change when the object has been modified. In contrast, components are black boxes to the developer because the software developer cannot see the implementation of the component, and the component cannot be modified to perform services outside of those described in the IDL. The behavior and properties of a component are specified by the IDL. If the developer invokes an interface as described by the IDL, the specified behavior from the component will be obtained. Interfaces of a component are like a contract. They ensure that the component will perform according to the IDL, regardless of how often it has been modified.

The biggest distinction between a component and an object is that an object is not compatible with another object developed using different programming languages. A programmer developing in Visual Basic could not use an object created using C++. For a company like Microsoft, which supports multiple programming languages, the incompatibility of objects built using different programming language presents a problem in maintaining object libraries. These libraries contain objects for commonly used services and functions that help programmers in their application development. Furthermore, the incompatibility of objects developed using different programming languages presents a barrier to reuse for enterprises that use multiple programming languages. Development efforts have to be spent recreating an object developed in one programming language if that object is needed for an application developed using another programming language. In contrast, component standards were developed specifically to address the issue that pieces of code developed using different languages cannot be made to interoperate with one another. Software programs that adhere to a component model will be able to interact with other components developed for the same model. The major component models are Common Object Model (COM) and its derivatives, such as COM+ and Microsoft XML Web Services platform (.NET), Enterprise JavaBeans (EJB), and Common Object Request Broker Architecture (CORBA).

The basic idea behind this bird's-eye view of distributed computing is that the concept of an object has moved from being a technique for determining and building the components of an application to a method for describing, encapsulating, assembling, and managing entire applications in a distributed computing environment. The original concept of an object could have been used to define a Binary Tree Abstract Data Type (ADT) and its associated methods and how it would perform in a given application. The distributed computing paradigm view of this object incorporates all of the original definitions as well as requiring the interfaces and techniques by which the Binary Tree ADT can have its methods and interfaces managed remotely, whether it be via a JavaBean, an Object Request Broker (ORB), or a COM component.

3.2.1.2 Case for Distributed Objects and Components

Distributed object computing is an architecture, a paradigm, and a set of technologies whereby objects can be distributed across a heterogeneous network in such a manner that arbitrary assemblies of these objects can operate as a single entity. Effectively, distributed objects extend the object-oriented analysis and design methodology to the network as a whole. The concept of an object refers to the nonincorporated software entity that is capable of being merged with other objects to form an assembly. Once integrated into the assembly, the object becomes a component.

The movement to distributed computing is because of five reasons:

1. Protection of expended investment
2. Gains in productivity
3. Protection of future investment
4. Enhanced utilization of assets
5. Effective interdepartmental data integration

The gains from realizing an effective means of creating, assembling, and managing a distributed set of objects are many. Extant software (i.e., legacy systems) can nonintrusively be incorporated into greater entities, thereby protecting existing investment. Order-of-magnitude gains in productivity can be realized as the developer moves away from the line-at-a-time paradigm and into the assembly, or industrial era of software development. For those objects that were created using the techniques of the earlier era, the possibilities of reuse are manifold, thus enabling investments future proof and optimizing future investment. Efficiencies in deployment and execution are also realized because the individual objects comprising an assembly (i.e., component) can be hosted on those machines that can best be utilized by that particular component. Finally, the various *islands* of technology can be bridged effectively and selectively for optimal efficiency and effectiveness. Thus, the accounts receivable, inventory, order, and accounts payable systems can be merged at the data flow level without replacement or costly rework of the individual applications.

3.2.2 Distributed Computing in the Enterprise

There are three primary approaches for realizing the distributed computing paradigm: CORBA, COM, and RMI. Before we discuss each of them in detail, here is an historical overview of their development.

Once networking became prevalent across academia and industry, it became necessary to share data and resources. In the early years of distributed computing, message passing (e.g., using sockets developed in the early 1980s) was the prevailing method of communication. This involved encoding the data into a message format and sending the encoded data over the wire. The socket interface allowed message passing using send and receive primitives on transmission control protocol (TCP) or user datagram protocol (UDP) transport protocols for low-level messaging over Internet protocol (IP) networks. Applications communicated by sending and receiving text messages. In most cases, the messages exchanged conformed to an application-level protocol defined by programmers. This worked well but was cumbersome due to the fact that the data had to be encoded and then decoded; also programmers developing a distributed application must have knowledge of what the others were doing to the data.

Programmers had to spend a significant amount of time specifying a messaging protocol and mapping the various data structures to and from the common transmission format. As the development of distributed computing applications increased, new mechanisms and approaches became necessary to facilitate the construction of distributed applications. The first distributed computing technology to gain widespread use was remote procedure call (RPC) developed in the 1980s by Sun Microsystems. RPC uses the client/server model and extends the capabilities of traditional procedure calls across a network. Remote procedure calls are designed to be similar to making local procedure calls. While in a traditional local procedure call paradigm the code segments of an application and the procedure it calls are in the same address space, in a remote procedure call the called procedure runs in another process and address space across the network on another processor.

RPC proved to be an adequate solution for the development of two-tier client/server architectures. As distributed computing became more widespread, the need to develop, for example, n-tier applications emerged and RPC could not provide the flexibility and functionality required. With such applications, multiple machines may need to operate simultaneously on the same set of data, and, hence, the state of that data became of great concern. Research in the area of distributed objects allowed overcoming this problem with the specification of two competing technologies: common object request broker architecture (CORBA) and distributed common object model (DCOM). Later, Java remote method invocation (RMI) was developed and also became a competitor.

The CORBA standard was developed by the Object Management Group (OMG) starting in the 1990s and defines an architecture that specifies interoperability between distributed objects on a network. With CORBA, distributed objects can communicate regardless of the operating system they are running on (e.g., Linux, Solaris, Microsoft Windows, or Mac OS). Another primary feature of CORBA is its interoperability between various programming languages. Distributed objects can be written in various languages (such as Java, C++, C, Ada, etc.). The main component of CORBA is the ORB (object request broker). Objects residing in a client make remote requests using an interface to the ORB running on the local machine. The local ORB sends the request to the remote ORB, which locates the appropriate object residing in a server and passes back an object reference to the requester. An object residing in a client can then make the remote method invocation of a remote object. When this happens the ORB marshals the arguments and sends the invocation over the network to the remote objects ORB that invokes the method locally and sends the results back to the client.

DCOM, developed by Microsoft, is a protocol that enables communication between two applications running on distributed computers in a reliable, secure, and efficient manner. DCOM is an extension of the Component Object Model (COM). COM is an object-based programming model and defines how components and their clients interact. COM allows the development of software components using a variety of languages and platforms to be easily deployed and integrated. The distributed COM protocol extends the programming model introduced by COM to work across the network by using proxies and stubs. Proxies and stubs allow remote objects to appear to be in the same address space as the requesting object. When a client instantiates a component that resides outside its address space, DCOM creates a proxy to marshal method calls and route them across the network. On the server side, DCOM creates a stub, which unmarshals method calls and routes them to an instance of the component.

Java RMI is a package for writing and executing distributed Java programs by facilitating object method calls between different Java Virtual Machines (JVM) across a network. Java RMI hides most of the aspects of the distribution and provides a conceptually uniform way by which local and distributed objects can be accessed. An RMI application consists of a server interface, a server implementation, a server skeleton, a client stub, and a client implementation. The server implementation creates remote objects that conform to the server interface. These objects are available for method invocation to clients. When a client wishes to make a remote method invocation it invokes a method on the local stub, which is responsible for carrying out the method call on the remote object. The stub acts as a local proxy. A server skeleton exists for each remote object and is responsible for handling incoming invocations from clients.

CORBA, DCOM, and Java RMI enjoyed considerable success, but they were beset by shortcomings and limitations when used in Web environments. For example,

- They tend to create tightly coupled distributed systems
- Some are vendor and platform specific (e.g., COM/DCOM only runs on Windows)
- The distributed systems developed run on closely administered environments
- Some use complex and proprietary protocols, and specific message formats and data representation

With the growth of the Web, the search soon started for a Web compliant replacement for this technology. In Chapter 8, we will see that Web Services are currently the most natural solution to develop distributed systems on the Web.

3.2.2.1 Component Object Request Broker Architecture (CORBA)

In 1989, a group of eight companies formed the Object Management Group (OMG) to promote the use of object technology and create standards for object interoperability. OMG introduced CORBA 1.0 in 1991 as the first vendor-independent object standard. What CORBA 1.0 brought was the IDL and a set of application programming interfaces (API) that allow objects to request and receive services from other objects. CORBA is programming language independent and platform neutral. As long as vendors develop programs following CORBA standards, their programs can interoperate with other CORBA-compliant programs. The platform neutral feature of CORBA means that CORBA-compliant programs can be executed by any platform with CORBA middleware. The CORBA standards have undergone several revisions to include a component model, support for transactions, a bridge to the component models (EJB and COM), and support for messaging service (such as Message Queuing [JMS]). Using CORBA standards, vendors have introduced middleware products to support CORBA components.

The central building blocks of CORBA are the object request broker or (ORB), IDL, dynamic invocation interface (DII), interface repository, and object adopter (OA). The ORB is the heart of the CORBA architecture. It serves as the middleware for object-based integration, similar to what RPC does for procedural programs. A client application can invoke a method through the ORB without worrying about the system platform, network connectivity, or object implementation details. A method in the object-oriented world is equivalent to a procedure in the procedural world. Methods are functions that an object has exposed to the outside world. In the component world, an object method could become a component interface. The client application sends a request to the ORB. The ORB delivers the request to the requested object, whether residing in a remote server or on the same server. Once the request has been processed, the ORB returns the results to the calling application. The ORB hides all of these concerns from the client application. To the client application, the method it is invoking appears to be implemented in the same platform and using the same programming language, even if the method might belong to a program executed on a different platform and constructed in a procedural language. The ORB abstracts all of these details to put the client and the invoked object on the same playing field. Every implementation of a CORBA ORB supports an interface repository. The client simply has to invoke an object using the interface definition stored in the interface repository.

CORBA, with its vendor-neutral position and set of *open* standards, has always provided the most hope for a truly unified way to incorporate any software object into a distributed assembly. However, the very *openness* of CORBA and its associated loose specifications has contrived to be the cause of its undoing. The CORBA products from most vendors—while compliant with the standards—in fact differ somewhat from one another and hence rely on vendor-specific protocols. This means that the key issue of standardization is illusory. The lack of a tight specification, while appealing to potential vendors in that it allows them to build their *own* CORBA implementation, means that the third-party market for CORBA-based applications does not exist because there is no standard.

3.2.2.1.1 How CORBA Works

CORBA uses an ORB for implementing its interobject invocations. The latest specification for the ORB specifies the Internet Inter-ORB Protocol (IIOP) as the protocol by which objects are *remoted*. The ORB acts much like a bus or backplane in a hardware device through which each CORBA object interacts with other CORBA objects.

For a CORBA client object to request service from a CORBA server object, the client must acquire a reference to the server component. The ORB will parse the available services (methods) of the server object and connect them with the request from the client object. The available methods are accessed by the ORB from Interface Definition Language (IDL) skeletons programmatically created for each server object. These IDL stubs are effectively the distributed object's interface made available for reading, distribution, and connection by the ORB. The IDL compiler provides type and method exposition information for each skeleton that is then stored in an Interface Repository. Effectively, the ORB is a message router and object invocation device that relies on IDL stubs and skeletons to resolve the various service requests.

The server-side skeleton will receive the invocation from the ORB and execute the requested method. Results from the method, as well as input arguments from the client proxy, will be routed via IIOP to the ORB that is responsible for providing them to the server stub. Hence, CORBA requires multiple ORBs if a multiplatform solution is being deployed. The essence of CORBA is the ORBs used to connect the various distributed objects. With the advent of IIOP, some of the vendor specificity problems have been eased, allowing ORBs from a variety of sources to interact successfully.

3.2.2.2 Microsoft Component Technologies

Microsoft developed the COM in the mid-1990s. COM provides a set of specifications for creating a COM component. A COM component can be created in several languages, including non-object-oriented languages, which are supported by Microsoft. Once the COM component is created, it is stored in binary code so programs developed using different programming languages can use it. COM is a component standard that provides black box reuse. A developer can access a COM component through interfaces provided by the component. Similar to CORBA, these interfaces are described using COM Interface Description Language (COM IDL). An interface is immutable in the sense that it is a contract for the service it provides. When a client application invokes an interface, it is guaranteed that the interface will always work the way it is defined in the COM IDL. A new interface has to be created if there are to be changes to an existing interface. As in CORBA interface, the immutability property is one of the core principles of component standards.

In COM, there is a Microsoft Interface Definition Language (MIDL) compiler that generates a stub and proxy objects from COM IDL for interaction with other components. A proxy object is a program of a target component that is local to the client application. If the client wants to invoke a method of a target component, the client sends the call to the proxy of the target component. The proxy relays the request to the stub object of the target component. The job of the stub is to receive the remote call on behalf of the target component and send the request to the target component for processing. When the request has been completed, the response is returned by the target component to the stub object, which relays the response to the client application via the proxy object. In comparison to CORBA, integration using COM does not require a broker. Similar to RPC, the client application communicates directly with the target component through proxy and stub.

To enable transparent communication of components across the network, Microsoft created Distributed COM (DCOM). DCOM does not alter how COM operates. It simply provides a mechanism for the client application to communicate with a remote target component using the network. When the client application and the target component reside on the same machine, there is no need to use DCOM. The request can be sent by COM through interprocess communication. The main benefit DCOM provides to the developer is location independence. When making the call, the developer does not have to worry about whether the request from the client application is local or remote. DCOM automatically handles the communication. To accomplish remote communication, DCOM uses Object Remote Procedure Call (ORPC). This protocol is similar to IIOP that CORBA uses to communicate between ORBs. The combination of COM and DCOM is a competing component model to CORBA. Microsoft has implemented COM on its Windows platform and Apple's Macintosh platform. COM has also been implemented on specific versions of UNIX platforms.

The COM technology has been developed further to COM+, .NET, and .NET Web Service.

3.2.2.2.1 How DCOM Works

DCOM is sometimes referred to as COM on the wire. DCOM works similarly to CORBA in that a client-side DCOM object creates a message and invokes a wire protocol similar to IIOP called ORPC (Object Remote Procedure Call) to communicate with the server DCOM object. Instead of an ORB, DCOM relies on a service control manager (SCM) to perform the various services of locating and activating an object implementation. As in CORBA, the server is responsible for invoking the method requested via interaction with the SCM. Once the client-side device has received a reference to the server-side object, it can access the exposed methods of the server.

In DCOM, the client side is called the proxy, while the server side is referenced as the stub. These stubs and proxies use an Interface Definition Language similar in purpose to that of the CORBA IDL. Instead of maintaining and using an *interface repository* as CORBA does, DCOM avails itself of the Microsoft registry services, thereby enhancing flexibility at the cost of increased complexity. Interestingly, DCOM server components can be built in a number of languages, including C++, Java, and COBOL.

3.2.2.3 Java Component Technologies

In 1995, Sun Microsystems introduced the Java development technology. The impetus behind Java was to create a development technology that could be run on any platform. Platform neutrality has always been the challenge software vendors faced. Prior to Java, software vendors had to develop several versions of their products for the different platforms their customers were using. The duplicated development efforts were a huge cost to the vendors. Java's motto is "Write Once, Run Anywhere." Instead of writing several versions of their programs, applications written in Java can be run on any platform that supports a JVM.

The main elements of J2EE technology are components, container services, and Web Services. J2EE offers two different component models, Servlet or Java Server Page (JSP) for the Web component and EJB for the server component. Servlet and JSP are primary J2EE component technology for Web development, while EJB is the component model for application logic. JSP is a text-based document that contains static content (i.e., images, text) and dynamic data. The JSP static data could be expressed in HyperText Markup Language (HTML) or XML code, while the dynamic data are controlled by Java code. JSP is an extension of Servlet. When JSP is being run, the JSP code is translated into a Java Servlet. Both Servlet and EJB operate in a separate container; Servlet operates in a Web container and EJB operates in an EJB container. A container is a runtime environment that hosts and manages a component. The container provides services that allow the component to operate, including transaction, messaging, remote access, security, and other services. The container will automatically use a container service during runtime depending on policies specified by the developer during design time. For instance, a policy might be to use messaging for remote communication rather than RPC.

The main difference between the Servlet and EJB component is that the Servlet specializes in communicating with Web browser. Thus, it takes requests using HTTP. An EJB communicates with other Java components using the Remote Method Invocation over Internet Inter–ORB Protocol (RMI–IIOP). According to J2EE 1.4 specification, the J2EE applications are required to access EJB through RMI–IIOP. Servlet can respond to HTTP commands, such as HTTP–POST and HTTP–GET. Following the J2EE specification, EJB should not interact directly with the Web browser. The one Web interaction allowed for EJB under J2EE specification is SOAP/HTTP protocol for Web Services. Because of the differences in the container, Servlet resides on the Web server that has the Servlet container, while EJB resides on the application server that has the EJB container. Other than that, both the Web and EJB containers provide similar services.

> Because of the similarities of these two component models, some software vendors developed their products purely using Servlet technology. The advantage of this approach is their products can be run on any Web server that supports a Servlet container. This minimizes the system footprint necessary to run their products. Web servers are originally designed to serve static content that does not require high-performance and distributed computing features. However, most Web servers that support Servlet containers are now offering high-performance features such as failover, clustering, and load balancing. Increasingly, the main difference between the Web server (that supports Servlet container) and the application server (that supports EJB container) is the support for integration with other systems and platforms.

In traditional Java programming paradigm, the Web components and EJB play distinct and complementary roles. How should the Web components interact with EJB components? JSP, Servlet, and EJB are created to work in conjunction with the Model–View–Controller (MVC) programming model. The MVC design paradigm separates an interactive application into three modules: model, view, and controller. The model module is the application logic and data. It is the heart of the application processing and corresponds to the application and data layer in the traditional three-layer application architecture. The view component is the presentation to the user. Typically, in a Web program, this component would contain the Web pages that are displayed to the users. The controller component is the link between the model and the view; it dispatches the requests from the views to the model and mediates the application flow. The controller functions similar to the ORB—both the ORB and the controller take requests from the client, dispatch the requests to the receiver, and return the results to the client. Because of the interactive nature of Web applications, following a request, the controller performs the additional task of selecting and synthesizing the view to display to the user. In this manner, the controller can be perceived as a specialized view that can display dynamic views.

According to the MVC model, the controller is typically performed by the controller Servlet. This Servlet takes the request from the user and invokes the appropriate model component(s). Once the request has been processed, the controller Servlet selects the appropriate HTML or JSP to display to the user. As mentioned earlier, JSP is translated into Servlet when it is run; JSP and Servlet usually work together. Therefore, the controller Servlet does not render the view per se; it sends the request to one or more objects to render the view to the user. As per the MVC model, EJB is an ideal candidate to serve that role.

Thus, in a typical flow, a request could be generated by the user from the Web browser. The controller Servlet takes the request and invokes the appropriate EJB component interface. When the result of the EJB component call has been received, the controller Servlet assembles the view to display to the user; the view is then rendered by JSP.

3.2.2.3.1 *How RMI Works*

Effectively, RMI uses Java language extensions to extend the Java Virtual Machine (JVM) address space so that it appears to include other virtual machines independent of where they might actually be hosted. RMI is effectively a JVM-to-JVM communication protocol allowing objects to be *passed*. Unlike IDL-based distributed models, RMI requires no mapping to common interface definition languages (IDLs). The syntax of RMI is such that it appears almost identical to that used for local invocations. Because the interobject communication relies on the JVM executable, objects can be distributed dynamically, thereby removing the requirement to provide installation on the client prior to implementation. This, of course, greatly eases the burden of distribution and maintenance. In addition, RMI benefits from the built-in security features of the JVM, thereby guaranteeing a secure distributed object environment.

JavaBean is effectively a technique and an instantiation of a software component model for Java. The JavaBean technique relies on the *serialization* of an object provided by the JVM and language constructs. Serialization allows the state of an object (and any objects it refers to) to be written to an output stream. Later, the serialized object can be recreated by reading from an input stream. This technique is used to transfer objects between a client and a server for RMI. Many Beans are provided not as class files but rather as preinitialized, serialized objects. JavaBean was designed for visual manipulation in a builder tool, much like an ActiveX component. JavaBeans are not an architecture for distributed computing but rather a technique

for building and distributing the components of a distributed computing environment. Effectively, the Bean technology provides a way
through Java *introspection* and *reflection* whereby a builder/developer
can determine the properties, methods, and events of a JavaBean.

3.3 Summary

This chapter looked into the characteristics of distributed systems, the
nature of n-tier architecture, and the enterprise component architecture.
The latter part of the chapter detailed the requirements of distributed systems that act as a reference for the remaining part of the book. The last part
discussed the three main component standards (CORBA, Microsoft, and
Java). Java and Microsoft are the most widely used component development
technologies.

4

Enterprise Application Integration (EAI)

In these times of market change and turbulence, enterprises are confronted with the increasing need to interconnect disparate systems to satisfy the need of the business. It is estimated that 35%–60% of an organization's IT resources are spent on integration. Enterprise application integration is the creation, maintenance, and enhancement of leading-edge competitive functionality of the enterprise's business solutions by combining the functionality of the existing legacy applications, commercial off-the-shelf (COTS) software packages, and newly developed custom applications via a common middleware.

4.1 Enterprise Applications

Enterprise applications are software applications developed to manage the business operations, assets, and resources of an enterprise. Their development process integrates the work of at least four groups, namely, GUI developers responsible for the design and development of widgets to ease human computer interaction; application programmers focusing on coding the business logic for the solution of a particular business problem; database managers building data models to structure and manage data storage, access, security, and consistency; and finally, application integrators for integrating existing applications and available technologies with the new applications.

In principle, there is no difference between enterprise applications and regular software applications other than the specific business purpose they are developed for. As the nature of business goals and processes vary, software solutions delivered for specific business problems vary as well. As a consequence, the number and the variety of applications delivered for each solution increase the complexity of managing the overall IT system. While having an automated solution to business problems increases effectiveness and efficiency and reduces cost, managing the complexity of the automation solution is a new business problem that companies have to deal with.

A high-level blueprint of a standard application template for a company can reduce that complexity. In response to this need, the design characteristics, limitations, interfaces, and rules of developing enterprise applications have been documented. This high-level description, the blueprint, of how an application should be developed to satisfy the business goals is known as Enterprise Application Architecture. This architecture defines an organizing structure for software application elements and the resources, their relationships and roles in an organization. Enterprise applications are usually developed independent of each other, and each of these applications manages their own data in their specific database system. This leads to data heterogeneity and inefficiency because the same data elements are stored multiple times in different databases. This creates the problem of managing the same logical data object stored in multiple data stores. Differences in data structures as well as in semantics are also possible. One of the challenges facing enterprises today is the task of integrating all these applications within the organization, even though they may use different operating systems and employ a variety of database solutions. Simplistic approaches soon become unmanageable as the number of applications to be integrated increases. Enterprise Application Integration (EAI) has the task of making independently developed applications that may also be geographically dispersed and may run on multiple platforms to work together in unison with the goal of enabling unrestricted sharing of data and business processes.

In order to accomplish this goal, middleware vendors provide solutions to transform, transport, and route the data among various enterprise applications. EAI faces significantly more management challenges than technical challenges, and its implementation is time consuming and needs substantial resources, particularly in upfront design. Among the software applications for managing company assets and resources, the most commonly used are Enterprise Resource Planning (ERP), Customer Relationship Management (CRM), Supply Chain Management (SCM), Business Intelligence Applications, and Human Resource (HR) Applications. ERP is, probably, the most general class of enterprise software that attempts to integrate all departments and functions across a company. ERP incorporates many different families of more specific enterprise applications. CRM solutions focus on strategies, processes, people, and technologies used by companies to successfully attract and retain customers for maximizing profitability, revenue, and customer satisfaction. Enterprise Content Management solutions provide technologies, tools, and methods used to capture, manage, store, preserve, and deliver content (document, voice and video recordings, etc.) related to organizational processes across an enterprise. SCM solutions focus on the process of planning, implementing, and controlling the operations of the supply chain, which includes the flow of materials, information, and finances as they move in a process from supplier to manufacturer, to wholesaler, to retailer, and to consumer. HR Management solutions provide a coherent approach to the recruitment and management of people working in organizations.

4.1.1 Management of Enterprise Applications

4.1.1.1 Manageability

Enterprise application systems are confronted by ever-changing requirements and environments and should be able to adapt to these changes dynamically. Consequently, better solutions have the following features:

- A high degree of adaptability and extendibility of the whole system built into the design
- Support for mass deployment
- Support for business-driven configuration scenarios
- A configurable security management system based on users, roles, and access control lists with distinct and configurable privileges
- Tracing capability for user actions

Another aspect of application complexity is the ability to manage software components that constitute the application. With the shrinking average lifetime of applications due to the increasingly competitive markets, incorporating updates to provide newer or latest features over time becomes an obstacle for managing software components. The problem is not new but became increasingly important with the growing number of two-tier applications at the end of the 1990s. The applications that are often built as rich clients and capable of directly accessing back-end databases are prime candidates for version conflicts and data integrity frauds, if the software update is not planned and enforced perfectly. All clients have to be updated before they attempt to interact with a new version of the back-end system. If this is not possible, a proper mitigation solution has to be deployed. Such a solution may sound promising upfront but can become very difficult to manage over time. Releasing a mitigation solution for every version deployed soon becomes a maintenance nightmare and may actually hinder future development, because every iteration of the software has to be aware of all previous combinations. The root cause of this problem is that client software has to keep track of the changes to remain up-to-date. This problem is avoided in three-tier applications as only the application deployed on the server needs to be updated, which is also the major reason for the success of typical three-tier Web applications.

Depending on the usage scenario, the necessary downtimes for updating software have to be minimized and made as transparent as possible for the user. This is one of the main reasons behind new concepts like the OSGi (Open Service Gateway initiative) that supports hot deployment with 24×7 availability (24 h/day and 7 days/week).

4.1.1.2 Maintainability

Given the inevitability of change, software components must be designed to support modification and extension over their lifetime. Maintainability enables utilization of software components over time. Developers use well-known software engineering techniques to achieve this goal. In addition to producing well-written documents for the source code and for the software itself, the software design and the strict usage of appropriate standards, best practices, patterns, and naming conventions are crucial. Analysis, design, and coding patterns help to develop a common language and encourage the use of best practices and simpler, more intuitive ways of understanding and maintaining applications.

4.1.1.3 Scalability

Enterprise applications are long-term investments and are meant to adapt to changing business needs. One of these needs is the capability to scale according to usage requirement. Scalability, in contrast to maintainability, does not imply an actual change of the features of certain software but rather addresses the need to sustain a running and most importantly usable system in terms of growing or fluctuating number of users and corresponding traffic of queries and reporting.

There are two major different techniques to achieve scalability: vertical scaling and horizontal scaling. With vertical scaling, an existing system is basically extended or updated to increase the capacity of existing hardware or software adding more processing power or, for instance, data storage. In contrast, horizontal scaling or clustering involves adding multiple entities of hardware and software together to act as one single logical unit. Vertical scaling is possible for most applications, whereas horizontal scaling may not always be possible or efficient enough. Mixtures of these two approaches are often referred to as diagonal scaling. Which of the aforementioned scaling techniques eventually is best for an application depends on many different factors, such as the programming language, architecture, deployed protocols, message size and frequency, requirements on the deployment environment, throughput, responsiveness, availability, or the total cost of ownership. As an example, consider a two-tier application with an old mainframe-based database. Porting the database to another hardware system, changing the entire database, or enhancing the existing database with clustering capabilities would be very expensive. In this case, vertical scaling is a better choice because it is less invasive than the horizontal approach and does not necessitate any internal changes on the application. In case of a typical three-tier application with a Web front-end and an application server and a database in the back, horizontal scaling is more flexible. New nodes of the Web front-end can be added dynamically as soon as the load increases, thus providing the flexibility by allocating resources dynamically where needed.

4.1.1.4 Interoperability

The ability of disparate applications working together is considered one of the most important characteristics of an Enterprise Application Architecture. Companies typically integrate business solutions from different software vendors, Starting from e-mail clients to intranet portals, finance management systems, supply chain management software, customer relationship management systems, and so on. Furthermore, companies are experiencing continuous changes (takeovers, mergers or restructures, and even splits). All this and the fact that the lifetime of software is limited force CIOs to plan for a transition phase in order to adapt to these changes. Software has to be built in a way to accommodate these inevitable changes and ensure interoperability with other applications.

Current software designs aim to address interoperability by providing flexible and extendable solutions. Monolithic software solutions and legacy applications were connected via proprietary communication protocols. This approach makes the enterprise depend on the solutions developed by a single vendor. Hence, enterprises are responding to the availability of open and flexible solutions. Many IT companies including IBM, Microsoft, Sun Microsystems, and BEA are now providing a whole arsenal of software based on standard interfaces enabling seamless integration.

4.1.1.5 Security

Security is playing an increasing role in current enterprises (see Section 19.1.2 "Security"). Companies are dedicating special resources to protect themselves from this growing threat. Mistakes done in this area often not only cause direct financial losses but can also have dramatic impact on the reputation and the value and trust of the company, especially for companies handling sensitive personal data, such as insurance, financial, and medical companies. Even the partial loss of sensitive data can put such companies out of business as soon as the customers start losing their trust. This is one reason why in the current Basel III Accord of the Bank of International Settlements, the security concept of a company is evaluated to assign a risk level and ultimately affect the credit ranking of the company.

An enterprise faces three different kinds of threats:

1. Attacks against personal- and customer-related data like medical records
2. Attacks against company goods like confidential contracts, licenses, or detailed business objectives
3. Attacks to abuse the company infrastructure in order to run illegal businesses like file sharing with pirate copies, for instance

In order to protect the business, the IT systems have to be up-to-date. Besides well-known mechanisms such as virus scanners, firewalls, and demilitarized zones, modern software architectures contain their own security layers. In Java, for instance, a Security Manager enforces policies to limit the rights to a distinct part of the source code. Even if an attacker manages to breach or compromise a certain part of the code, he or she only acquires the minimal subset of rights the code was granted to that part of the code in order to fulfill its tasks.

This helps to minimize the impact of vulnerabilities. In spite of all the efforts, data breaches are bound to occur. Depending on the domain a company is working in, the company may be even obliged to comply with rules and regulations and may face noncompliance charges for not following the regulations such as the Payment Card Industry Data Security Standard, Visa Member rules, or the Health Insurance Portability and Accountability Act. Hence, companies must adopt solutions to protect their data and provide forensic evidence of attacks that can be used in courtrooms. To minimize the risks, a plan has to be developed to detect security breaches by monitoring IT systems for unusual behavior. The system must respond to such anomalies. Once a breach has been detected, a previously established response plan has to be invoked to respond and handle the incident appropriately. The response plan has to be comprehensive and should not only cover steps to be taken in order to fix the breach in security, but also define notification agency and the nature of the forensic evidence to be gathered and recorded for admissibility as evidence in a court of law.

4.1.1.6 Reliability

Reliability is defined as the ability to perform and maintain distinct functionalities within predefined parameters for a specified period of time. The importance of this nonfunctional requirement may vary drastically depending on the usage scenario. An intranet portal will most likely have different availability requirements than a stock-trading portal during business hours.

Reliability can be differentiated into a data-centric part and a service-centric part. Data-centric reliability concerns the data the application is working with. Service-centric reliability focuses on the guaranteed availability of services. In online shopping, credit card validation is a vital service for the business, whereas a service to verify zip codes for the order address has a noticeably lower priority. A service to provide currency exchange rates have to be as accurate as possible and its data reliability is essential. Service reliability is characterized by the number, date, time, and time span of scheduled downtimes, what is or what is not considered force majeure, and how much latency is acceptable. Some of the questions to be asked before the actual software is developed or chosen concern its latency, tolerance to failure, and the fallback mechanisms available in case of a failure.

4.1.1.7 Usability

Usability is defined as the ease of understanding and use for availing a service. While well-designed and ergonomic software is better accepted by users, the usability is also affected by familiarity, continuity, and strength of acquired habits. How can we decide if the Microsoft Windows GUI or the GUI from Apple Macintosh is better? Objectively, even though it is not possible to define an absolute standard for usability, there are well-accepted norms, patterns, and best practices for enhancing usability. There are standards to evaluate and measure the accessibility and the usability for different user groups, for example, as defined in Section 508 of the Rehabilitation Act or the Web Content Accessibility Guidelines from the World Wide Web Consortium (W3C). However, rules, regulations, and laws change from region to region, which makes it almost impossible to find one standard for all situations. In short, while accessibility and usability are critical for the acceptance of applications, which standards are applicable depends on the context of usage.

4.1.2 Systems Heterogeneity in Enterprises

This problem refers to the fact that in a large enterprise or an inter-enterprise system consisting of an enterprise and its partners, more than one technology is generally used to integrate applications. Therefore, it is literally impossible to impose enterprise-wide standards in this respect.

Various kinds of technological heterogeneity can exist in a large enterprise, including the following:

a. *Middleware heterogeneity*: In a large enterprise, more than one type of middleware is generally used. The two most common types are application servers and message-oriented middleware (MOM). In addition, brand (vendor) heterogeneity requires support for different brands of application servers and MOMs.

b. *Protocol heterogeneity*: This heterogeneity refers to the different transport protocols being used to access the services offered by various applications. Examples of such protocols include IIOP, JRMP, HTTP, and HTTPS.

c. *Synchrony heterogeneity*: There is almost always a need to support both synchronous and asynchronous interactions between applications. In addition, there is sometimes a need for callback methods and publish and subscribe. Therefore, many times a situation arises where the styles of interaction supported by two applications that wish to interact do not match. Hence, these applications cannot interact with each other.

d. *Protocol mismatch*: Related to the heterogeneity of communication protocols is the problem that arises when different applications want to communicate with each other using incompatible protocols. For example, Application A might want to communicate with Application B using HTTP. However, for Application B, the suitable protocol might be IIOP. In such cases, a protocol transformation is needed so that Application A can communicate with Application B.

e. *Diversity of data formats*: A problem arises when there is diversity in the data format being exchanged. Most of the time, the data are dependent on the middleware being used. This diversity of data can also cause a problem if two applications that wish to interact support different data formats.

f. *Diversity of interface declarations*: A problem arises when there are large differences in the way the service interfaces are being declared and used to invoke the services. For example, the way interfaces are declared in CORBA and Java RMI is different.

g. *No common place for service lookup*: A problem arises when there is no common place to look up services to deal with the diversity of the services in a large enterprise.

Another common problem is that as soon as a new version of provider software becomes available, consumer applications must be modified to account for the change in the provider application. The solution to this problem requires that methods be found that allow the services to be extended—for example, by adding more parameters—without breaking the previous versions of the consumer application.

This diversity and extendibility have been partly dealt with by developing standards and by further development in technology. We discuss these standards in Chapter 7, whereas the further development in technology is discussed in Chapter 9.

4.2 Integration of Enterprise Applications

EAI provides components for integrating applications with external applications and technologies within the enterprise and is designed to work with third-party products. By employing EAI effectively, an enterprise can leverage its existing enterprise-wide information assets, that is, customer relationships:

- To provide new products and services easily and quickly
- To streamline its internal process and operations
- To strengthen supply relationships
- To enhance customer relationships

As EAI enables enterprise-wide integration of diverse applications across various products and divisions, it provides the enterprise with a 360° view of its customer relationships across multiple channels of interaction. Every customer perceives the enterprise as a whole and also expects to be recognized and valued by the enterprise as a whole; familiarity with customers' earlier interactions and purchases helps frontline members of the enterprise to create opportunities for selling other products or additional add-ons and services to the earlier purchases.

 One of the important objectives of application integration is to achieve the integration between applications with as reduced a level of coupling or interdependency as possible so that the impact of changes in one application does not affect the other application(s).

4.2.1 Basics of Integration

The basic concepts related with EAI are described here. A robust and flexible EAI provides a combination of the methods of integration and modes of communication that are embodied into the various models of integration that are deployed within the EAI architecture as discussed next.

4.2.1.1 Methods of Integration

Methods of integration are the approaches used to guide a request from a sender to a receiver. The two primary methods of integration are as follows:

a. Messaging—In this approach, the sender constructs a message that contains information on the actions desired as well as the data required to perform these actions—the message contains both the control information and data. Messages provide a lot of flexibility because the control information can be easily changed and extended; they are independent of any of the applications. However, to function correctly, the integration messages must be predefined precisely so that the messages can be coded and decoded in exactly the same way by all senders and receivers.

b. Interface—In this approach, the sender communicates through an interface, which defines explicitly the actions that can be invoked by an application; the interface is self-describing in terms of the actions that can be taken. Interfaces make the application look like a procedure, or a method, or an object. Interfaces are difficult to change and extend; they are associated with a particular application.

c. Connector—In this approach, the application provides an access point that allows either a message or invocation on an interface to be passed into the application; a connector is more than an interface providing additional capabilities like error handling and validation checking, conversion and transformation of data into appropriate format, and managing state information enabling guaranteed delivery or graceful recovery. Many applications do not have a predefined or prebuilt entry point into the application. In such cases, one may need to use data files, databases, user interfaces, or memory as the entry point for the injection of the request. It is at this point that the correct integration model must be selected—presentation, data, or functional—to build the right connector based on the internal structures of the application.

4.2.1.2 Modes of Communication

The flexibility of systems is critically dependent on the modes of communications that are utilized by the systems. Assuming that a request refers to a communication from a sender to a receiver, the two basic options for communications are as follows:

1. Synchronous communication—This requires the sender of a request to wait until a reply, which is the result of the request, is received before continuing the processing. Synchronous communication between systems implies a high degree of coupling and requires the sender and the receiver to coordinate the communications with their internal processing. A reliable network infrastructure is essential for this kind of communication. It is used when the sender requires a notification of the receipt or needs the result of the processing from the receiver. For instance, interactive systems need a synchronous type of communication.

   There are three popular types of synchronous communications:
   a. Request/response
   b. Transmit
   c. Polling

2. Asynchronous communication—This allows the sender to continue processing after sending the request without waiting for a reply to this request. The sender does not concern itself with whether or when the request has been received, how it is processed, or the results returned from the receiver. Asynchronous communications does not demand a high degree of coupling and also does not require the sender to coordinate the communications with its internal processing. It is used when the communication of information is required without the need to coordinate activities or responses.

There are three popular types of asynchronous communications:

a. Message passing—This is used in situations where information needs to be transmitted but a reply is not required. This needs a reliable network for guaranteed delivery.

b. Publish/subscribe—This is used in situations where a reply is not required, but unlike all other cases, the recipient is determined based on the content of the request and the predeclared interest of the receiver application. This type of communication is useful for STP type of functional integration (see Section 4.2.2.2.3, "Straight-Through Processing").

c. Broadcast—This is used in situations where again a reply is not needed but the request is sent to all the applications and each receiver decides if it is interested in the request/message and accordingly processes that request/message in accordance with the business and functional logic programmed into each of the receiver systems.

4.2.1.3 Middleware Options

Middleware is software that enables disparate applications to interact with each other—it facilitates the communication of requests between software components through the use of predefined interfaces or messages. The five basic types of middleware are as follows:

1. Remote procedure call (RPC)—This is based on the notion of developing distributed applications that integrate at the procedure level but across a network

2. Database access middleware—This is based on the notion of developing distributed applications that integrate at the distributed data level whether in files or databases but across the network

3. Message Oriented Middleware (MOM)—This is based on the notion of developing distributed applications that integrate at the message level but across the network

4. Distributed object technology (DOT)—This is based on the notion of developing distributed applications that integrate at the interface level but those that make the application look like an object

5. Transaction processing monitor (TPM)—This is based on the notion of developing distributed applications that integrate at the distributed transaction level but across the network

4.2.2 Models of Integration

An integration model defines the approach and configurations used to integrate software applications depending on the nature and methods of the envisaged integration. There are three possible points of integration, namely, presentation, functional, and data integration.

4.2.2.1 Presentation Integration

In this model, the integration is accomplished by deploying a new and uniform application user interface—the new application appears to be a single application although it may be accessing several legacy and other applications at the back–end. The integration logic, the instructions on where to direct the user interactions, communicates the interaction of the user to the corresponding application using their existing presentations as a point of integration. It then integrates back any results generated from the various constituent applications. Thus, a single presentation could replace a set of terminal-based interfaces and might incorporate additional features, functions, and workflow for the user. For instance, a mainframe application can be integrated into a new Microsoft Windows application at the front-end using the screen-scraping technology that effectively copies, maps, and imports data from specific locations on character-based screens of the mainframe application onto the new schemas and data structures of the new system.

Presentation integration is the easiest to achieve and can be automated almost 100%; however, it is also the most limiting of the three models.

4.2.2.2 Functional Integration

In this model, the integration is accomplished by invoking from other applications functionality or from the business logic of the existing applications by using code level interfaces to the existing applications. This might be achieved at the level of an object or a procedure or via application programming interface (API) if it exists for each of the corresponding applications. The business logic includes the processes and workflow as well as the data manipulation and rules of interpretation. For instance, to change the customer's address in an enterprise application, the functionality of the existing customer order and billing application can be accessed if it is functionally integrated with these later applications. Rather than re-create the logic in the new application, it is more efficient and less error prone to reuse the existing logic.

Traditionally, remote procedure calls (RPCs), which have been employed for this kind of integration, have provided the definitions for access and basic communications facilities. However, lately, distributed processing middleware has become the preferred method of integration as it not only provides a more robust approach to the interface definitions and communications but also enables runtime support for intercomponent requests. The three categories of distributed processing are as follows:

1. Message oriented Middleware (MOM)—This achieves integration by providing for the communication of messages between applications by means of the messages placed in MOM, which itself is implemented in a variety of configurations, including message queuing and message passing. MOM is then responsible for delivering to the target system. Microsoft's MSMQ, BizTalk, IBM's MQSeries, and Talarian's SmartSockets are examples of MOM.

2. Distributed object technology (DOT)—This achieves integration by providing object interfaces that make applications look like objects. The application can then be accessed by other applications across a network through the object interfaces. OMG's CORBA, Microsoft's COM+, and Sun's Java 2 Enterprise Edition (J2EE) are examples of DOT.

3. Transaction processing monitors (TPMs)—These achieve integration by providing critical support for integrity of distributed resources such as databases, files, and message queues across distributed architectures by allowing various types of transactions to be managed using a variety of concepts including two-phase commit. BEA's Tuxedo is an example of TPM.

Functional integration that is more flexible than the other two integration models can be applied in three different forms as described later.

4.2.2.2.1 Synchronization

This corresponds to the coordination of data updates from multiple sources across integrated applications that may have been developed and enhanced over a long period of time. It provides integration that is loosely coupled and predominantly asynchronous. These applications may represent various relationships that a customer may have had with the enterprise or manage employee- or product-related information. When an update is made into any of the systems, the update needs to be propagated across all of these systems. Typically, synchronization is implemented by propagating a request that describes the intended action and the corresponding data to each of the relevant systems.

4.2.2.2.2 Component Integration

Component integration is the integration of applications where a well-defined interface exists that allows a component to be accessed via requests from other components without modifications. The interfaces for each component must identify the specific functions that the component supports. It provides integration that is tightly coupled and predominantly synchronous.

4.2.2.2.3 Straight-Through Processing (STP)

This corresponds to a coordinated set of automated actions executed across all relevant applications in the correct order of precedence automatically, that is, without human intervention. It provides integration that is tightly coupled and can be both synchronous and asynchronous. This kind of process is commonly associated with workflow though it does not involve decision making or complicated scheduling. For instance, an order for a product is placed on a Website; the Order Processing System (OPS) creates the order and notifies the logistics and shipping system to ship the product. When the order is completed, the OPS is notified of the change of status and the billing system triggers a bill for payment. Once the payment is received, the OPS is notified to close the order.

4.2.2.3 Data Integration

In this model, integration is accomplished by bypassing the existing application business logic and directly accessing the data created, processed, and stored by each of the corresponding applications. For instance, an Oracle-based billing system can be integrated with an IBM-based customer order system using the database gateway technology that integrates the DB2 database with the Oracle database.

This has been one of the earliest models applied for accessing information from databases, including

- Batch file transfer
- Open database connectivity (ODBC)
- Database access middleware
- Data transformation

The data integration model provides greater flexibility than the presentation integration model; it simplifies access to data from multiple sources and also allows the data to be reused across other applications. However, integrating at the data level necessitates rewriting of any functionality required by each of the applications, which implies greater effort for avoiding inconsistencies, standardizing, testing, and debugging for each of the applications on an ongoing basis. Since this model is highly sensitive to changes in the data models for each of the applications, this integration model is not very amenable for change and maintenance.

4.2.2.4 Business Process Integration

Achieving business process integration is often connected with business process reengineering and is not a sole technical problem. It, however, requires the implementation of several technical layers as the foundation and integrates applications at a higher level of abstraction. SOA, BPEL, and related technologies today provide new opportunities for making integrated information systems more flexible and adaptable to business process changes. This way, our information systems can get more agile, provide better support for changing requirements, and align closer to business needs (see Chapter 10, "Service Composition").

4.2.2.5 Business-to-Business Integration

There is a growing need to enable inter-enterprise integration, often referred to as business-to-business (B2B) integration, or e-business. The requirements today for online, up-to-date information, delivered with efficiency, reliability, and quality, are very high and gone are the days where a company could just publish offline catalogs on their Web pages.

Customers today expect immediate response and are not satisfied with batch processing and several days of delay in confirming orders. However, these delays are often the case when e-business is not backed by an efficiently integrated enterprise information system. Immediate responsiveness, achieved by the highly coupled integration of the back-end (enterprise information systems) and the front-end (presentation) systems, is a key success factor. Nonintegrated systems fail to meet business expectations; the primary reason for this is the lack of enterprise integration. In an e-business scenario, Applications from one company are invoking operations on front-end applications belonging to other companies. Only if these front-end systems are satisfactorily connected with back-end systems can the other company be able to provide an immediate and accurate response, which is an essential prerequisite for successful B2B collaboration.

4.2.3 Patterns of Integration

Integration patterns can be grouped into point-to-point integration and hub-and-spoke integration based on the way applications are connected. In the first approach, the applications are directly connected, while in the second, message exchanges go through a third party before being delivered to the final destination.

4.2.3.1 Point-to-Point Integration

Point-to-point integration is the simplest way to integrate independently developed application silos and do not require significant upfront investment. Each application is connected directly and explicitly to others. To link each application to another directly, an interface needs to be developed. This style may work well if the number of applications to be integrated is not large and there is no intention to scale out. Otherwise, it may quickly become unmanageable as the number of applications silos increases: if there are N applications to be integrated, then the number of interfaces to be developed becomes $N \times N$; that is, the number of interfaces to be developed grows on the order of N^2.

Another problem with point-to-point integration is the inability to respond to changes quickly. This is because the interfaces are hardwired to the application pairs, and changes within the enterprise information infrastructure may require rewiring the interfaces.

4.2.3.2 Message-Oriented Integration

In message-oriented integration solutions, the applications communicate with each other by sending and receiving messages through a middleware that manages the message queue associated with each application. Integration of two applications is by sending and receiving messages to the appropriate queue and the middleware ensures that the messages are delivered. However, point-to-point aspect of the integration is not eliminated, since applications are required to specify the recipients of the messages.

4.2.3.3 Hub–Spoke Integration

Spoke–hub integration eliminates the need to encode the address of the recipient. A centralized enterprise application middleware routes messages to their destinations based on the content and the format of the message. All applications are connected to the central integration hub like the spokes on a bicycle wheel. For this reason, this integration style is called spoke–hub integration. The concept is effectively used in many industries, such as transportation and telecommunication.

 A collection of point-to-point integrations with one common end results effectively in a hub–spoke integration pattern.

Spoke–hub integration reduces the number of connections from order of N^2 to N. This means that only as many interfaces as the number of applications need to be developed. In practice, a centralized integration hub provides a place for the adapters and it is the responsibility of the application developer to provide an adapter for each hub they connect to. Without standardization of adapters, this requires some development and testing resources (see Chapter 5, Section 5.12.1, "Replacing a Point-to-Point Integration Architecture with a Broker").

The spoke–hub integration is effectively implemented by message broker software that translates messages from one protocol to another, making sure that the data structures are compatible. Message brokers allow the rules of communication between applications to be defined outside the applications so that application developers do not need to worry about designing adaptors for every other application. When a message is received, the message broker runs the rule over the received message, transforms the data if needed, and inserts it into the appropriate queue. The rules are defined in a declarative way based on the communication protocols used by the applications. The message broker uses these rules to identify the message queues where the messages should be relayed. Publish/Subscribe software architecture style can be used to implement message brokers. Accordingly, applications publish their messages that are then relayed to the receiving applications that subscribe to them.

The major drawback of the message broker approach is the difficulty in managing and configuring the rules when the dependencies between applications are complex. Also, because message-based communications are inherently asynchronous, the solution may not be well suited for synchronous communication requirements, such as real-time computing or near-real-time computing.

4.3 Summary

This chapter focuses on enterprise applications and aspects related to the integration of applications. It describes the basics and models of integration including presentation, functional, data, business process, and bussiness-to-bussiness integration. The last part of the chapter explains the various patterns of integration, which will be used in Chapter 5.

5

Integration Technologies

Comprehensive enterprise-wide integration infrastructure usually requires more than one technology. Typically, also, because of the existing technologies, we will have to use a mixture of technologies. When selecting and mixing different technologies, we have to focus on their interoperability. Interoperability between technologies will be crucial because we will use them to implement the integration infrastructure. Achieving interoperability between technologies can be difficult even for technologies based on open standards. Small deviations from standards in products can deny the *on-paper* interoperability. For proprietary solutions, interoperability is even more difficult. It is not only the question of if we can achieve interoperability but also how much effort we have to put in to achieve it. Technologies used for integration are often referred to as middleware.

5.1 Middleware

Middleware is system services software that executes between the operating system layer and the application layer and provides services. It connects two or more applications, thus providing connectivity and interoperability to the applications. Middleware is not a silver bullet that will solve all integration problems. Due to overhyping in the 1980s and early 1990s, the term *middleware* has lost popularity but is coming back in the last few years. The middleware concept, however, is today even more important for integration, and all integration projects will have to use one or many different middleware solutions. Middleware is mainly used to denote products that provide glue between applications, which is distinct from simple data import and export functions that might be built into the applications themselves.

All forms of middleware are helpful in easing the communication between different software applications. The selection of middleware influences the application architecture, because middleware centralizes the software infrastructure and its deployment. Middleware introduces an abstraction layer in the system architecture and thus reduces the complexity considerably. On the other hand, each middleware product introduces a certain communication overhead into the system, which can influence performance, scalability, throughput, and other efficiency factors. This is important to consider when designing the integration architecture, particularly if our systems are mission critical and are used by a large number of concurrent clients.

Middleware is connectivity software that is designed to help manage the complexity and heterogeneity inherent in distributed systems by building a bridge between different systems, thereby enabling communication and transfer of data. Middleware could be defined as a layer of enabling software services that allow application elements to interoperate across network links, despite differences in underlying communications protocols, system architectures, operating systems, databases, and other application services. The role of middleware is to ease the task of designing, programming, and managing distributed applications by providing a simple, consistent, and integrated distributed programming environment. Essentially, middleware is a distributed software layer, or *platform*, that lives above the operating system and abstracts over the complexity and heterogeneity of the underlying distributed environment with its multitude of network technologies, machine architectures, operating systems, and programming languages.

The middleware layers are interposed between applications and Internet transport protocols. The middleware abstraction comprises two layers. The bottom layer is concerned with the characteristics of protocols for communicating between processes in a distributed system and how the data objects, for example, a customer order, and data structures used in application programs can be translated into a suitable form for sending messages over a communications network, taking into account that different computers may rely on heterogeneous representations for simple data items. The layer above is concerned with interprocess communication mechanisms, while the layer above that is concerned with non-message- and message-based forms of middleware. Message-based forms of middleware provide asynchronous messaging and event notification mechanisms to exchange messages or react to events over electronic networks. Non-message-based forms of middleware provide synchronous communication mechanisms designed to support client–server communication.

Middleware uses two basic modes of message communication:

1. *Synchronous* or time dependent: The defining characteristic of a synchronous form of execution is that message communication is synchronized between two communicating application systems, which must both be up and running, and that execution flow at the client's side is interrupted to execute the call. Both sending and receiving applications must be ready to communicate with each other at all times. A sending application initiates a request (sends a message) to a receiving application. The sending application then blocks its processing until it receives a response from the receiving application. The receiving application continues its processing after it receives the response.

2. *Asynchronous* or time independent: With asynchronous communication, an application sends (requestor or sender) a request to another while it continues its own processing activities. The sending application does not have to wait for the receiving application to complete and for its reply to come back. Instead, it can continue processing other requests. Unlike the synchronous mode, both application systems (sender and receiver) do not have to be active at the same time for processing to occur.

The basic messaging processes inherently utilize asynchronous communication. There are several benefits to asynchronous messaging:

1. Asynchronous messaging clients can proceed with application processing independently of other applications. Loose coupling of senders and receivers optimizes system processing by not having to block sending client processing while waiting for the receiving client to complete the request.

2. Asynchronous messaging allows batch and parallel processing of messages. The sending client can send as many messages to receiving clients without having to wait for the receiving clients to process previously sent messages. On the receiving end, different receiving clients can process the messages at their own speed and timing.

3. There is less demand on the communication network because the messaging clients do not have to be connected to each other or the MOM while messages are processed. Connections are active only to put messages to the MOM and get messages from the MOM.

4. The network does not have to be available at all times because of timing independence of client processing. Messages can wait in the queue of the receiving client if the network is not available. MOM implements asynchronous message queues at its core. It can concurrently service many sending and receiving applications.

Despite the performance drawbacks, synchronous messaging has several benefits over asynchronous messaging. The tightly coupled nature of synchronous messaging means the sending client can better handle application errors in the receiving client. If an error occurs in the receiving client, the sending client can try to compensate for the error. This is especially important when the sending client requests a transaction to be performed in the receiving client. The better error handling ability of synchronous messaging means it is easier for programmers to develop synchronous messaging solutions. Since both the sending and receiving clients are online and connected, it is easier for programmers to debug errors that might occur during the development stage. Since most developers are also more familiar with programming using synchronous processing, this also facilities the development of synchronous messaging solutions over asynchronous messaging solutions.

When speaking of middleware products, we encompass a large variety of technologies. The most common forms of middleware are as follows:

1. Database access technologies
2. Asynchronous Middleware
3. Synchronous Middleware
4. Message-oriented Middleware
5. Request/Reply Messaging Middleware
6. Transaction Processing Monitors
7. Object Request Brokers
8. Application Servers
9. Web Services
10. Enterprise Service Buses
11. Enterprise Systems

We discuss these in detail in the following sections.

5.2 Database Access Technologies

Database access technologies provide access to the database through an abstraction layer, which enables us to change the actual Database Management System (DBMS) without modifying the application source code. In other words, it enables us to use the same or similar code to access different database sources. Therefore, database access technologies are useful for extracting data from different DBMSs. The technologies differ in the form of interfaces to the database they provide. They can offer function-oriented or object-oriented access to databases. The best known representatives are Java Database Connectivity (JDBC) and Java Data Objects (JDO) on the Java platform and Open Database Connectivity (ODBC) and Active Data Objects (ADO.NET) on the Microsoft platform.

To expose data to the outside world, data source applications (i.e., databases) could incorporate a standard-based data access component for a remote application to perform functions on the database. The generic name for this type of standard is call level interface (CLI). The concept of CLI was originally created by the Structured Query Language (SQL) Access Group, an industry group created to define industry SQL standards. CLI shields the developer from the individual database. As long as the database is CLI compliant, the developer can use the same code to access different SQL databases. There are two main varieties of data integration technologies using CLI concept. The first variety is the one championed by Microsoft called the Open Database Connectivity (ODBC) standard and other standards like Object Link and Embedding database (OLE DB). The other variety is the JDBC standard from Javasoft.

5.2.1 Microsoft Open Database Connectivity (ODBC)

ODBC is a CLI and has a set of standard function calls to a database. Using ODBC, an application can remotely access a database—developers do not care what database or platform is used to store application data. The same code for data access can work for any database that has an ODBC driver, as long as the right ODBC Driver Manager exists for that platform. This eases the task of data access. Initially, ODBC has encountered database performance issues. However, it has evolved into a high-performance database access mechanism.

In order for ODBC to work, an operating system-specific driver manager needs to be utilized. A driver manager dynamically determines which ODBC driver to use for a program to access a database that is ODBC compliant. The ODBC driver takes the request from the calling program and translates it to a native format that the database can understand, and the database performs the request. Microsoft provides the ODBC Driver Manager for its operating systems; there are other ODBC Driver Managers for other operating systems from other vendors. As long as an ODBC driver exists for a database, an application can ask the database to perform a request that is supported natively in the database. Therefore, if a function supported by ODBC does not exist in the database, the ODBC driver for that database cannot support that function. Conversely, if a database has a function that is not supported in the ODBC standard, then the ODBC driver cannot support that function.

5.2.2 Java Database Connectivity (JDBC)

Like ODBC, JDBC is a CLI and it has its own set of functions. It enables a Java program to access a database with a JDBC driver. The architecture of JDBC is similar to ODBC. There is the JDBC Driver Manager, which is supplied in the Java Development Toolkit. When using the JDBC Driver Manager, the developer has to register the driver with the Driver Manager in the Java program.

There are four types of JDBC drivers:

- Type 1 is the JDBC–ODBC bridge driver.
- Type 2 is native application programming interfaces (APIs) partly Java-technology-enabled driver.
- Type 3 is the net-protocol fully Java-technology-enabled driver.
- Type 4 is the native-protocol fully Java-technology-enabled driver.

5.3 Asynchronous Middleware

In an environment where multiple applications and Web Services need to interact with each other, it is not practical to expect that each application knows the signature characteristics of every other application's methods. Instead, the intricacies of the service interface should not necessarily be known to all interacting applications. Asynchronous communication promotes a loosely coupled environment in which an application does not need to know the intimate details of how to reach and interface with other applications. Each participant in a multistep business process flow need only be concerned with ensuring that it can send a message to the messaging system. In general, asynchronous communication is often the preferred solution for EAI and cross-enterprise computing, especially when applications want to transfer data between internal enterprise information systems, for example, databases and ERP packages, or between their systems and those of their partners.

5.3.1 Store and Forward Messaging

With the store and forward queuing mechanism, messages are exchanged through a queue, which is the destination to which senders send messages and a source from which receivers receive messages. Messages are placed on a virtual channel called a message queue by a sending application and are retrieved by the receiving application as needed—the queue is a container that can keep hold of a message until the recipient collects it. The message queue is independent of both the sender and receiver applications and acts as a buffer between the communicating applications. The physical location of the queue or the physical details of the host platform are immaterial, all that is required is that an application is in some way registered or connected to the message queue subsystem. This provides a useful form of abstraction that enables physical implementations to be changed on either platform, without affecting the rest of the implementation.

The store and forward queuing mechanism is typical of a many-to-one messaging paradigm where multiple applications can send messages to a single application. The same application can be sender, receiver, or both sender and receiver. Message queuing provides a highly reliable, although not always timely, means of ensuring that application operations are completed.

5.3.2 Publish/Subscribe Messaging

Publish/subscribe messaging is a slightly more scalable form of messaging when compared to the store and forward mechanism. With this type of asynchronous communication, the application that produces information publishes it and all other applications that need this type of information subscribe to it. Messages containing new information are placed in a queue for each subscriber by the publishing application. Each application in this scheme may have a dual role: it may act as a publisher or subscriber of different types of information. The subscription list can be easily modified, on the fly, providing a highly flexible communications system that can run on different systems and networks. The publish/subscribe messaging mode usually includes the ability to transform messages, acting as an interpreter, which enables applications that were not designed to work together to do so.

The message server takes the responsibility of delivering the published messages to the subscribing applications based on the subscribed topic. Every message has an expiration time that specifies the maximum amount of time that it can live from the time of its publication in a topic. The message server first delivers messages to its associated active subscribers and then checks to make sure if there are any nonactive durable subscribers subscribed to the published topic. If, after the initial delivery, any of the durable subscribers did not acknowledge receipt of the message, the message is retained in the message server for the period of the expiration time, in the hope that the durable subscribers, if any, will connect to the message server and accept delivery of the message.

All subscribers have a message event listener that takes delivery of the message from the topic and delivers it to the messaging client application for further processing. Subscribers can also filter the messages that they receive by qualifying their subscriptions with a message selector. Message selectors evaluate a message's headers and properties (not their bodies) with the provided filter expression strings.

5.3.3 Point-to-Point Messaging

Point-to-point model is a pull-based or polling-based model, where messages are requested from a queue instead of being pushed to the client automatically as is the case with publish/subscribe model. Many large systems are divided into several separate units; the point-to-point messaging model provides reliable communication for such multistaged applications. The point-to-point messaging model allows clients to send and receive messages both synchronously and asynchronously via queues. One important difference between the publish/subscribe messaging and the point-to-point messaging is that point-to-point messages are always delivered, regardless of the current connection status of a receiver.

5.3.4 Event-Driven Processing Mechanism

The asynchrony, heterogeneity, and inherent loose coupling that characterize modern applications in a wide area network promote event interaction as a natural design abstraction for a growing class of software systems.

Such systems are based on a technical infrastructure known as an event notification service. An event notification service complements other general-purpose middleware services, such as point-to-point and multicast communication mechanisms, by offering a many-to-many communication and integration facility. Clients in an event notification scheme are of two kinds: objects of interest, which are the producers of notifications, and interested parties, which are the consumers of notifications. It is noteworthy that a client can act as both an object of interest and an interested party. An event notification service typically emulates the publish/subscribe asynchronous messaging scheme wherein clients publish event (or notification) messages with highly structured content, and other clients make available a filter (a kind of pattern) specifying the subscription: the content of events to be received at that client. Event message distribution is handled by an underlying content-based routing network, which is a set of server nodes interconnected as a peer-to-peer network. The content-based router is responsible for sending copies of event messages to all clients whose filters match that message.

In order to achieve scalability in a wide area network, the event notification service by necessity must be implemented as a distributed network of servers. It is the responsibility of the event notification service to route each notification through the network of servers to all subscribers that registered matching subscriptions and to keep track of the identity of the subscriber that registered each subscription. The event notification scheme is particularly appealing for developing service-based applications. The fact that notifications are delivered based on their content rather than on an explicit destination address adds a level of indirection that provides a great deal of flexibility and expressive power to clients of the service.

5.4 Synchronous Middleware

Programming models for synchronous forms of middleware are composed of cooperating programs running in several interacting distributed processes. Such programs need to be able to invoke operations synchronously in other processes, which frequently run in different computing systems. The most familiar approaches to non-message-based forms of middleware are typified by the remote procedure call (RPC) and the remote method invocation (RMI).

5.4.1 Remote Procedural Call (RPC)

RPC is based on the function call technique in traditional programming. With RPC, the client application passes the arguments for the function call to a local stub. Stubs are code within the local system that handles communication and passing data to and from the remote system. When a local stub receives the arguments, it establishes communication with the server and passes the arguments for the procedure call to the stub on the server. The client application blocks processing while it waits for the response to its remote procedure calls. When the server stub receives the arguments, it calls the procedure. After the server executes the procedure call, it passes the results to the server stub, which sends the results to the client stub. The client stub passes the results to the calling application. The connection between the two systems is closed only after the server returns the results of the request or the connection reaches a preset time limit. The client proceeds with its processing after it has received the response. Figure 5.1 describes the basic operation of RPC. In this example, the calling application in system A calls the RPC client stub. The stub then communicates with the RPC server stub in system B. After the target application in system B finishes processing, the result is sent back to the calling application via the client and server stubs.

RPC utilizes synchronous communication, which is different from the asynchronous nature of messaging. Using RPC, an application can invoke functions on another system as if they are on the same system. This is a tremendous help for application development. In a client–server development environment, developers can build distributed applications that span multiple computers using RPC without having to worry about network interface details. Unlike message-oriented middleware (MOM), RPC is not a discrete middleware layer. It requires stub codes on the client and the server. As long as the RPC stubs are available on the client and the server, the communication can be established directly without a discrete middleware intermediary. RPC technology comes with its set of specifications; when applications on different platforms follow the same set of specifications, these applications can interact with one another. Thus, RPC standards are platform independent.

RPCs work well for smaller, simple applications where communication is primarily point to point (rather than one system to many). RPCs do not scale well to large, mission-critical applications, as they leave many crucial details to the programmer, including handling network and system failures, handling multiple connections, and synchronization between processes. RPC-style programming leads to tight coupling of interfaces and applications. In an RPC environment, each application needs to know the intimate details of the interface of every other application—the number of methods it exposes and the details of each method signature it exposes. This figure clearly shows that the synchronized nature of RPC tightly couples the client to the server. The client cannot proceed—it is blocked—until the server responds, and the client fails if the server fails or is unable to complete. This can be understood by the fact that when performing a synchronous operation across multiple processes, the success of one RPC call depends on the success of all downstream RPC-style calls that are part of the same synchronous request/response cycle. This makes the invocation a whole-or-nothing proposition. If one operation is unable to complete for any reason, all other dependent operations will fail with it.

Remote procedure calls are also a client/server infrastructure intended to enable and increase interoperability of applications over heterogeneous platforms. Similar to MOM, it enables communication between software on different platforms and hides almost all the details of communication. RPC is based on procedural concepts—developers use remote procedure or function calls. The first implementations date back to the early 1980s. The main difference between MOM and RPC is the manner of communication. While MOM supports asynchronous communication, RPC promotes synchronous, request–reply communication (sometimes referred to as *call/wait*), which blocks the client until the server fulfills its requests. To achieve remote communication, applications use procedure calls; RPC middleware hides all communication details, which makes using remote procedure calls very similar to local procedure calls.

RPC guards against overloading a network, unlike the asynchronous mechanism, MOM. There are a few asynchronous implementations available, but they are more the exception than the rule. RPC increases the flexibility of architecture by allowing a client of an application to employ a function call to access a server on a remote system. RPC allows the remote access without knowledge of the network address or any other lower-level information. The semantics of a remote call is the same whether or not the client and server are collocated. RPC is appropriate for client/server applications in which the client can issue a request and wait for the server to return a response before continuing with its own processing. On the other hand, RPC requires that the recipient be online to accept the remote call. If the recipient fails, the remote calls will not succeed, because the calls will not be temporarily stored and then forwarded to the recipient when it is available again, as is the case with MOM.

RPC is often connected with the Distributed Computing Environment (DCE), developed by the Open Systems Foundation (OSF). DCE is a set of integrating services that expand the functionality of RPC. In addition to RPC, the DCE provides directory, time, security, and thread services. Over these fundamental services, it places a layer of data-sharing services, including distributed file system and diskless support. Technologies that predominantly use RPC-style communication include the Common Object Request Broker Architecture (CORBA), the Java Remote Method Invocation (RMI), DCOM, Active X, Sun RPC, Java API for XML-RPC (JAX-RPC), and the Simple Object Access Protocol (SOAP) v1.0 and v1.1. Component-based architectures such as EJB are also built on top of this model. However, due to their synchronous nature, RPCs are not a good choice to use as the building blocks for enterprise-wide applications where high performance and high reliability are needed. The synchronous tightly coupled nature of RPCs is a severe hindrance in system-to-system processing where applications need to be integrated together. Under synchronous solutions, applications are integrated by connecting APIs together on a point-to-point basis; consequently, it results in a lot of integration points between applications.

5.4.2 Remote Method Invocation (RMI)

Traditional RPC systems are language neutral and therefore cannot provide functionality that is not available on all possible target platforms. The Java RMI provides a simple and direct model for distributed computation with Java objects on the basis of the RPC mechanism. The Java RMI establishes interobject communication. If the particular method happens to be on a remote machine, Java provides the capability to make the RMI appear to the programmer to be the same as if the method is on the local machine. Thus, Java makes RMI transparent to the user. RMI applications comprise two separate programs: a server and a client. RMI provides the mechanism by which the server and the client communicate and pass information back and forth.

There are two different kinds of classes that can be used in RMI: *remote* and *serializable* classes. A remote object is an instance of a remote class. When a remote object is used in the same address space, it can be treated just like an ordinary object. But if it is used externally to the address space, the object must be referenced by an object handle. Correspondingly, a serializable object is an instance of a serializable class. A serializable object can be copied from one address space to another. This means that a serializable object can be a parameter or a return value. Note that if a remote object is returned, it is the object handle being returned.

5.5 Messaging-Oriented Middleware (MOM)

Message-oriented middleware (MOM) is a client/server infrastructure that enables and increases interoperability, flexibility, and portability of applications. It enables communication between applications over distributed and heterogeneous platforms. It reduces complexity because it hides the communication details and the details of platforms and protocols involved. The functionality of MOM is accessed via APIs. It typically resides on both ends, the client and the server side. It provides asynchronous communication and uses message queues to store the messages temporarily. The applications can thus exchange messages without taking care of the details of other applications, architectures, and platforms involved. The messages can contain almost any type of data; asynchronous nature of communication enables the communication to continue even if the receiver is temporarily not available. The message waits in the queue and is delivered as soon as the receiver is able to accept it. But asynchronous communication has its disadvantages as well. Because the server side does not block the clients, they can continue to accept requests even if they cannot keep pace with them, thus risking an overload situation.

MOM products are proprietary products and have been available from the mid-1980s. Therefore, they are incompatible with each other. Using a single product results in dependence on a specific vendor; this can have negative influence on flexibility, maintainability, portability, and interoperability. MOM product must specifically run on each and every platform being integrated. Not all MOM products support all platforms, operating systems, and protocols. However, Java platform provides ways to achieve relatively high independence from a specific vendor through a common interface, used to access all middleware products—the Java Message Service (JMS).

MOM is particularly suitable for integrating applications by reason of features like the following:

- *Transparent cooperation of heterogeneous systems*: The integration broker provides transformation software to transform application data running under diverse programming environments, operating systems, and hardware platforms.

- *Prioritization of requests*: All messages in an MOM environment may have priority attached to them. This forces higher-priority messages to be processed before lower-priority messages sent at an earlier time.

- *Automatic message buffering and flow control*: A distributed application often will need to read messages from diverse applications and programs. To support this undertaking, each application can have message queues that transparently buffer the messages when there are variable traffic rates, providing automatic flow control.

- *Persistent messaging*: This enables reliability and ensures that messages are guaranteed to be delivered at most once to their subscribers.

- *Flexibility and reliability*: Flexibility with MOM is achieved because an application can send its messages whenever it decides, independently of the recipient's availability. Senders and recipients are independent and thus unattached. Reliability is achieved because a persistent message is never lost.

- *Load balancing*: The asynchronous nature of MOM enables flexibility for load balancing. Load balancing is achieved as messages can be forwarded from a relatively busy application system to a less busy one. Dynamic load balancing can be designed into an MOM environment using selected algorithms including *least busy* and *round-robin*. This results in a cost-effective use of network facilities. In addition, the load balancing facility generates highly available message queues and enables an effective peak-hours network management. With this configuration, a low-bandwidth system may still attain acceptable performance, whereas it would have collapsed in a synchronous messaging environment.

- *Scalability and optimal use of resources*: When process volumes increase, MOM brokers employ *dynamic routing* and *multiplexing* techniques. Dynamic routing allows clients and servers that are not preprogrammed to communicate as the MOM automatically connect the requestor to the necessary service (without developer intervention). In addition, in case of a server breakdown, the MOM platform can dispatch a message to another backup server. Multiplexing is a function offered by MOM brokers that enables several applications to share a message queue.

MOM has demonstrated an ability to deliver the benefits of asynchronous messaging for applications and process-to-process interoperability, distributed transaction processing (such as banking, brokerage, airline reservations), distributed enterprise workflow (such as process manufacturing, insurance claims processing), real-time automation (such as utility and process control), and systems management (such as distributed backup, software distribution), among others.

5.5.1 Integration Brokers

Integration brokers perform necessary content and format transformation to translate incoming messages into a format that the subscribing system(s) can understand and utilize. Integration brokers are usually built on top of some MOM implementations and so the general principles of MOM also apply to them. An integration broker is usually built on a queue manager and routes messages to applications. The integration broker allows multiple applications to implement a published service with the broker providing application integration. In addition to these functions, integration brokers account for any differences at the structural level of applications; integration brokers take care of structural mismatches by keeping track of message schemas and by changing accordingly the content of messages to the semantics of a specific application. These unique capabilities of integration brokers enable them to broker not only between applications but also between types of middleware.

An *integration broker* is an application-to-application middleware service that is capable of one-to-many, many-to-one, and many-to-many message distribution. An integration broker is a software hub that records and manages the contracts between publishers and subscribers of messages. When a business event takes place, the application will publish the message(s) corresponding to that event. The broker reviews its lists of subscriptions and activates delivery to each subscriber for this type of message so that subscribers receive only the data to which they subscribe.

Integration brokers consist of components that provide the following functions:

- Message transformation: The message transformation functionality *understands* the format of all messages transmitted among applications. This is possible since the integration broker holds a repository of schemas of interchanged messages. Using this knowledge, the broker can translate between schemas by restructuring the data of these messages. In this way, subscribing applications can make sense of received messages.

- Business rules processing: The integration broker allows the application of business rules to messages so that new application logic can reside within the integration broker.

- Routing services: The routing functionality takes care of the flow control of messages.

- Naming services: The directory services functionality is needed since integration brokers function in a distributed environment and need a way to locate and use network resources. Applications using the integration broker are able to find other applications or hardware on the network.

- Adapter services: Many integration brokers use adapters as layers between the broker and large enterprise's back-end information systems to convert the data formats and application semantics from the source application to the target application. Adapters map the differences between two distinct interfaces: the integration broker interface and the native interface of the source or target application. For instance, an integration broker vendor may have adapters for several different source and target applications (such as packaged ERP applications), adapters for certain types of databases (such as Oracle, Sybase, or DB2), or even adapters for specific brands of middleware.

- Repository services: Repository houses information on rules, logic, objects, and metadata on target and source applications. The repository keeps track of input/output to the applications, its data elements, interrelationships between applications, and all the metadata from the other subsystems of the broker like the rules processing component. Metadata is one of the key elements of any integration solution as it is used to describe the structure of information held in disparate systems and processes.

- Events and alerts: Messages passing through the integration broker may trigger events or alerts based on specified conditions. Such conditions may be used for tracking business processes that move outside the range of given parameters and create a new message, run a special-purpose application, or send an alert.

5.5.2 Java Message Service (JMS)

As indicated earlier, applications that were designed for use with one middleware product could not be used with another middleware product. Middleware vendors attempted to standardize programming interfaces to MOM packages, but with little success. In 1999, Sun launched the Java Message Service (JMS), a framework that specified a set of programming interfaces by which Java programs could access MOM software. JMS is not a messaging system itself; it is an abstraction of the interfaces and classes needed by messaging clients when communicating with different messaging systems. JMS is a vendor-independent API for enterprise messaging that can be used with many different MOM vendors. JMS acts as a wrapper around different messaging products, allowing developers to focus on actual application development and integration, rather on the particulars of each other's APIs; application developers use the same API to access many different systems. JMS not only provides a Java API for connectivity to MOM systems but also supports messaging as a first-class Java distributed computing paradigm on the same footing as RPC. JMS-based communication is a potential solution in any distributed computing scenario that needs to pass data either synchronously or asynchronously between application elements, for instance; interfacing Enterprise Java Beans (EJB) with legacy applications and sending legacy-related data between the two.

JMS provides two principal models of MOM messaging:

1. The JMS point-to-point messaging model allows JMS clients to send and receive messages both asynchronously and synchronously via queues. A given queue may have multiple receivers, but only one receiver may consume each message. This guarantees that, for example, if a packaging order is sent to multiple warehouses, a single warehouse receives the message and processes the order.

2. The JMS publish/subscribe messaging model allows publishers to send messages to a named topic; all subscribers to this topic receive all messages sent to this topic. There may be multiple message listeners subscribed to each topic and an application can be both sender and receiver. JMS supports different message-sending configurations, including one-to-one messages, one-to-many messages, and many-to-many messages. One-to-one messages allow one message to be sent from one publisher (sender) to one subscriber (receiver), that is, point-to-point messaging; one-to-many messages allow one message to be sent from one publisher to numerous subscribers; and many-to-many messages allow many messages to be sent from many publishers to numerous subscribers.

JMS supports two types of message delivery: reliable message delivery and guaranteed message delivery. With reliable message delivery, the messaging server will deliver a message to its subscribing client as long as there are no application or network failures—delivery would fail if some disruption were to occur during delivery. With guaranteed message delivery, the message server will deliver a message even if there are application or network failures. The messaging server will store the message in its persistent store and then forward the message to its subscribing clients. After the client processes the message, it sends an acknowledgment to the messaging server and verifies the receipt of the message.

5.6 Request/Reply Messaging Middleware

Most of the asynchronous messaging mechanisms that we have examined so far follow the *fire-and-forget* messaging principle. On many occasions, applications require that request/reply messaging operations be performed. Here, we can distinguish between two types of request/reply messaging operations: synchronous request/reply messaging and asynchronous request/ reply messaging operations. Synchronous request/reply messaging is often necessary when trying to integrate with a Web Service client that blocks and waits for a synchronous response to return to it. In the asynchronous version of request/reply messaging, the requestor (sender) expects the reply to arrive at a later time and continue its work unaffected.

To perform a request/reply operation, the sender must use two channels: one for the request and one for the reply. The request message needs to contain reference to the receiver's end point, along with a correlation identifier that is needed to correlate the request with the response message. The requestor needs to poll a reply channel for the reply message. Both request/reply messaging modes can be layered on top of message-oriented middleware. Some Message-oriented middleware systems can further automate this process by managing the contents of the request/reply message.

5.7 Transaction Processing Monitors

Most messaging systems allow for transactional messaging that entail four ACID properties, namely, atomicity, consistency, isolation, and durability. In a synchronous request using distributed objects, the request by the calling object and the response from the receiving object use one process. This ensures the calling object receives the response to the request in the same connection to the object-based middleware. In contrast, transactional messaging uses one process for sending the message and a separate process for receiving a response to the message. Because the processes are distinct, it is also not possible for most MOM products to encapsulate the sending of a message and the receiving of that same message into one transaction—the transactional context is lost once the message has been sent to the destination. This is a result of the asynchronous, loosely coupled nature of messaging.

An approach to implementing transactional scope across the sending and receiving messaging clients is for the programmer to include a compensating transaction in the application logic. For instance, if the transaction in the receiving client fails, a message has to be sent to the sending client to inform it of the failure. When the sending client receives the message, it has to perform a compensating transaction to reverse the original database transaction it has performed. Due to the asynchronous nature of message queuing, it is difficult for MOM to maintain ACID properties of transactions. A basic MOM product does not provide this mechanism, and it has to be custom developed.

 A transaction represents a sequence of database operations (insert, update, delete, select) for which the system guarantees four properties also known as Atomicity, Consistency, Isolation, and Durability (ACID):

- *Atomicity*: A transaction is executed completely or not at all, thus exhibiting the characteristics of atomicity. As a consequence, all changes to the data made by this transaction become visible only if the transaction reaches commit successfully. Otherwise, if the transaction was terminated abnormally before reaching a commit, the original state of the data from the beginning is restored.
- *Consistency*: The property of consistency guarantees that all defined integrity or consistency constraints are preserved at the end of a transaction, that is, a transaction always moves the database from one consistent state to another consistent state. This has two consequences: In case a consistency constraint is violated, the transaction may be abnormally terminated and secondly, constraints can be temporarily violated during transaction execution but must be preserved upon the commit.
- *Isolation*: A transaction behaves as if it runs alone on the database without any concurrent operations. Furthermore, it only sees effects from previously committed transactions.
- *Durability*: When a transaction reaches the commit, it is guaranteed that all changes made by this transaction will survive subsequent system and disk failures.

In an application program, the boundaries of a transaction are specified by two commands:

- Begin-of-Transaction (BOT) denotes the beginning of the operation sequence – in some systems (e.g. in SQL database systems), this command is implicitly performed after the end of the previous transaction.
- Commit and rollback denote the end of the transaction. Commit is the successful end and requires that all updates must be made permanent, while rollback is for aborting the entire sequence, that is, undoing all effects.

The ACID properties are usually implemented by different components of a data management solution. Maintaining isolation of transactions is achieved by a concurrency control component implementing the concept of serializibility. Atomicity and durability are guaranteed by providing recovery strategies coping with possible failures. Finally, consistency is either explicitly supported by checking integrity rules or only implicitly by allowing rollbacks of transactions.

The transaction is the mechanism that binds the client to one or more servers and is the fundamental unit of recovery, consistency, and concurrency in a client–server system. From the perspective of application integration, transactions are more than just business events. They have become an important vehicle of guaranteeing consistency and robustness in distributed systems.

Transaction processing (TP) monitors enable building online TP by coordinating and monitoring the efforts of separate applications. TP monitor technology provides the distributed client–server environment with the capability and capacity to efficiently and reliably develop, execute, and manage transaction applications. TP monitors reside between front-end applications and back-end applications and databases to manage operations on transactional data; they manage processes and orchestrate applications by breaking complex applications into a set of transactions. Under the control of a TP monitor, a transaction can be managed from its point of origin—typically on a client—across one or more servers and back to the originating client. When a transaction ends, all parties involved agree that it either succeeded or failed. Transaction models define when a transaction starts, when it ends, and what the appropriate units of recovery are in case of failure. A TP monitor is needed for transactions requiring guaranteed completion of multiple discrete functions on multiple application systems.

Transaction processing (TP) monitors are important middleware technology in mission-critical applications. They represent the first generation of application servers. TP monitors are based on the concept of transactions. They monitor and coordinate transactions among different resources. Although the name suggests that this is their only task, they have at least two very important additional roles: providing performance management and security services. They provide performance management with load balancing and resource pooling techniques, which enable efficient use of computing resources and thus a larger number of concurrent clients. TP monitors map client requests through application service stateless routines to improve system performance. They can also take some application transition logic from the client. They also provide security management where they enable or disable access of clients to certain data and resources. TP monitors can be viewed as middle-tier technology, and this is why they are predecessors of today's application servers.

TP monitors have been traditionally used in legacy information systems. They are based on the procedural model, use remote procedure calls for communication between applications, and are difficult to program because of complex APIs through which they provide the functionality. In spite of that, they have been successfully used for more than 25 years. TP monitors are proprietary products, which makes migration from one product to another very difficult.

A TP monitor can manage transactional resources on a single server or across multiple servers, and it can also cooperate with other TP monitors in federated arrangements. TP monitors are primarily designed to run applications that serve large numbers of clients. By interjecting themselves between clients and servers, TP monitors can manage transactions, route them across systems, load balance their execution, and restart them after failures. The router subsystem of a TP monitor mediates the client request to one or more server processes. Each server in turn executes the request and responds. Typically, the server manages a file system, database, or other mission-critical resources shared among several clients.

TP monitors have several drawbacks to contend with; namely, TP monitors are much more intrusive than MOM—they demand more modification of the applications themselves in order take advantage of the TP monitor's specific services. TP can adversely affect application performance because, from the point of view of the transaction's requestor, the processing of a transaction is synchronous. The requestor must wait until all the processing of the transaction has completed before it can proceed with further computations. Moreover, during the processing of a transaction, all the resources used by it are locked until the transaction completes—no other application can use these resources during the execution of a transaction.

5.8 Object Request Brokers

Object request brokers (ORBs) are a middleware technology that manages and supports the communication between distributed objects or components. ORBs enable seamless interoperability between distributed objects and components without the need to worry about the details of communication. The implementation details of ORB are not visible to the components. ORBs provide location transparency, programming language transparency, protocol transparency, and operating system transparency.

The communication between distributed objects and components is based on interfaces. This enhances maintainability because the implementation details are hidden. The communication is usually synchronous, although it can also be deferred synchronous or asynchronous. ORBs are often connected with location services that enable locating the components in the network. ORBs are complex products but they manage to hide almost all complexity. More specifically, they provide the illusion of locality—they make all the components appear to be local, while in reality, they may be deployed anywhere in the network. This simplifies the development considerably but can have negative influence on performance. A basic outline of ORB architecture is shown in the next figure.

The differences between messages and interfaces as the method of integrating are subtle but important. Interface-based integration requires the specification and implementation of a well-defined interface that describes the actions that an application can perform. The interface is associated with an application. Messages are not associated with any application. In addition, with interfaces, the actions that can be processed by any application are easy to read and explicitly stated. Messages, as described previously, inherently hide the applications that use them. The nature of interfaces usually requires less processing to decode than a message, and errors are discovered earlier in the development process. The bottom line is that messages provide a lower degree of coupling than interfaces at the cost of greater potential of errors and inability to reuse solutions.

Over the long term, interface-based integration should be easier to reuse and maintain because the interface is explicitly defined and visible to the developer while messages can be hidden in the applications. It does not require looking at the code to see if an application will respond to a request. Interfaces are self-describing in terms of the actions that can be taken. With messages, either documentation or code must be read unless a directory service is provided that maps message use by application. Interfaces, however, may be more difficult to change and extend depending on implementation. For example, depending on the design of a message, control information can be modified without having to change application code. Changes to an interface may require a compilation of any and all applications to make them effective. Finally, interfaces require some discipline in the definition of the interface to allow plug and play.

ORB products may choose different scenarios as to how and where to implement their functionality. They can move some functions to the client and server components, or they can provide them as a separate process or integrate them into the operating system kernel.

There are three major standards of ORBs:

1. OMG CORBA ORB compliant

2. Java RMI and RMI-IIOP (Internet Inter-ORB Protocol)

3. Microsoft COM/DCOM/COM+/.NET Remoting/WCF

There are many ORB products compliant with the CORBA ORB specifications and various implementations of RMI and RMI-IIOP. Particularly, RMI-IIOP is important, because it uses the same protocol for communication between components as the CORBA ORB, namely, the IIOP. This makes RMI-IIOP (Internet Inter-ORB Protocol) interoperable with CORBA.

5.9 Application Servers

Application servers handle all or the majority of interactions between the client tier and the data persistence tier. They provide a collection of already mentioned middleware services, together with the concept of a management environment in which we deploy business logic components—the container. In the majority of application servers, we can find support for Web Services, ORBs, MOM, transaction management, security, load balancing, and resource management. Application servers provide a comprehensive solution to enterprise information needs. They are also an excellent platform for integration. Today, vendors often position their application servers as integration engines or specialize their common purpose application servers by adding additional functionality, like connections to back-end and legacy systems, and position their products as integration servers. Although such servers can considerably ease the configuration of different middleware products, it is still worth thinking of what is underneath.

Whether used for integration or new application development, application servers are software platforms. A software platform is a combination of software technologies necessary to run applications. In this sense, application servers, or more precisely the software platforms that they support, define the infrastructure of all applications developed and executed on them. Application servers can implement some custom platform, making them the proprietary solution of a specific vendor (these are sometimes referred to as proprietary frameworks). Such application servers are more and more rare. On the other hand, application servers can support a standardized, open, and generally accepted platform, such as Java Enterprise Edition.

The following lists the most important aspects of a platform:

- Technical aspects define the software technologies that are included in the software platform, the architecture of the applications developed for that platform, interoperability, scalability, portability, availability, reliability, security, client contracts, possibilities to grow and accommodate new solutions, and so on. In terms of integration, a very important aspect is the interoperability with other systems.

- Openness enables the application server vendors and third-party companies to have some possibility of influencing the development of the platform. Different solutions exist, from fully closed platforms that bind us to a certain vendor to the fully open platforms, for example, the open-source initiative, where everything, even source code, is free and can also be freely modified. Open platforms are often defined with specifications. These are documents that strictly define the technologies included in the platform and enable different vendors to implement the platform (e.g., as application servers). A tight specification guarantees consistency, and a platform defined in terms of specifications can also have a reference implementation and a set of compatibility tests.

- Interoperability among platform implementations is crucial for the adoption of a certain platform. Particularly, the way the platform regulates additions and modifications is crucial. The stricter a platform is with the implementation of the core specification, the better chances it has to be successful and to gain a large market share. Each platform, however, has to provide ways for application servers to differentiate their product, possibly through implementing some additional functionality.

- The Cost of the platform is also an important factor, and it is probably the most difficult to assess because it includes the cost of the application server and other development software, the cost of hardware, the training, and the cost of the maintenance of the applications through their lifecycle.

- The Last factor is maturity, from which we can predict how stable the platform is. The more mature the platform is, the more it has been tested and the more has been proved that it is suitable for large-scale applications.

5.10 Web Services

Web Services are the latest distributed technology. They provide the technological foundation for achieving interoperability between applications using different software platforms, operating systems, and programming languages. From the technological perspective, Web Services are the next evolutionary step in distributed architectures. Web Services are similar to their predecessors but also differ from them in several aspects.

Web Services are the first distributed technology to be supported by all major software vendors. Therefore, it is the first technology that fulfills the universal interoperability promise between applications running on disparate platforms. The fundamental specifications that Web Services are based on are SOAP (Simple Object Access Protocol), WSDL (Web Services Description Language), and UDDI (Universal Description, Discovery, and Integration). SOAP, WSDL, and UDDI are XML based, making Web Services protocol messages and descriptions human readable (see Chapter 8, "Web Services").

From the architectural perspective, Web Services introduce several important changes compared to earlier distributed architectures. They support loose coupling through operations that exchange data only. This differs from component and distributed object models, where behavior can also be exchanged.

Operations in Web Services are based on the exchange of XML-formatted payloads. They are a collection of input, output, and fault messages. The combination of messages defines the type of operation (one way, request/response, solicit response, or notification). This differs from previous distributed technologies. For more information, please refer to WSDL and XML schema specifications.

Web Services provide support for asynchronous as well as synchronous interactions. They introduce the notion of end points and intermediaries. This allows new approaches to message processing. Web Services are stateless and utilize standard Internet protocols such as HTTP (hypertext transfer protocol), SMTP (Simple Mail Transfer Protocol), FTP (File Transfer Protocol), and MIME (Multipurpose Internet Mail Extensions). So, connectivity through standard Internet connections, even those secured with firewalls, is less problematic.

In addition to several advantages, Web Services also have a few disadvantages. One of them is performance, which is not as good as distributed architectures that use binary protocols for communication. The other is that plain Web Services do not offer infrastructure and quality of service (QoS) features, such as security and transactions, which have been provided by component models for several years. Web Services fill this important gap by introducing additional specifications:

- WS-Security: This addresses authentication and message-level security and enables secure communication with Web Services.
- WS-Coordination: This defines a coordination framework for Web Services and is the foundation for WS-Atomic Transaction and WS-Business Activity.

- Transaction specifications (WS-Atomic Transaction and WS-Business Activity): These specify support for distributed transactions with Web Services. Atomic Transaction specifies short duration, ACID transactions, and Business Activity specifies longer running business transactions, also called compensating transactions.

- WS-Reliable Messaging: This provides support for reliable communication and message delivery between Web Services over various transport protocols.

- WS-Addressing: This specifies message coordination and routing.

- WS-Inspection: This provides support for dynamic introspection of Web Service descriptions.

- WS-Policy: This specifies how policies are declared and exchanged between collaborating Web Services.

- WS-Eventing: This defines an event model for asynchronous notification of interested parties for Web Services.

5.11 Enterprise Service Bus (ESB)

An Enterprise Service Bus (ESB) is a software infrastructure acting as an intermediary layer of middleware that addresses the extended requirements that usually cannot be fulfilled by Web Services, such as integration between Web Services and other middleware technologies and products, higher level of dependency, robustness, and security, management, and control of services and their communication. An ESB addresses these requirements and adds flexibility to communication between services and simplifies the integration and reuse of services. An ESB makes it possible to connect services implemented in different technologies (such as EJBs, messaging systems, CORBA components, and legacy applications) in an easy way. An ESB can act as a mediator between different, often incompatible, protocols and middleware products (see Chapter 9, "Enterprise Service Bus (EBS)").

The ESB provides a robust, dependable, secure, and scalable communication infrastructure between services. It also provides control over the communication and the use of services. It has message interception capabilities, which allow us to intercept requests to services and responses from services and apply additional processing to them. In this manner, the ESB acts as an intermediary. An ESB usually provides routing capability to route the messages to different services based on their content, origin, or other attributes and transformation capability to transform messages before they are delivered to services. For XML-formatted messages, such transformations are usually done using XSTLT (Extensible Style sheet Language for Transformations) or XQuery engines.

An ESB also provides control over the deployment, usage, and maintenance of services. This allows logging, profiling, load balancing, performance tuning, charging for service use, distributed deployment, on-the-fly reconfiguration, etc. Other important management features include the definition of correlation between messages, definition of reliable communication paths, and definition of security constraints related to messages and services.

An ESB should make services broadly available. This means that it should be easy to find, connect, and use a service irrespective of the technology it is implemented in. With broad availability of services, an ESB can increase reuse and can make the composition of services easier. Finally, an ESB should provide management capabilities, such as message routing, interaction, and transformation, which we have already described.

5.12 Enterprise Systems

There are many issues to deal with in enterprise integration, but at the core is an architectural problem concerning modifiability. Consider enterprise has "n" number of different business applications that need integrating to support some new business processes that may entail communicating between these "n" applications using their published messaging interfaces. Assuming one-way messages only, this means the business process under consideration must be able to transform its source data into the remaining $(n - 1)$ different message formats (this direct dependency creates a tight coupling between these applications); but this does not end here; similar changes may be necessitated in other applications leading to requirement for corresponding number of interfaces. In the general case, the number of interfaces between N applications is $N \times (N - 1)/2$. So as N grows, the number of possible interfaces grows exponentially, making such point-to-point architectures nonscalable in terms of modifiability. Considering that many interfaces between two applications are two way, requiring two transformations, and most applications have more than one interface, so in reality, the number of interfaces between N tightly coupled applications can be considerably greater than $N(N - 1)/2$. This results in a highly complex architecture.

5.12.1 Replacing a Point-to-Point Integration Architecture with a Broker

An introduction of a message broker as an intermediatory between the applications can reduce the complexity dramatically. Complexity in the integration end points, namely, the business applications, is greatly reduced as they just send messages using their native formats to the broker, and these are transformed inside the broker to the required destination format. If there is a need to change an end point, then one just needs to modify the message transformations within the broker that are dependent on that end point. No other business applications have to be made aware of the change or be modified. Thus, the number of interfaces from applications to the broker is reduced to N plus the number of interfaces required to transform from the broker to these very applications would be N, resulting in a total of $2 \times N$ interfaces. Thus, by introduction of the broker, the problem complexity reduces broadly from $N \times N$ to $N + N$—a dramatic reduction in the reduction in complexity (see Chapter 4, Section 4.2.3.3, "Hub–Spoke Integration").

There is a downside here in that the brokers are potentially a performance bottleneck, as all the messages between applications must pass through the broker. Good brokers support replication and clustered deployments to scale their performance. But of course, this increases deployment and management complexity and, more than likely, the license costs associated with a solution.

5.12.2 Enterprise Systems with an Enterprise Model

So message brokers are very useful, but not a panacea by any means for integration architectures. There is however a design approach that can be utilized that possesses the scalability of a point-to-point architecture with the modifiability characteristics of broker-based solution. The solution is to define an enterprise data model (also known as a canonical data model) that becomes the target format for all message transformations between applications. For instance, a common issue is that all your business systems have different data formats to define customer information. When one application integrates with another, it (or a message broker) must transform its customer message format to the target message format (say) the canonical customer information format.

Using this canonical message format, a message exchange is now reduced to the following steps:

1. Source application transforms local customer data into canonical customer information format.
2. Source sends message to target with canonical customer information format as payload.
3. Target receives message and transforms the canonical customer information format into its own local customer data representation.

This implies that each end point, that is, business application must know

- How to transform all messages it receives from the canonical format to its local format
- How to transform all messages it sends from its local format to the canonical format

Thus, by using the enterprise data model to exchange messages, we get the best of both worlds. The number of transformations is reduced to $2 \times N$ (assuming a single interface between each end point). This gives us much better modifiability characteristics. Also, as there are now considerably fewer and less complex transformations to build, the transformations can be executed in the end points themselves. We have no need for a centralized, broker-style architecture. This scales well, as there's inherently no bottleneck in the design. And there's no need for additional hardware for the broker and additional license costs for our chosen broker solution.

However, the strategy of using an enterprise data model for obviating the need of enterprise integration issues is not prevalent—the primary reason for this is existing enterprises have to confront the reality of myriad of enterprise applications and the corresponding enterprise data models and the sheer impossibility of converting or migrating to a uniform canonical enterprise data model.

5.13 Summary

Most of the enterprises have a host of existing enterprise applications and infrastructure technologies including integration technologies themselves. We looked at a spectrum of technologies that have been employed toward obtaining integration between enterprise applications. These include database access technologies, asynchronous and synchronous middleware, message-oriented middleware, request/reply messaging middleware, transaction processing monitors, object request brokers, application servers, Web Services, enterprise service buses, and enterprise systems. The inability of enforcing a uniform enterprise system/architecture/data model within the enterprise gave a fillip to the service-oriented architecture, which we discuss in Chapter 7.

6

J2EE for Enterprise Integration

6.1 Choosing an Enterprise Application Integration Platform

To implement the integration architecture requires an enterprise software platform that gathers together all of the necessary technologies and middleware solutions needed for building enterprise information systems. The prime choices are the J2EE platform, COBRA (Common Object Request Broker Architecture), and the Microsoft.NET architecture.

6.1.1 CORBA

CORBA is a standardized open architecture managed by OMG (Object Management Group). CORBA technology can be considered a generalization of Remote Procedure Call (RPC) technology and includes several improvements on the data objects and on the data primitives. The purpose of this technology and architecture was to enable the development of distributed applications and services that can interoperably communicate with other disparate applications over the network. The CORBA was essentially developed to bring about a discipline to implement portability and interoperability of applications across different hardware platforms, operating environments, and disparate hardware implementations. CORBA technology uses a binary protocol called Internet Inter-ORB Protocol (IIOP) for communicating with remote objects.

6.1.2 DCOM

In the mid-1990s, Microsoft Corporation introduced a technology called COM that enabled the development of software modules called components for integrating applications over the client–server architecture. To build these components, developers adhered to the COM specification so that the components could operate interoperably within the network. The DCOM technology, introduced sometime in the late 1990s, enabled interaction among network-based components to bring in the Distributed Computing Environment (DCE). DCOM technology is essentially built on an object

135

RPC layer, which in turn is on top of DCE RPC to enable the communication among remote objects. DCOM technology uses a binary protocol, termed Object Remote Procedure Call (ORPC), for distributed communication among remote objects. Technologies such as Object Linking and Embedding (OLE), ActiveX, and Microsoft Transaction Server (MTS) are some of Microsoft's technological advancements built on COM and DCOM technologies.

6.1.3 J2EE

Developed by Dr. James Gosling of Sun Microsystems, Java technology was introduced in 1995. It was based on a simple idea that a Java Virtual Machine (JVM) would behave the same way on any platform, and therefore, applications developed using Java programming language would behave reliably and consistently on any platform. The Java programming environment provided unique features unlike any other programming language, namely, portability and platform independence. The core feature is the Java Runtime Environment (JRE) that can be made available on any hardware or operating environment. The application is developed using the Java programming language and compiled into platform-independent bytecodes. This bytecode can then be deployed to run on JRE that is installed on any compatible system.

Java EE is the server-side extension of Java. The applications are not just Java objects but are also appropriate server-side components. For creating Web applications, components such as Java Servlets and JavaServer Pages (JSP) are used and deployed on Web servers, and these Web servers run on JRE. Likewise, for creating enterprise applications, components such as Enterprise JavaBeans (EJB) are developed and deployed, optionally with Web applications, in application servers. Again, these application servers also run in JRE.

The J2EE platform is controlled by the Java Community Process (JCP). The JCP is responsible for the development of the whole Java technology. Anyone can join the JCP and influence the evolution of the Java platform. Modifications to the platform require consensus among the members of the JCP process. This guarantees that there will be no rapid changes that would cut the compatibility with existing software. On the other hand, it gives the members the possibility to influence its development and direction. Sun Microsystems (now part of Oracle Corp.) still has the rights on the J2EE trademark and requires licenses to be paid, but it is the JCP, not Oracle, who controls the platform.

The J2EE specification, controlled by the JCP, is then implemented by many different application server vendors who compete in a very large market. This means that the customer is free to choose the application server and the vendor and, at a later date, to switch to a different vendor if needed, but any existing software should be portable between different vendors' implementations.

6.1.4 .NET

The .NET product suite is largely a rewrite of Windows DNA, which constitutes Microsoft's previous platform constituents for developing enterprise applications. The .NET technologies offer language independence and language interoperability. A .NET component can be written partially in different programming languages. The .NET technology converts this composite language component into an intermediary neutral language called Microsoft Intermediate Language (MSIL). This MSIL code is then interpreted and compiled to a native executable file. The .NET Framework also includes a runtime environment called the Common Language Runtime (CLR). This environment is analogous to the Sun Microsystems' Java Runtime Environment (JRE).

Microsoft has packed a number of servers as part of the .NET platform called the .NET Enterprise Servers. These servers provide vital services for hosting enterprise-class applications. Some important servers included as part of the .NET Servers are SQL Server, Exchange Server, Commerce Server, Cluster Server, Host Integration Server, and BizTalk Server.

If .NET becomes an open platform and is ported to operating systems other than Windows, it will be a valuable contender to Sun's J2EE.

6.2 Enterprise Application Integration (EAI) Using J2EE

J2EE is the result of Sun's effort to integrate the assortment of Java technologies and API together into a cohesive Java development platform for developing complex distributed Java applications. Sun's enhancement of the n-tier development model for Java, combined with the introduction of specific functionalities to permit easier development of the server-side scalable Web-based enterprise applications, has led to a wide adoption of Java for Web-centric application development.

Enterprise application development entails expertise in a host of areas like interprocess communications, memory management, security issues, and database-specific access queries. J2EE provides built-in support for services in all these areas, enabling developers to focus on implementing business logic rather than intricate code that supports basic application support infrastructure.

There are numerous advantages of application development in the J2EE area:

- J2EE offers support for componentization of enterprise applications that enable higher productivity via reusability of components, rapid development of functioning applications via prebuilt functional components, higher-quality test-driven development via pretested components, and easier maintenance via cost-effective upgrades to individual components.

- J2EE offers support for hardware and Operating systems (OS) independence by enabling system services to be accessed via Java and J2EE rather than directly via APIs specific to the underlying systems.
- J2EE offers a wide range of APIs to access and integrate with third-party products in a consistent manner, including databases, mail systems, and messaging platforms.
- J2EE offers clear-cut segregation between system development, deployment, and execution, thus enabling independent development, integration, and upgradation of components.
- J2EE offers specialized components that are optimized for specific types of roles in an enterprise application like Entity Beans for handling persistent data and Session Beans for handling processing.

All the aforementioned features make possible rapid development of complex, distributed applications by enabling developers to focus on developing business logic, implementing the system without being impacted by prior knowledge of the target execution environment(s), and creating systems that can be ported more easily between different hardware platforms and Operating systems (OS).

6.2.1 Reference Architecture

The objective of the flexibility and reusability can be achieved primarily at two levels: application architecture level and the application component design level. The reference architecture is the vision of the application architecture that integrates common elements into a component structure modeling the current business and also positioning it to meet the challenges of the future. From a technical point of view, the architecture positions the development organization to automatically meet the benchmark requirements on time to market, flexibility, and performance.

A set of key elements drive the definition of the reference architecture that is comprised of three layers, namely, business objects, process-oriented or service-based objects, and user interface layer.

The defining elements of enterprise applications are as follows:

- *Business entities* are the foci of the enterprise applications. These range from top-level entities such as a customer or a supplier down to bottom-level entities such as purchase orders, sales orders, or even individual level line items of these orders. Entities participate in the business processes, have attributes or properties, have methods for responding to requests for information, and have different sets of enforceable policies or rules applicable to them. The latter include the requirement for persistence of the state of the entities as reflected in the snapshot of all attributes.

- *Business processes* carry out the tasks of the enterprise. They have some kind of specified workflow and essentially involve one or more business entities. They must be executed in a secure manner and also be accessible via a host of user interfaces or devices or clients.

- *User interactions* carry out the access and display of information related to business entities as an outcome of some business processes for scrutiny by the users of the enterprise application. This essentially involves some kind of screen flow or page navigation, attributes for presentation, user requests, or generated responses, that is, static or dynamic content, form-oriented processing, and error handling. The user interaction could be via a host of user interfaces or devices or clients.

Each of these elements gives rise to the three primary architecture layers of the reference architecture. These layers could reside on the same physical layer or be distributed across a network. Figure 6.1 presents the three architecture layers constituting the reference architecture.

6.2.1.1 User Interaction Architecture

User interactions are modeled by user interface components that comprise the User Interaction Architecture. In J2EE platform, this is typically implemented as a combination of servlets and Java Server Pages (JSP). In a Web-based application, this layer would process HTML form submissions, manage state within an application, generate Web-page content, and control navigation between pages. Many of the functions within this layer can be automated through configurable foundation components.

6.2.1.2 Service-Based Architecture

Business processes are modeled by service components that comprise the Service-Based Architecture. In J2EE platform, this is typically implemented as a process-oriented object wrapped with a stateless Session Bean. The concept of services allows the front end to be decoupled from the back-end business object components. The service-based layer adds tremendous value in terms of flexibility, reusability, and component design.

6.2.1.3 Business Object Architecture

Business entities are modeled by object components that comprise the Business Object Architecture. Each of these components manages the data

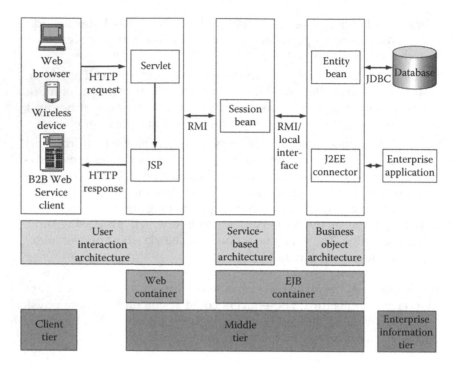

FIGURE 6.1
Enterprise application in J2EE.

and business logic associated with a particular entity, including persistence of that data into a relational database. In J2EE platform, this is typically implemented as a combination of regular Java classes and Entity beans in J2EE application. The database access can be implemented by the container in the case of CMP (Container-Managed Persistence) Entity Beans or by the developer in the case of BMP (Bean-Managed Persistence) Entity Beans or regular Java classes. The persistence of each business object is abstracted out to the extent possible so that separate data objects, persistent frameworks, or CMP services can be used to affect the data object persistence in the database.

A major portion of the reference architecture is a generic and configurable implementation of the Model 2 architecture discussed later in the following section.

6.2.2 Realization of the Reference Architecture in J2EE

The Java Enterprise Edition (J2EE) platform provides a component-based approach to implement n-tier distributed enterprise applications. Figure 6.1 shows how the J2EE components provide the implementations for the different layers of the reference architecture.

The components that make up the application are executed in runtime environments called containers. Containers are used to provide infrastructure-type services such as life-cycle management, distribution, and security. Containers and components in the J2EE application are broadly divided into three tiers. The client tier is typically a Web browser or alternatively Java application client. The middle tier contains two primary containers of the J2EE application, namely, Web container and EJB container. The function of the Web container is to process client requests and generate corresponding responses, while the function of the EJB container is to implement the business logic of the application. The EIS tier primarily consists of data sources and a number of interfaces and APIs to access the resources and other existing or legacy applications.

6.2.2.1 JavaServer Pages and Java Servlets as the User Interaction Components

JSP and Java Servlets are meant to process and respond to Web user request. Servlet provides a Java-centric programming approach for implementing Web tier functionality. The Servlet API provides an easy-to-use set of objects that process HTTP requests and generate HTML/XML responses. JSPs provide an HTML-centric version of the Java Servlets. JSP components are document based rather than object based and possess built-in access to Servlet API request and response objects as also the user session object. JSPs also provide a powerful custom tag mechanism, enabling the encapsulation of reusable Java presentation code that can be placed directly into the JSP document.

6.2.2.2 Session Bean EJBs as Service-Based Components

Session Beans are meant for representing services provided to a client. Unlike Entity Beans, Session Beans do not share data across multiple clients—each user requesting a service or executing a transaction invokes a separate Session Bean to process the request. A stateless Session Bean after processing a request goes on to the next request or next client without maintaining or sharing any data. On the other hand, stateful Session Beans are often constructed for a particular client and maintain a state across method invocations for a single client until the component is removed.

6.2.2.3 Entity Bean EJBs as the Business Object Components

Entity Beans are meant for representing persistent data entities within an enterprise application. One of the major component services that are provided to the Entity Beans is that of Container Managed Persistence (CMP). However, in EJB 2.0 specification, CMP persistence is limited to one table only. Any object-relational mapping involving more than a one-to-one table-object mapping is supported only through Bean-Managed Persistence (BMP) (see "Entity Beans").

6.2.2.4 Distributed Java Components

Java Naming and Directory Interface (JNDI) enables naming and distribution of Java components within the reference architecture. JNDI can be used to store and retrieve any Java object. However, JNDI is usually used to look up for component (home or remote) interfaces to enterprise beans. The client uses JNDI to look up the corresponding EJB Home interface, which enables creation, access, or removal of instances of Session and Entity Beans. In case of local Entity Bean, a method invocation is proxied directly to the bean's implementation. While in case of remote Entity Beans, the Home interface is used to obtain access to the remote interface to invoke the exposed methods using RMI. The remote interface takes the local method call, serializes the objects that will be passed as arguments, and invokes the corresponding remote method on the distributed object. These serialized objects are converted back into normal objects to invoke the method to return the resulting value upon which the process is reversed to revert the value back to the remote interface client.

6.2.2.5 J2EE Access to the EIS (Enterprise Information Systems) Tier

J2EE provides a number of interfaces and APIs to access resources in the EIS tier. The use of JDBC API is encapsulated primarily in the data access layer or within the CMP classes of the Entity Bean. Data sources that map to a database are defined in JDBC, which can be looked up by a client searching for a resource using the JNDI. This enables the J2EE application server to provide connection pooling to different data resources, which should appropriately be closed as soon as the task is over to prevent bottlenecks.

The various J2EE interfaces and APIs available are as follows:

- Java Connector Architecture provides a standard way to build adapters to access existing enterprise applications.
- JavaMail API provides a standard way to access mail server applications.
- Java Message Service (JMS) provides a standard interface to enterprise messaging systems. JMS enables reliable asynchronous communication with other distributed components. JMS is used by Message-Driven Beans (MDBs) to perform asynchronous or parallel processing of messages.

6.2.3 Model–View–Controller Architecture

The Model 2 architecture is based on the Model–View–Controller (MVC) design pattern. A generic MVC implementation is a vital element of the reference architecture as it provides a flexible and reusable foundation for very rapid Web-based application development.

The components of the MVC architecture are as follows:

- *View* deals with the display on the screens presented to the user.
- *Controller* deals with the flow and processing of user actions.
- *Model* deals with the business logic.

MVC architecture modularizes and isolates screen logic, control logic, and business logic in order to achieve greater flexibility and opportunity for reuse. A critical isolation point is between the presentation objects and the application back-end objects that manage the business logic and data. This enables the user interface to affect major changes on the display screens without impacting the business logic and data components.

View does not contain the source of data and relies on the model to furnish the relevant data. When the model updates the data, it notifies as also furnishes the changed data to the view so that it can re-render the display to the user with the up-to-date data and correct data.

The controller channels information from the view on the user actions for processing by the business logic in the model. Controller enables an application design to flexibly handle things such as page navigation and access to the functionality provided by the application model in case of form submissions. Thus, controller provides an isolation point between the model and the view, resulting in a more loosely coupled front end and back end.

Figure 6.2 gives a complete picture of how objects in the MVC architecture are mapped to the reference architecture in J2EE.

6.2.4 Overview of J2EE Platform Technologies

These can be subdivided into three main categories, namely, component services, horizontal services, and communication services.

6.2.4.1 Component Services

These services assist in expediting and simplifying the development of the enterprise applications. They insulate the resulting applications from the underlying J2EE APIs.

6.2.4.1.1 JavaServer Pages (JSP)

As discussed in MVC architecture previously, the objective is to separate the presentation and content from the application logic, with the presentation and content contained in the JSP. JSP is similar to server-side scripting technology, except that JSP is compiled, whereas scripts are interpreted. JSP utilizes the Java Servlet technology to achieve server-side processing.

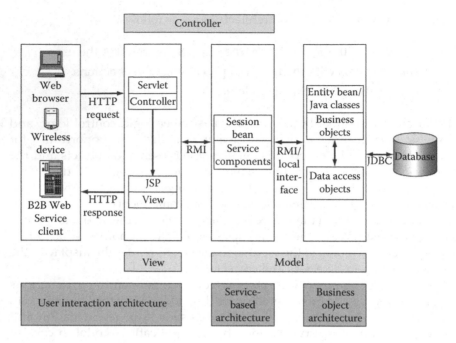

FIGURE 6.2
MVC and enterprise application architecture.

A JSP consists of Java code embedded within a structured document such as HTML or XML. The basic idea is to use the markup language for the static portion of the presentation and embed special tags within the page to mark up the dynamic content. The tags are used to process incoming requests from a client and consequently generate a response. As JSP uses additional system resources, wherever the presentation content is static, a plain HTML page should be used. The use of JSP allows the presentation code to be easily maintained as regular HTML code and shields the Web developer from having to deal with unfamiliar language and tools.

Java Scriptlets can be embedded in a JSP file, though their usage should be kept to the minimum. Sun recommends the use of JSP where there is a significant amount of dynamic content.

6.2.4.1.2 Servlets

Servlets are primarily used as a conduit for passing data back and forth between a Web client and an enterprise application running on a server. Servlets are server-side programs that execute in a servlet engine, which often forms a part of the HTTP server, but may also run standalone as well. Servlets run inside the servlet engine or container hosted on a Web server; the servlet container manages the life cycle of a servlet and translates the

Web client's requests into object-based requests and, in reverse, the object-based responses back to the Web client via HTTP.

Servlets provide a more effective alternate mechanism to the traditional CGI scripts for interaction between the server-based business logic and the Web-based clients. Servlets are usually employed to handle preliminary tasks like gathering and checking for valid inputs from the entry fields of a Web page. Once the basic checks are completed, the data are then passed on to more suitable component(s) for actual processing.

JSP specification provides the JSP with the same capabilities as the servlet. The basic idea is to leverage JSP for presentation-centric tasks and utilize the servlets for business logic processing-centric tasks. Servlets are preferred for more logical tasks as they are also comparatively easier to debug. Since Java code is embedded within the JSP, it may seem that the separation of presentation from business logic is not realistic. JSP should primarily be focused on presentation, and any Java code embedded within the JSP should primarily be for communication with servlets, other control or data entities, and so on.

JSP development usually adopts the Model 2 architecture based on the Model-View-Controller (MVC) architecture discussed earlier. It uses one or more servlets as controllers; requests received by frontline servlet(s) are redirected to the concerned JSPs. Usually JavaBeans is used as the model that acts as the conduit to pass information between the controller servlet(s) and the JSPs. The controller fills in the JavaBean based on the request, and the JSP in turn composes the actual page using values from the JavaBean.

6.2.4.1.3 *Enterprise JavaBeans (EJB)*

EJB components encapsulate business logic. EJB defines a comprehensive component model for building scalable, distributed server-based enterprise Java application component.

EJB components have four parts:

1. An implementation class that contains the business logic.
2. Home and remote interfaces that present the EJB's methods to the outside world.
3. A deployment descriptor: An XML file that is used to configure the EJB component being deployed in a J2EE server. For example, a deployment descriptor can define the security properties or transaction properties of EJB methods.

EJBs are container-managed components; that is, the container manages their life cycle and, based on the configuration specified in the deployment descriptor, interacts on behalf of the EJBs with various J2EE services.

The success of EJBs is based on a set of key concepts. First, EJBs are deployed within a container hosted by an application server, rather than deploying directly onto the application server. A container provides the environment for execution of EJBs, management of their life cycle, and provisioning of additional services. Second, EJBs take an approach based on proxy pattern rather than a monolithic component, which effectively separates out the component into client objects and remote objects. While the EJB user only sees the client object represented by the EJB interfaces, the remote object is free to change in terms of implementation details like location on the network and underlying transport mechanism. Third, EJBs use the concept of deployment descriptors that decouples the development from the deployment aspects.

There are three types of EJBs:

1. Entity Beans: Entity beans are EJBs designed specifically to represent data in a persistent store, which is typically a database. They encapsulate persistent data in a data store, which is typically a row or record of data in a database table. Apart from the built-in database access and synchronization capabilities, entity beans automatically provide the ability to share both state and behavior across multiple clients concurrently, disaster recovery facilities, and so on.

    An entity bean consists of a Home interface, a Remote interface, an implementation class, and a primary key class. The Home interface defines create, finder, remove, and home methods. The remote interface defines business methods. It also has a primary key class that contains methods for operating on the primary key for a single or a compounded database table. The implementation class implements all of the life cycle, finder, select, and business methods. Like all EJBs, entity beans also make use of the deployment descriptor to hold additional information pertaining to the component including transaction settings on business methods, relationships with other entity beans, and persistent filed settings.

2. Session Beans: Session beans are the most popular of the EJBs and are used primarily to manage transactions or client sessions. In an enterprise application, they are often used as the main controller connecting servlets or JSPs to entity beans or other components. Apart from the built-in transaction management and state management capabilities, EJB container also provides additional services such as automated resource management, concurrency, and security.

    These are used mainly for transient activities. They are nonpersistent and often encapsulate bulk of the business logic. While stateful session beans retain client state between successive interactions with the client, stateless session beans do not do so. In the case of a stateless session bean, each successive invocation of the bean is treated as an independent activity.

Entity beans have the option to use two different kinds of persistence mechanisms. First, container-managed persistence (CMP) entails using entity beans, while all database access and synchronization is handled automatically by the EJB container. Second, bean-managed persistence (BMP) entails using entity beans, while all data access and synchronization is handled by hand-crafted custom code.

3. Message-Driven Beans (MDBs): Message-driven beans are EJBs designed to be asynchronous consumers and processors of JMS messages. These were introduced newly in J2EE 1.3 and are useful for situations where synchronicity is not essential, for example, integrating loosely coupled systems. Session beans employ Remote Procedure Call (RPC)-based communication, which has the disadvantage that the sender must wait for a response before it can undertake the next activity. Message-driven beans are stateless, and, unlike session and entity beans, Message-driven beans do not have published interfaces.

Message-driven beans are useful for achieving the following:

- Efficiency: Messaging can be used for separating out those elements of the business logic that can be processed independent of the main thread of processing. This enables the main thread to obviate the need to expend resources and time on nonessential operations and move on to the next requests.

- Decoupling: Different subsystems are developed so that they are not tightly integrated with each other.

- Flexible integration: Loosely coupled systems can be composed by using Message-driven beans to *wrap* existing systems.

6.2.4.2 Horizontal Services

These are general services that are required across multiple tiers in enterprise application.

6.2.4.2.1 Java Database Connectivity (JDBC)

JDBC enables via database neutral APIs to perform a host of operations like obtaining database connections, execute SQL queries and updates via these connections, and process the results of such queries. The J2EE extensions also provide support for connection pooling and distributed transactions.

The database driver modules in the JDBC are responsible for mapping a database neutral request onto the request expected by a specific RDBMS.

6.2.4.2.2 *Java Naming and Directory Interface (JNDI)*

In J2EE application servers, JNDI provides via neutral APIs a mechanism that is used by the clients, Web components, and EJBs to find J2EE resources using a symbolic naming scheme. Directory and naming services are used to map symbolic names or a set of search attributes onto a resource. For example, the Domain Name System maps symbolic host names onto their Internet addresses. Similarly, a File System maps a symbolic pathname onto a system file identifier like an inode in Unix. Individually, all of these naming and directory services have specific APIs and may be written in a variety of languages.

6.2.4.2.3 *Java Connector Architecture (JCA)*

JTS is a comprehensive service that supports distributed transactions and consequently two-phase commit protocol.

JCA, which is a subset of JTS, is made available as a resource to J2EE application.

6.2.4.2.4 *XML Processing APIs*

XML is a widely accepted way of representing data in a standard format that can be validated against a Document Type Definition (DTD) or schema. These data can then be transmitted between various systems that convert the neutral format to system-specific formats, for example, to a relational form. Just as Java provides code portability, XML provides data portability. XML is also used as the configuration language in J2EE.

6.2.4.3 **Communication Services**

6.2.4.3.1 *HTTP/HTTPS*

Hypertext Transfer Protocol (HTTP) is a text-based protocol used for communication across the Internet and supports Web browser interactions with the HTTP servers listening on server machines. The protocol is stateless in that the server does not maintain any client state—every request made by an HTTP client will have to provide all essential information needed to process the request like

- The nature of the client (e.g., kind of browser)
- The kind of request
- The resource target on the server (e.g., a particular HTML page or servlet)
- The data to be sent to the server

Client requests are matched with server responses that provide information about the request like

- The status of the request
- The MIME type of the response data
- The response data

HTTPS is the secure form of the HTTP in that the HTTP communications are transmitted over the Secure Socket Layer (SSL).

6.2.4.3.2 *Remote Method Invocation (RMI)*

RMI enables communication between distributed objects transparent of their remote locations on the network by communicating with a local proxy or a stub that is generated automatically to communicate with the corresponding remote objects. The code for the proxy is generated automatically and communicates using sockets with the remote object—there is helper code at the remote end that reads from the socket, processes the bytes, and makes the method call on the remote object.

Initially, RMI allowed communication only between Java objects. Subsequently, RMI began to support communication with non-Java objects using RMI-IIOP that is a CORBA transport protocol on top of TCP/IP.

RMI enables developers to effectively concentrate on developing the business logic rather than worrying about the details of the distribution.

6.2.4.3.3 *Java Message Service (JMS)*

JMS enables asynchronous communication between producers and consumers in that a producer sends a message to a queue or topic, and, rather than waiting for a response, it moves on to undertake other tasks. As and when they are ready, consumers read messages from queues and topics.

JMS employs two models of communication:

1. Point to point: This involves the uses of FIFO queues and supports one-to-one and many-to-one interactions between producers and consumers. Message objects are created by the producers and sent to a named queue. Consumers who wish to read messages from the head of the queue obtain a reference to the queue head and then listen or wait for messages to be placed on the queue. As and when a message is placed on the queue, it will be read and removed from the queue by the listening consumer.
2. Publish and subscribe: This involves the use of topics and supports many-to-many communication between producers and consumers. Topics are analogous to newsgroups, and consumers subscribe to one or more topics. As and when producers publish messages to topics, a separate copy of a message is sent to each consumer.

6.2.4.3.4 *JavaMail*

JavaMail enables the sending and receiving of e-mail from within a Java program. It provides APIs that enable the creation of MIME message objects that can be sent and received using the underlying mail protocols like SMTP, POP3, and IMAP4. It is a form of asynchronous mail though slower than JMS and is used mainly for interaction between end users.

6.3 Summary

This chapter describes the platforms for the realization of the Enterprise Application Integration, namely, CORBA, DCOM, J2EE, and .NET. Then we sketch the reference architecture and detail the realization of this reference architecture in J2EE. In the end, the chapter describes the services constituting the J2EE platform.

Section II

Road to Cloudware

Enterprises require much more agility and flexibility than what could be provided by EAI solutions. SOA exposes the fundamental business capabilities as flexible and reusable services—the services support a layer of agile and flexible business processes that can be easily changed to provide new products and services to keep ahead of the competition.

Chapter 7 presents the basic concepts and characteristics of SOA. Chapter 8 presents the defining architecture of Web Services. Chapter 9 explains the basic design of an enterprise service bus (ESB), and Chapter 10 introduces the principles of service composition and the related business process execution language (BPEL). Chapters 11 and 12 respectively, present two different service delivery models, namely, applications service providers (ASP) and grid computing.

In the final analysis, cloud computing is an extension of the *network is computer* vision, namely, *network is service provider*.

7

Service-Oriented Architecture

Integration seems to be one of the most important strategic priorities mainly because new innovative business solutions demand integration of different business units, business systems, enterprise data, and applications. Integrated information systems improve the competitive advantage with unified and efficient access to the information. Integrated applications make it much easier to access relevant, coordinated information from a variety of sources. It is clear that replacing existing systems with new solutions will often not be a viable proposition. Companies soon realize that the replacement is more complicated, more complex, more time consuming, and more costly than even their worst-case scenarios could have predicted: too much time and money have been invested in them, and there is too much knowledge incorporated in these systems. Therefore, standard ways to reuse existing systems and integrate them into the global, enterprise-wide information system must be defined.

The modern answer to application integration is a SOA with Web Services; SOA is a style of organizing (services), and Web Services are its realization. An SOA with Web Services is a combination of architecture and technology for consistently delivering robust, reusable services that support today's business needs and that can be easily adapted to satisfy changing business requirements. An SOA enables easy integration of IT systems, provides multichannel access to your systems, and automates business processes. When an SOA with its corresponding services is in place, developers can easily reuse existing services in developing new applications and business processes.

A service differs from an object or a procedure because it is defined by the messages that it exchanges with other services. A service's loose coupling to the applications that host it gives it the ability to more easily share data across the department, enterprise, or Internet. An SOA defines the way in which services are deployed and managed. Companies need IT systems with the flexibility to implement specialized operations, to change quickly with the changes in business operations, to respond quickly to internal as well as external changes in conditions, and consequently gain a competitive edge. Using an SOA increases reuse, lowers overall costs, and improves the ability to rapidly change and evolve IT systems, whether old or new.

An SOA also maps IT systems easily and directly to a business's operational processes and supports a better division of labor between the business and technical staff. One of the great potential advantages of solutions created using an SOA with SOAP or REST Web Services is that they can help resolve this perennial problem by providing better separation of concerns between business analysts and service developers. Analysts can take responsibility for defining how services fit together to implement business processes, while the service developers can take responsibility for implementing services that meet business requirements. This will ensure that the business issues are well enough understood to be implemented in technology and the technology issues are well enough understood to meet the business requirements.

Integrating existing and new applications using an SOA involves defining the basic Web Service interoperability layer to bridge features and functions used in current applications such as security, reliability, transactions, metadata management, and orchestration; it also involves the ability to define automated business process execution flows across the Web Services after an SOA is in place. An SOA with Web Services enables the development of services that encapsulate business functions and that are easily accessible from any other service; composite services allow a wide range of options for combining Web Services and creating new application functionality.

As a prerequisite, one will have to deal with a plethora of legacy technologies in order to service-enable them. But the beauty of services and SOA is that the services are developed to achieve interoperability and to hide the details of the execution environments behind them. In particular, for Web Services, this means the ability to emit and consume data represented as XML, regardless of development platform, middleware, operating system, or hardware type. Thus, an SOA is a way to define and provision an IT infrastructure to allow different applications to exchange data and participate in business processes, regardless of the operating systems or programming languages underlying these applications.

7.1 Defining SOA

SOA provides an agile technical architecture that can be quickly and easily reconfigured as business requirements change. The promise of SAO is that it will break down the barriers in IT to implement business process flows in a cost-effective and agile manner that would combine the best of custom solutions as well as packaged applications while simultaneously reducing lock-in to any single IT vendor.

A Service-oriented architecture (SOA) is a style of organization that guides all aspects of creating and using business services throughout their life cycle (from conception to retirement), as well as defining and provisioning the IT infrastructure that allows different applications to exchange data and participate in business processes regardless of the operating systems or programming languages underlying these applications. The key organizing concept of an SOA itself is a service. The processes, principles, and methods defined by SOA are oriented toward services; the development tools selected by an SOA are oriented toward creating and deploying services; and the runtime infrastructure provided by an SOA is oriented to executing and managing services.

A service is a sum of constituting parts including a description, the implementation, and the mapping layer (termed as transformation layer) between the two. The service implementation, termed as the executable agent, can be any environment for which Web Service support is available. The description is separated from its executable agent; one description might have multiple different executable agents associated with it and vice versa. The executable agent is responsible for implementing the Web Service processing model as per the various Web Service specifications and runs within the execution environment, which is typically a software system or programming language environment. The description is separated from the execution environment using a mapping or transformation layer often implemented using proxies and stubs. The mapping layer is responsible for accepting the message, transforming the XML data to be native format, and dispatching the data to the executable agent.

Web Service roles include requester and provider; a requester can be a provider and vice versa. The service requester initiates the execution of a service by sending a message to a service provider, which executes the service and returns the results, if any specified, to the requester.

7.1.1 Services

Services are coarse-grained, reusable IT assets that have well-defined interfaces (or service contracts) that clearly separate the service accessible interface from the service technical implementation. This separation of interface and implementation serves to decouple the service requesters from the service providers so that both can evolve independently as long as the service contract remains unchanged.

A service is a location on the network that has a machine-readable description of the messages it receives and optionally returns. A service is therefore defined in terms of the message exchange patterns it supports. A schema for the data contained in the message is used as the main part of the contract between a service requester and a service provider; other items of metadata describe the network address for the service, the operations it supports, and its requirements for reliability, security, and transactional integrity. However, developing a service is quite different from developing an object because a service is defined by the message it exchanges with other services, rather than a method signature. A service usually defines a coarse-grained interface that accepts more data in a single invocation than an object because of the need to map to an execution environment, process the XML, and often access it remotely. Services are executed by exchanging messages according to one or more supported message exchange patterns (MEPs), such as request/response, one-way asynchronous, or publish/subscribe. Services are meant to solve interoperability problems between applications and for use in composing new applications or application systems, but not meant like objects to create the detailed business logic for the applications.

From a business perspective, services are IT assets that correspond to real-world business activities or identifiable business functions that can be accessed according to the service policies related to

- Who is authorized to access the service
- When can the service be accessed
- What is the cost of using the service
- What are the reliability levels of using the service
- What are the performance levels for the service

A service is normally defined at a higher level of abstraction than an object because it is possible to map a service definition to a procedure-oriented language, such as COBOL or PL/1, or to a message queuing system such as JMS or MSMQ, as well as to an object-oriented system such as J2EE or the .NET Framework. Whether the service's execution environment is a stored procedure, message queue, or object does not matter; the data are seen through the filter of a Web Service, which includes a layer that maps the Web Service to whatever execution environment is implementing the service. The use of XML in Web Services provides a clear separation between the definition of a service and its execution so that Web Services can work with any software system. The XML representation of the data types and structures of a service via the XML schema allows the developer to think solely in terms of the data being passed between the services without having to consider the details of the service's implementation. This is quite in contrast to the traditional nature of the integration problem that involves figuring out the implementation of the service in order to be able to talk to it.

One of the greatest benefits of service abstraction is its ability to easily access a variety of service types, including newly developed services, wrapped legacy applications, and applications composed of other newer and legacy services.

7.2 SOA Benefits

SOA delivers the following business benefits:

a. *Increased business agility*: SOA improves throughput by dramatically reducing the amount of time required to assemble new business applications, from existing services and IT assets. SOA also makes IT significantly easier and less expensive to reconfigure and adapt services and IT assets to meet new and unanticipated requirements. Thus, the business adapts quickly to new opportunities and competitive threats, while IT quickly changes existing systems.

b. *Better business alignment*: As SOA services directly support the services that the organization provides to customers.

c. *Improved customer satisfaction*: As SOA services are independent of specific technology, they can readily work with an array of customer-facing systems across all customer touch points that effectively reduce development time, increase customer engagement time, and, hence, increase customer solutioning, enabling enhanced customer satisfaction.

d. *Improved ROI of existing assets*: SOA dramatically improves the ROI of existing IT assets by reusing them as services in the SOA by identifying the key business capabilities of existing systems and using them as the basis for new services as part of the SOA.

e. *Reduced vendor lock-in and switching costs*: As SOA is based on loosely coupled services with well-defined, platform-neutral service contracts, it avoids vendor and technology lock-in at all levels, namely, application platform and middleware platform.

f. *Reduced integration costs*: SOA projects can focus on composing, publishing, and developing Web Services independently of their execution environments, thus obviating the need to deal with avoidable complexity. Web Services and XML simplify integration because they focus on the data being exchanged instead of the underlying programs and execution environments.

Technical benefits of SOA include the following:

a. *More reuse*: Service reuse lowers development costs and speed.

b. *Efficient development*: As services are loosely coupled, SOA promotes modularity that enables easier and faster development of composite applications. Once service contracts have been defined, developers can separately and independently design and implement each of the various services. Similarly, service requestors too can be designed and implemented based solely with reference to the published service contracts without any need to contact the concerned developers or without any access to the developers of the service providers.

   c. *Simplified maintenance*: As services are modular and loosely coupled, they simplify maintenance and reduce maintenance costs.

   d. *Incremental adoption*: As services are modular and loosely coupled, they can be developed and deployed in incremental steps.

7.3 Characteristics of SOA

7.3.1 Dynamic, Discoverable, Metadata Driven

Services should be published in a manner by which they can be discovered and consumed without intervention of the provider. Service contracts should use metadata to define service capabilities and constraints and should be machine readable so that they can be registered and discovered dynamically to lower the cost of locating and using services, reduce corresponding errors, and improve management of services.

7.3.2 Designed for Multiple Invocation Styles

Design and implement service operations that implement business logic that supports multiple invocation styles, including asynchronous queuing, request/response, request/callback, request/polling, batch processing, and event-driven publish/subscribe.

7.3.3 Loosely Coupled

This implies loose coupling of interface and technology; interface coupling implies that the interface should encapsulate all implementation details and make them nontransparent to service requesters, while technology coupling measures the extent to which a service depends on a particular technology, product, or development platform (operating systems, application servers, packaged applications, and middleware).

7.3.4 Well-Defined Service Contracts

Service contracts are more important than the service implementations because it defines the service capabilities and how to invoke the service in an interoperable manner. A service contract clearly separates the service's externally accessible interface from the service's technical implementation;

consequently, the service contract is independent of any single service implementation. The service contract is defined based on the knowledge of the business domain and not so much on the service implementation. The service contract is defined and managed as a separate artifact, is the basis for sharing and reuse, and is the primary mechanism for reducing interface coupling.

Changing a service contract is generally more expensive than modifying the implementation of a service because while changing a service implementation is relatively a localized effort, changing a service contract may entail changing hundreds or thousands of service requesters.

7.3.5 Standard Based

Services should be based on open standards as much as possible to the following benefits:

- Minimizing vendor lock-in by isolating from proprietary, vendor-specific technologies and interfaces
- Increasing the choice of service requesters for alternate service providers and vice versa

It is important to base the service-level data and process models on mature business domain standards as and when they become available. This is in addition to complying with technology standards like SOAP, WSDL, UDDI, and the WS* specification.

7.3.6 Granularity of Services and Service Contracts

Services and service contracts must be defined at a level of abstraction that makes sense to service requesters as also service providers. To achieve this, services should perform discrete tasks and provide simple interfaces to encourage reuse and loose coupling.

An abstract interface at the appropriate level of granularity promotes ready substitutability, which enables any of the existing service providers to be replaced by a new service provider without affecting any of the service requesters.

7.3.7 Stateless

Services should be stateless because they scale more efficiently as any service request can be routed to any service instance. In contrast, stateful interactions do not scale efficiently because the server needs to track which service is serving which client and cannot reuse a service until the conversation is finished or a time-out has occurred.

Thus, services should be implemented so that each invocation is independent and does not depend on the service maintaining client-specific conversations in memory or in persistent state between the invocations.

7.3.8 Predictable Service-Level Agreements (SLAs)

Service delivery platform must provide service-level management capabilities for defining, monitoring, incident logging, and metering of SLAs for service usage. SLAs should be established early because they affect service design, implementation, and management. There should also be provision for fine-tuning of SLAs based on the feedback of ongoing operations.

Typically, SLAs define metrics for services such as response time, throughput, availability, and meantime between failures. Above all, SLAs are usually tied up to a business model whereby service requesters pay more for higher or more stringent SLAs but charge a penalty when service providers default on their SLA commitments.

7.3.9 Design Services with Performance in Mind

Service invocation should not be modeled on local function calls since local transparency may result in a service that is on another machine on the same LAN or another LAN or WAN.

7.4 SOA Ingredients

Web Services are new standards for creating and delivering cooperative applications over the Internet. Web Services allow applications to communicate irrespective of the platform or the operating system.

7.4.1 Objects, Services, and Resources

Any distributed system involves sending messages to some remote entity. Underlying the differences between many systems are the abstractions used to model these entities; they define the architectural qualities of the system. Three abstractions—in particular, object, resource, and service—are commonly used to describe remote entities; their definitions, however, are not always clearly distinguished. Yet the nature of these abstractions has a profound effect on the distributed communication paradigms that result from their use. One approach to identifying the similarities and differences between them is to understand them in terms of their relationship to two properties: state and behavior.

7.4.1.1 Objects

Objects have both state and behavior. The state is maintained through the internal data, and the behavior of the object is defined through the public operations on that data. A primary issue in these systems is the management of object identifiers, which are global pointers to instances of objects. It has been argued that architectures based on global pointers lead to brittle systems if they scale to Internet size because of the proliferation of references and the need to maintain the integrity of the pointers. As a result, these systems are considered to be best suited to medium-sized networks within a single domain, with known latencies and static addressing and intimate knowledge of the middleware models used.

7.4.1.2 Services

A service is a view of some resource, usually a software asset. Implementation detail is hidden behind the service interface. The interface has well-defined boundaries, providing encapsulation of the resource behind it. Services communicate using messages. The structure of the message and the schema, or form, of its contents are defined by the interface. Services are stateless. This means all the information needed by a service to perform its function is contained in the messages used to communicate with it.

The service abstraction shares commonalities with the object abstraction but displays crucial differences:

- Like an object, a service can have an arbitrary interface.
- Like distributed object systems that use an IDL, services usually describe this interface in a description language.
- Unlike objects, services use a message-oriented model for communication. This has quite different semantics and implications to invoking a procedure on a remote object. In the latter, what the remote entity is plays a part. In the case of objects, the class of an object must be known. Once this class is known, behavior based on the class can be inferred by the consumer. Services, however, do not share class. Instead, they share contracts and schema. Therefore, what an entity is has no bearing on communication, and nothing is inferred. Furthermore, communication with an object involves addressing an instance. This is not the case with services as is discussed in the next item.
- Unlike objects, services do not have state. Object orientation teaches us that data and the logic that acts on that data should be combined, while service orientation suggests that these two things should be separate. Therefore, a service acts upon state but does not expose its own state. Put another way, services do not have instances. You cannot create and destroy a service in the way you can to an object.

7.4.1.3 Resources

The term *resource* is used here to specifically refer to the abstraction used by the Web and related initiative such as the Semantic Web. Such a resource is different from a distributed object in a number of ways:

Resource state is not hidden from a client as it is in object systems. Instead, standard representations of state are exposed. In object systems, the public interface of an object gives access to hidden state:

- Unlike distributed objects, resources do not have operations associated with them. Instead, manipulation and retrieval of resource representations rely on the transfer protocol used to dereference the uniform resource identifier (URI).
- As a consequence, a resource can be viewed as an entity that has state, but not the logic to manipulate that state, that is, no behavior.

Because resources have no behavior, they do not define how their state can be manipulated. While this could be viewed as limiting and potentially leading to ad hoc, underspecified interactions, in the case of the Web, the opposite is actually true. While an object-oriented system defines proprietary behavioral interfaces for every object, leading to a proliferation of means of manipulating objects, the Web uses a single, shared interface: HTTP. The few methods defined by HTTP allow arbitrary resources to be exchanged and manipulated, making interactions between entities far simpler and hence scalable. Imagine, for example, that every Web server defined its own interface to accessing the resources in its charge. This would require a browser to digest a new service interface and generate client-side code every time you clicked on a link, a process that would severely influence the scalability of the system as a whole.

7.4.2 SOA and Web Services

Web Services are new standards for creating and delivering cooperative applications over the Internet. The basic Web Service architecture consists of specifications (SOAP, WSDL, and UDDI) that support the interaction of a Web Service requester with a Web Service provider and the potential discovery of the Web Service description. The provider is typically publishing a WSDL description of its Web Service, and the requester accesses the description using a UDDI or other type of registry and requests the execution of the provider's service by sending a SOAP message to it. The basic Web Service standards are good for some SOA-based applications but not adequate for many others.

SOAP, originally defined as Simple Object Access Protocol is a protocol specification for exchanging structured information in the implementation of Web Services in computer networks. It relies on XML as its message format and usually relies on other application-layer protocols, most notably Remote Procedure Call (RPC) and HTTP for message negotiation and transmission. SOAP can form the foundation layer of a Web Service protocol stack, providing a basic messaging framework on which Web Services can be built.

As a simple example of how SOAP procedures can be used, a SOAP message can be sent to a Web Service–enabled website—for example, a house price database—with the parameters needed for a search. The site returns an XML-formatted document with the resulting data (prices, location, features, etc.). Because the data are returned in a standardized machine-parseable format, it may be integrated directly into a third-party site.

The SOAP architecture consists of several layers of specifications for MEPs, underlying transport protocol bindings, message processing models, and protocol extensibility. SOAP is the successor of XML-RPC. SOAP makes use of an Internet application-layer protocol as a transport protocol. Critics have argued that this is an abuse of such protocols, as it is not their intended purpose and therefore not a role they fulfill well. Proponents of SOAP have drawn analogies to successful uses of protocols at various levels for tunneling other protocols.

Both SMTP and HTTP are valid application-layer protocols used as transport for SOAP, but HTTP has gained wider acceptance because it works well with today's Internet infrastructure; specifically, HTTP works well with network firewalls. SOAP may also be used over HTTPS (which is the same protocol as HTTP at the application level but uses an encrypted transport protocol underneath) with either simple or mutual authentication; this is the advocated WS-I method to provide Web Service security as stated in the WS-I Basic Profile 1.1. This is a major advantage over other distributed protocols such as GIOP/IIOP or DCOM, which are normally filtered by firewalls. XML was chosen as the standard message format because of its widespread use by major corporations and open-source development efforts. Additionally, a wide variety of freely available tools significantly ease the transition to a SOAP-based implementation.

The advantages of using SOAP over HTTP is that SOAP allows for easier communication through proxies and firewalls than previous remote execution technology and is versatile enough to allow for the use of different transport protocols. The standard stacks use HTTP as a transport protocol, but other protocols are also usable (e.g., SMTP). SOAP is platform independent and language independent, and it is simple and extensible.

Because of the verbose XML format, SOAP can be considerably slower than competing middleware technologies such as CORBA (Common Object Request Broker Architecture). This may not be an issue when only small messages are sent. To improve performance for the special case of XML with embedded binary objects, Message Transmission Optimization Mechanism was introduced. When relying on HTTP as a transport protocol and not using WS-Addressing or an ESB, the roles of the interacting parties are fixed. Only one party (the client) can use the services of the other. Developers must use polling instead of notification in these common cases.

Most uses of HTTP as a transport protocol are made in ignorance of how the operation is accomplished. As a result, there is no way to know whether the method used is appropriate to the operation. The REST architecture has become a Web Service alternative that makes appropriate use of HTTP's defined methods.

7.4.3 SOA and RESTful Web Services

REpresentational State Transfer (REST) is a style of software architecture for distributed hypermedia systems such as the World Wide Web. As such, it is not strictly a method for building *Web Services*. The terms *representational state transfer* and *REST* were introduced in 2000 in the doctoral dissertation of Roy Fielding, one of the principal authors of the Hypertext Transfer Protocol (HTTP) specifications 1.0 and 1.1.

REST refers to a collection of network architecture principles, which outline how resources are defined and addressed. The term is often used in a looser sense to describe any simple interface that transmits domain-specific data over HTTP without an additional messaging layer such as SOAP or session tracking via HTTP cookies. These two meanings can conflict as well as overlap. It is possible to design a software system in accordance with Fielding's REST architectural style without using HTTP and without interacting with the World Wide Web. It is also possible to design simple XML+HTTP interfaces that do not conform to REST principles but instead follow a model of remote procedure call. Systems that follow Fielding's REST principles are often referred to as *RESTful*.

Proponents of REST argue that the Web's scalability and growth are a direct result of a few key design principles. Application state and functionality are abstracted into resources. Every resource is uniquely addressable using a universal syntax for use in hypermedia links, and all resources share a uniform interface for the transfer of state between client and resource. This transfer state consists of a constrained set of well-defined operations and a constrained set of content types, optionally supporting code on demand. State transfer uses a protocol that is client–server based, stateless and cacheable, and layered.

An important concept in REST is the existence of resources, each of which is referenced with a global identifier (e.g., a URI in HTTP). In order to manipulate these resources, components of the network (user agents and origin servers) communicate via a standardized interface (e.g., HTTP) and exchange representations of these resources (the actual documents conveying the information). For example, a resource, which is a circle, may accept and return a representation that specifies a center point and radius, formatted in Scalable Vector Graphics (SVG), but may also accept and return a representation that specifies any three distinct points along the curve as a comma-separated list.

Any number of connectors (clients, servers, caches, tunnels, etc.) can mediate the request, but each does so without *seeing past* its own request (referred to as *layering,* another constraint of REST and a common principle in many other parts of information and networking architecture). Thus, an application can interact with a resource by knowing two things: the identifier of the resource and the action required—it does not need to know whether there are caches, proxies, gateways, firewalls, tunnels, or anything else between it and the server actually holding the information. The application does, however, need to understand the format of the information (representation) returned, which is typically an HTML, XML, or JSON document of some kind, although it may be an image, plain text, or any other content.

RESTful Web Services rely on HTTP as a sufficiently rich protocol to completely meet the needs of Web Service applications. In the REST model, the HTTP GET, POST, PUT, and DELETE verbs are used to transfer data (often in the form of XML documents) between client and services. These documents are *representations* of *resources* that are identified by normal Web URIs (Uniform Resource Identifiers). This use of standard HTTP and Web technologies means that RESTful Web Services can leverage the full Web infrastructure, such as caching and indexing. The transactional and database integrity requirements of CRUD (Create, Retrieve, Update, and Delete) correspond to HTTP's POST, GET, PUT, and DELETE.

One benefit that should be obvious with regard to web-based applications is that a RESTful implementation allows a user to bookmark specific *queries* (or requests) and allows those to be conveyed to others across e-mail and instant messages or to be injected into wikis, etc. Thus, this *representation* of a path or entry point into an application state becomes highly portable. A RESTFul Web Service is a simple Web Service implemented using HTTP and the principles of REST. Such a Web Service can be thought of as a collection of resources comprising three aspects:

1. The URI for the Web Service
2. The MIME type of the data supported by the Web Service (often JSON, XML, or YAML but can be anything)
3. The set of operations supported by the Web Service using HTTP methods, including but not limited to POST, GET, PUT, and DELETE

REST provides improved response time and reduced server load due to its support for the caching of representations. REST improves server scalability by reducing the need to maintain session state. This means that different servers can be used to handle different requests in a session.

REST requires less client-side software to be written than other approaches, because a single browser can access any application and any resource. REST depends less on vendor software and mechanisms, which layer additional messaging frameworks on top of HTTP. It provides equivalent functionality when compared to alternative approaches to communication, and it does not require a separate resource discovery mechanism, because of the use of hyperlinks in representations. REST also provides better long-term compatibility because of the capability of document types such as HTML to evolve without breaking backward or forward compatibility and the ability of resources to add support for new content types as they are defined without dropping or reducing support for older content types.

SOAs can be built using REST services—an approach sometimes referred to as (ROA) REST-oriented architecture. The main advantage of ROA is ease of implementation, agility of the design, and the lightweight approach. The latest version of WSDL now contains HTTP verbs and is considered an acceptable method of documenting REST services. There is also an alternative known as WADL (Web Application Description Language).

7.4.3.1 Web Application Description Language (WADL)

Most RESTful services are described using free text. Although the number of RESTful services is increasing, there is currently no standard approach to describe this kind of services. If, for WS-* Web Services, the Web Service Description Language is the de facto standard, in the RESTful service domain, all approaches to describe services are still in an early stage. More structured approaches such as the Web Application Description Language (WADL) are currently emerging. The Web Application Description Language is an XML-based language for the description of RESTful services. It is designed to provide a machine-processable protocol description format for use with HTTP-based Web applications, especially those using XML to communicate.

A WADL document is defined using the following elements:

- *Application* is a top-level element that contains the overall description of the service. It might contain grammars, resources, methods, representation, and fault elements.
- *Grammars* acts as a container for definitions of any XML structures exchanged during the execution of the protocol described by the WADL document. Using the subelement *include*, one or more structures can be included.
- *Resources* act as a container for the resources provided by the Web application.
- *Resource* describes a single resource provided by the Web application. Each resource is identified by a URI, and the associated resources parent element. It can contain the following subelements: *path_variable* that is used to parameterize the identifiers of the

parent resource, zero or more *method* elements, and zero or more *resource* elements.

- *Method* describes the input to and output from an HTTP protocol method that may be applied to a resource. A method element might have two child elements: a request element that describes the input to be included when applying an HTTP method to a resource and a response element that describes the output that results from performing an HTTP method on a resource. A request element might contain query variable elements.

- *Representation* describes a representation of the state of a resource and can either be declared globally as a child of the application element, embedded locally as a child of a request or response element, or referenced externally.

- *Fault* is similar to a representation element in structure but differs in that it denotes an error condition.

7.4.3.2 Data Exchange for RESTful Services

Data exchange is one important aspect in any communication between distinct applications. Just as any other kind of service, RESTful services can be seen as applications or software entities that can be accessed over the Internet. There are two existing approaches to exchange data to and from RESTful services, namely, JSON, a lightweight computer data interchange format, and XML, a general language for sharing structured data between information systems.

JSON, short for JavaScript Object Notation, is a lightweight computer data interchange format, easy for humans to read and write, and, at the same time, easy for machines to parse and generate. Another key feature of JSON is that it is a completely language independent. Programming in JSON does not raise any challenge for programmers experienced with the C family of languages.

Though XML is a widely adopted data exchange format, there are several reasons for preferring JSON to XML:

- Data entities exchanged with JSON are typed, while XML data are typeless (some of the built-in data types available in JSON are string, number, array, and Boolean); XML data, on the other hand, are all strings.

- JSON is lighter and faster than XML as on-the-wire data format.

- JSON integrates natively with JavaScript, a very popular programming language used to develop applications on the client side; consequently, JSON objects can be directly interpreted in JavaScript code, while XML data need to be parsed and assigned to variables through the tedious usage of DOM APIs (Document Object Model APIs).

JSON is built on two structures:

1. A collection of name/value pairs. In various languages, this is realized as an object, record, structure, dictionary, hash table, keyed list, or associative array.
2. An ordered list of values. In most languages, this is realized as an array, vector, list, or sequence.

7.5 SOA Applications

An SOA can be thought of as an approach to building IT systems in which business services are the key organizing principle to align IT systems with the needs of the business. Any business that can implement an IT infrastructure that allows it to change more rapidly than its competitors has an advantage over them. The use of an SOA for integration, business process management, and multichannel access allows any enterprise to create a more strategic environment, one that more closely aligns with the operational characteristics of the business. Earlier approaches to building IT systems resulted in systems that were tied to the features and functions of a particular environment technology (such as CORBA, J2EE, and COM/DCOM) since they employed environment-specific characteristics like procedure or object or message orientation to provide solutions to business problems. The way in which services are developed aligns them better with the needs of the business than was the case with previous generations of technology. What is new in the concept of SOA is the clear separation of the service interface from execution technology, enabling choice of the best execution environment for any job and tying all of these executional agents together using a consistent architectural approach.

7.5.1 Rapid Application Integration

The combination of Web Services and SOA provides a rapid integration solution that readily aligns IT investments and corporate strategies by focusing on shared data and reusable services rather than proprietary integration products. These enterprise application integration (EAI) products proved to be expensive, consumed considerable time and effort, and were prone to higher rates of failure. Applications can more easily exchange data by using a Web Service defined at the business logic layer than by using a different integration technology because Web Services represent a common standard across all types of software. XML can be used to independently define the data types and structures. Creating a common Web Service layer or *overlay* of services into the business logic tiers of application also allows you to use a common service repository in which to store and retrieve service descriptions. If a new application seeks to use an existing service into one of these applications, it can query the repository to obtain the service description to quickly generate (say) SOAP messages to interact with it. Finally, the development of service-oriented entry points at the business logic tier allows a business process management engine to drive an automatic flow of execution across the multiple services.

7.5.2 Multichannel Access

Enterprises often use many channels to ensure good service and maintain customer loyalty; therefore, they benefit from being able to deliver customer services over a mixture of access channels. In the past, enterprises often developed monolithic applications that were tied to single access channel, such as a 3270 terminal, a PC interface, or a Web browser. The proliferation of access channels represented a significant challenge to IT departments to convert monolithic applications to allow multichannel access. The basic solution is to service-enable these using an SOA with Web Services that are good for enabling multichannel access because they are accessible from a broad range of clients, including Web, Java, C#, and mobile devices. In general, business services change much less frequently than the delivery channels through which they are accessed. Business services refer to operational functions such as vendor management, purchase order management, and billing, which do not vary very often, whereas client devices and access channels are based on new technologies, which tend to change.

7.5.3 Business Process Management

A business process is a real-world activity that consists of a set of logically related tasks that, when performed in an appropriate sequence and in conformity with applicable rules, produce a business outcome. Business process management (BPM) is the name for a set of software systems, tools, and methodologies that enable enterprises to identify, model, develop, deploy, and manage such business processes. BPM systems are designed to help align business processes with desirable business outcomes and ensure that the IT systems support those business processes. BPM systems let business users model their business processes graphically in a way that the IT department can implement; the graphical depiction of a business process can be used to generate an executable specification of the process. Unlike traditional forms of system development where the process logic is deeply embedded in the application code, BPM explicitly separates the business process logic from other application code. Separating business process logic from other application code renders increased productivity, reduced operational costs, and improved agility. When implemented correctly, enterprises can quickly respond to changing market conditions and seize opportunities for gaining competitive advantage.

SOA with Web Services can better automate business processes because Web Services help achieve the goals of BPM more quickly and easily.

7.6 Summary

Large-scale enterprise applications are increasingly being woven together from applications, packages, and components that were never designed to work together and may even run on incompatible platforms. This gives rise to a critical need for interoperability, one that becomes even more important as organizations start building a new generation of wide-area integrated applications that directly incorporate functions hosted by business partners and specialist service providers. Services and service-oriented architectures are pragmatic responses to the complexity and interoperability problems encountered by the builders of previous generations of large-scale integrated applications. Although it is possible to design and build *service-oriented systems* using any distributed computing or integration middleware, only Web Services technologies can today meet the critical requirement for seamless interoperability that is such an important part of the service-oriented vision. Just as J2EE middleware lets Java client applications call methods provided by J2EE components, the main purpose of SOA and Web Services is to enable applications to invoke functionality provided by other applications (developed in different languages and platforms). This chapter presented the definition and characteristics of service-oriented architectures along with alternate approaches to realizing the vision of service-oriented systems, namely, Web Services and RESTful services.

8

Web Services

Web Services are new standards for creating and delivering cooperative applications over the Internet. They allow applications to communicate irrespective of the platform or the operating system. By using Web Services, developers can eliminate major porting and quality testing efforts, potentially saving millions of dollars. They will radically change the way that applications are built and deployed in future.

A developer can create an application out of reusable components. But what good is it to have a large library of reusable components if nobody knows that they exist, where they are located, and how to link to and communicate with such programmatic components? Web Services are standards for finding and integrating object components over the Internet. They enable a development environment where it is no longer necessary to build complete and monolithic applications for every project. Instead, the core components can be combined from other standard components available on the Web to build the complete applications that run as services to the core applications.

Some of the past approaches for enabling program-to-program communications included combinations of program-to-program protocols such as Remote Procedure Call (RPC) and Application Programming Interfaces (APIs) coupled with architectures such as Common Object Model (COM), the Distributed Common Object Model (DCOM), and the Common Object Request Broker Architecture (CORBA). But, without a common underlying network, common protocols for program-to-program communication, and a common architecture to help applications to declare their availability and services, it has proven difficult to implement cross platform program-to-program communication between application modules. These previous attempts to set up standards for accomplishing these objectives were not very successful because

- They were not functionally rich enough and are difficult to maintain as best of breed
- They were vendor specific as opposed to using open and cross vendor standards
- They were too complex to deploy and use

The use of Web Service standards holds the potential for correcting each of these deficiencies. This new approach presents applications as services to each other and enables applications to be rapidly assembled by linking application objects together.

With the advent of the Internet and its protocols, most vendors and enterprises have graduated to a common communication and network protocol—the Internet's TCP/IP. And with the availability of Web standards such as Extensible Markup Language (XML); Simple Object Access Protocol (SOAP); Universal Description, Discovery and Integration (UDDI); and Web Services Description Language (WSDL), vendors enable customers to

1. Publish specifications about application modules via WSDL
2. Find those modules (either on the internal intranet or on the Internet) via the UDDI
3. Bind the applications together to work seamlessly and cooperatively and deliver the holistic functionality of composite application via SOAP and XML

Major hardware vendors and certain key software vendors are looking at these new Web standards for providing solutions for program-to-program communication. IBM's WebSphere Server environment, Sun Microsystems's Open Network Environment (ONE) comprised of various Sun technologies and third-party products, and Microsoft's .NET initiatives deliver Web Service–based solutions.

The significance of Web Services for the future is by reason of the following:

- Web Services will enable enterprises to reduce development costs and expand application functionality at a fraction of the cost per traditional application development and deployment method.
- Web Services will enable independent software vendors (ISVs) to bring products to market more quickly and respond to competitive threats with more flexibility.
- Web Services will enable enterprises to reuse existing legacy application functionality with the latest applications and technologies.
- Web Services will obviate the need of porting applications to different hardware platforms and operating systems at great expense.
- Web Services enable applications to communicate irrespective of platform or operating system.
- Web Services will have the effect of leveling the playing field because it will enable even specialized boutique application firms to compete

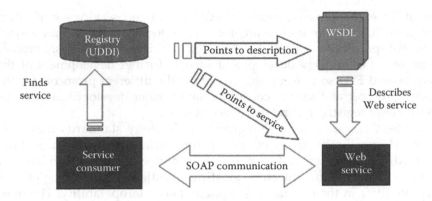

FIGURE 8.1
Web services usage model.

easily with well-established and resourceful original equipment manufacturers (OEMs).

- Web Services will enable only those OEMs to flourish that focus on providing comprehensive implementations and highly productive application development environments for Web Services.
- Web Services will enable applications to be packaged not only as licenses but also as services; this will give a big fillip to ASP services (as discussed in the earlier section) and will consequently expand the overall market size tremendously.
- Web Services will enable value-added resellers (VARs) to rapidly add new functionality to current product offerings or to customize the existing applications of customers.
- Web Services will enable enterprises to adapt better to changing market conditions or competitive threats

Figure 8.1 shows the Web Services usage model.

8.1 Web Service Standards

The standards are a collection of specifications, rules, and guidelines formulated and accepted by the leading market participants and are independent of implementation details. Standards establish a base for commonality and enable wide acceptance through interoperability. Examples of standards include a common communication language (XML), a common format for exchanging messages (SOAP), a common service specification format (WSDL), a common means for service lookup (UDDI), and a standard that specifically deals with interoperability issues (WS-I Basic Profile). Examples of technology development include further development of the ideas behind ESB, so as to be able to handle the different protocols for the service provider and service consumer, and further development of registries for easy registration and discovery of services.

Web Services are a set of integration technology standards that were designed specifically to meet the requirements arising from service-oriented architectures and systems. In many ways, Web Services are really not much different from existing middleware technologies, but they do differ in their focus on simplicity and interoperability. The most important feature offered by Web Services is that all major software vendors have agreed to support them. Interoperability is still not, of course, guaranteed to be painless but at least the problems encountered will be bugs and misinterpretations of common standards, not intentionally introduced incompatibilities between similar but different proprietary technologies.

All application integration technologies, including Web Services, really only provide four basic functions that let developers (and programs) do the following:

1. Find suitable services (using UDDI or another directory)
2. Find out about a service (using WSDL)
3. Ask a service to do something (using SOAP)
4. Make use of services such as security (using WS-* standards)

SOAP, WSDL, and UDDI were the first Web Service standards to be published, but they only meet the most basic requirements for application integration. They lack support for security, transactions, reliability, and many other important functions. This gap is being progressively filled by a series of standards (commonly called WS-*) first outlined by IBM and Microsoft at a W3C workshop in 2001. The task of creating these additional standards and getting industry-wide agreement is a confusing work in progress, with specifications in varying degrees of maturity and supported by various standards bodies. Some specifications complement, overlap, and compete with each other. There are now however production-ready implementations available for many of them. See http://www.w3.org/2002/ws/ for some insights into these specifications.

Web Services are XML standards. Services are defined using XML, and applications request services by sending XML messages and the Web Service standards make extensive use of other existing XML standards wherever possible. There are multiple Web Service standards, and these can be organized into various categories. This number of standards may suggest complexity rather than the desired simplicity, and in many applications, only a few core standards are actually in use. There is also increasingly good tool and library/framework support for these standards, so developers only have to understand the capabilities offered rather than the detailed XML syntax; in the following is a simple Web Service definition using the Java API for XML Web Services (JAX-WS), part of the JEE platform. Creating a Web Service is very simple:

```
package brokerservice.endpoint;
import javax.jws.WebService;
@WebService
public class Broker {
    @WebMethod
    public String viewStock(String name) {
    //code omitted
    }
}
```

So, with toolkits like JAX-WS, the service developer does not need to create or understand XML messages formatted as SOAP. The JAX-WS runtime system simply converts the API calls and responses to and from underlying SOAP message formats.

One of the simplifying principles underlying Web Services is that the various message fields and attributes used to support functions such as security and reliability are totally independent of each other. Applications only need to include just those few fields and attributes needed for their specific purposes and can ignore all the other standards. For example, a SOAP request might identify the requestor of a service by including a username and password in the form specified in the WS-Security UsernameToken profile. This user/password-related information is the only security-related header element included in the message. WS-Security supports other forms of user authentication, as well as encryption and digital signatures, but as these are not used by the service, they do not appear at all in the SOAP message request.

Another aim of the Web Service standards is to provide good support for system architectures that make use of *intermediaries*. Rather than assuming that clients always send requests directly to service providers, the intermediary model assumes that these messages can (transparently) pass along a chain of other applications on their way to their final destination. These intermediaries can do anything with the messages they receive, including routing them, auditing, logging, checking security, or even adding or subtracting bits of the message's content. Web services provide support for intermediary-based architectures in a number of ways. These include tagging header elements with the role of their intended recipient and supporting the *end-to-end* principle for functions such as security, so ensuring that they continue to work even if messages pass through intermediaries rather than traveling directly from client to service. For example, the client can use mechanisms provided by WS-Security to protect sensitive information intended only for the credit card application, hiding it from the router that the message must pass through on its journey.

We start our discussion of standards with the Extensible Markup Language (XML) because XML forms the basis on which most of the other standards are built.

8.2 XML

John Bosak of Sun Microsystems is credited with the revolutionary work on eXtensible Markup Language (XML). The idea of XML essentially emerged from the other nonexpendable markup languages such as Generalized Markup Language (GML) from IBM, Standardized Generalized Markup Language (SGML) from ISO, and Hypertext Markup Language (HTML) from ECRN. XML's popularity essentially stems out of its extensible capability. One of the biggest contributions of XML is its capability of interoperability. The development of XML resulted in its adoption by a variety of industries—both vertical and horizontal. This has resulted in the creation of a large number of XML vocabularies that cater to the interoperability needs of different industries. But the biggest impact of XML for enterprise solution has as a part of the SOAP, WSDL, and UDDI technologies.

XML is probably the most important of the standards on which Web Services are built. XML documents are often used as a means for passing information between the service provider and the service consumer. XML also forms the basis for WSDL (Web Services Description Language), which is used to declare the interface that a Web Service exposes to the consumer of the service. Additionally, XML underlies the SOAP protocol for accessing a Web Service. Lastly, UDDI (Universal Description, Discovery, and Integration), which is used to publish and discover a Web Service, is also based on XML. Similar to HTML, XML uses tags. However, unlike HTML, where tags are used to indicate how the data should be presented or displayed, in XML, tags are used to describe what the data are. Another difference from HTML is that tags are not fixed but can be invented whenever there is a need for a new one.

XML has been adopted as a popular middleware-independent standard format for the exchange of data and documents. XML is basically the lowest common denominator upon which the IT industry can agree. Unlike CORBA, IDL, and Java interfaces, XML is not bound to any particular technology or middleware standard and is often used today as an ad hoc format for processing data across different, largely incompatible middleware platforms. XML is free and comes with a large number of tools on many different platforms, including different open source parsing APIs such as SAX, StAX, and DOM.

These tools enable the processing and management of XML documents. Another advantage of XML is that it retains the data's structure in transit. In addition, XML is very flexible, and this flexibility positions XML as the most suitable standard for solving middleware and application heterogeneity problems. XML also solves the data format problem mentioned previously.

A basic XML document consists of a top element. This top element may consist of data (the payload), an attribute, and any number of other elements in a recursive manner. A sample portion of a simple XML document is shown in Listing 8.1. This document contains a top element named address, which has a single attribute that is used to specify the country. This top element has also four child elements, which provide information on the name of the person, the street address, the city, and the postal code. Each of these child elements has data (that is, a payload) contained in them. For example, the data for the name element are *Sachin Tendulkar*.

Listing 8.1: Basic XML document structure

```
1 <address country = "India">
2      <name>Sachin Tendulkar</name>
3      <street>Dadar Gymkhana</street>
4      <city>Mumbai</city>
5      <state>Maharashtra</state>
6      <postal-code>45561</postal-code>
7 </address>
```

The grammar and structure of an XML type document is defined in a schema. Another important concept used in XML is namespaces, which are used to avoid the collision of names in different spaces and to extend the use of vocabulary defined in one specific domain to other domains. The discussion will include schemas, namespaces, and various models to use for XML parsing, processing, creating, and editing.

8.3 WSDL

Web Services Description Language (WSDL) is an XML-based language for describing the interface and other characteristics of a Web Service. This is the second application of XML to solve the heterogeneity problems mentioned earlier in this chapter. WSDL offers the following advantages in the description of the services as compared to previously described approaches:

- Unlike CORBA's IDL and RPC's specification files, WSDL is more completely agnostic toward programming languages and middleware technologies. This feature of WSDL is the direct result of it being based on XML, thus making WSDL suitable to describe almost any type of service.

- WSDL provides a method of specifying a communication protocol for invoking a service. Therefore, a service is free to choose any protocol it can conveniently implement.
- WSDL also provides a way to specify a message format for communicating with a given service. Therefore, a service is free to choose any convenient message format. An example of a message format is SOAP.
- WSDL also provides wide latitude for the service provider to specify the type of service operations they offer. In general, four different types of service operations can be specified, including synchronous operations and asynchronous operations.
- Finally, WSDL has a method for specifying a service endpoint. A service endpoint is the network address at which the service is available for invocation.

WSDL is used to describe Web Services, including their interfaces, methods, and parameters. WSDL characterizes the interface, which consists of two parts:

1. An abstract interface description containing the supported operations, the operation parameters, and their types
2. A binding and implementation description containing a binding of the abstract description to a concrete transport protocol, message format, and network address

The WSDL description of a service called StockQuoteService that provides a single operation named GetLastTradePrice is depicted in Listing 8.2.

Listing 8.2 WSDL for the GetLastTradePrice service

```
<?xml version = "1.0"?>
<definitions name = "StockQuote"
            targetNamespace = "http://myCompany.com/
              stockquote.wsdl"
            xmlns:tns = "http://myCompany.com/stockquote.
              wsdl"
            xmlns:soap = "http://schemas.xmlsoap.org/wsdl/
              soap/"
            xmlns:xsd = "http://www.w3.org/2001/XMLSchema"
            xmlns = "http://schemas.xmlsoap.org/wsdl/">
[Abstract data type definitions]
            <message name = "GetLastTradePrice">
```

[Data that is sent]
```
                <part name = "body" type = "xsd:string"/>
            </message>
            <message name = "LastTradePrice">
```
[Data that is returned]
```
                <part name = "body" type = "xsd:float"/>
            </message>
            <portType name = "StockQuotePortType">
```
[Port type containing one operation]
```
            <operation name = "GetLastTradePrice">
```
[An operation with Input and Output messages]
```
                    <input message =
                        "tns:GetLastTradePrice"/>
                    <output message =
                        "tns:LastTradePrice"/>
                </operation>
            </portType>

            <binding name = "StockQuoteBinding"
                type = "tns:StockQuotePortType">
                <soap:binding style = "document"

            transport = "http://schemas.xmlsoap.org/
                soap/http"/>
            <operation name = "GetLastTradePrice">
                <soap:operation soapAction =
                            "http://myCompany.com/
                                GetLastTradePrice"/>
                    <input>
                            <soap:body use = "literal"/>
                    </input>
                    <output>
                            <soap:body use = "literal"/>
                    </output>
                </operation>
```
[Binding to a specific protocol]
```
            </binding>

        <service name = "StockQuoteService">
            <documentation>Stock quote service</
                documentation>
```
[Binding to a specific service]
```
                <port name = "StockQuotePort"
                        binding = "tns:StockQuoteBinding">
                <soap:address location =
                        "http://myCompany.com/
                            stockServices"/>
            </port>
        </service>
        </definitions>
```

This operation takes one parameter symbol of type string that names the stock of interest and returns a float that holds the most recently traded price. As this listing shows, a complete WSDL document consists of a set of definitions, starting with a root *definition* element, followed by six individual element definitions—types, message, portType, binding, service and port—which describe a service.

Here is a brief description of the top six elements:

1. *types*: This element defines the data types contained in messages exchanged as part of the service. Data types can be simple, complex, derived, or array types. Types (either schema definitions or references) that are referred to in a WSDL document's message element are defined in the WSDL document's type element.

2. *message*: This element defines the messages the service exchanges. A WSDL document has a message element for each message that is exchanged, and the message element contains the data types associated with the message. For example, in the Listing 8.2, the first message contains a single part that is of the string type.

3. *portType*: This element specifies, in an abstract manner, operations and messages that are part of the service. A WSDL document has one or more portType definitions for each service it defines. In Listing 8.2, only one port type, StockQuotePortType, is defined.

4. *binding*: This element binds the abstract port type, and its messages and operations, to a transport protocol and to a message format. In Listing 8.2, one operation, GetLastTradePrice, is defined, which has an input message and an output message. Both of these messages are exchanged in SOAP body formats. The binding transport protocol is HTTP.

5. *service*: This element together with the port defines the name of an actual service and, by providing a single address for binding, assigns an individual endpoint for the service.

6. *port*: A port can have only one address. The service element groups related ports together and, through its name attribute, provides a logical name for the service. In Listing 8.2, one service (stockServices) is defined that has a single port (or endpoint) with the address http://mycompany.com/stockServices.

WSDL is well supported by development environments such as Visual Studio, Eclipse, and WebSphere. These tools can generate WSDL automatically from program method and interface definitions, and they take in WSDL service definitions and make it easy for developers to write code that calls these services. One adverse side effect of this tool support is that it tends to encourage developers to think of services as remote methods, rather than moving to the preferable and richer message-based model provided by Web Services.

8.4 SOAP and Messaging

Although adopting XML is an important step forward in dealing with heterogeneity and extensibility requirements, XML by itself is not sufficient for two parties (the service provider and service consumer applications) to properly communicate. For effective communications, the parties must be able to exchange messages according to an agreed-upon format. Simple Object Access Protocol (SOAP) is such a protocol, providing a common message format for services. SOAP originally stood for Simple Object Access Protocol, but it is now officially no longer an acronym. SOAP clients send XML request messages to service providers over any transport and can get XML response messages back in return.

SOAP is a text-based messaging format that uses an XML-based data encoding format. SOAP is independent of both the programming language and the operational platform, and it does not require any specific technology at the endpoints, thus making it completely agnostic toward vendors, platforms, and technologies. It specifies a simple but extensible XML-based application-to-application communication protocol, roughly equivalent to DCE's RPC or Java's RMI, but much less complex and far easier to implement as a result. This simplicity comes from deliberately staying well away from complex problems, such as distributed garbage collection and passing objects by reference. Its text format also makes SOAP a firewall-friendly protocol. Although SOAP was originally designed to work only with HTTP, any transport protocol or messaging middleware can be used to carry a SOAP message. All that the SOAP standard does is define a simple but extensible message-oriented protocol for invoking remote services, using HTTP, SMTP, UDP, or other protocols as the transport layer and XML for formatting data.

The SOAP message is a complete (or valid) XML document, with the top element being the envelope element. The envelope element contains a body element and an optional header element. The body element usually carries the actual message, which is consumed by the recipient. The header element is generally used for advanced features for intermediate processors.

Listing 8.3 An example of SOAP message

```
<soap:envelope xmlns:soap = "http://schemas.xmlsoap.org/soap/
    envelope/"
soap:encodingStyle = "http:/schemas.xmlsoap.org/soap/
    encoding/"/>
        <soap:header>
        </soap:header>
```

```
<soap:body>
        <m:GetLastTradePrice xmlns:m = "http://example.
            org/Tradeprice" >
        <tickerSymbol> COMPANY </tickerSymbol>
        </m:GetLastTradePrice>
        </soap:body>
</soap:envelope>
```

The listing shows how a SOAP message is encoded using XML and illustrates some SOAP elements and attributes. As the listing shows, the top element in SOAP must be the envelope element, which must contain two namespaces. The namespace SOAP:encodingStyle indicates the SOAP encoding, and the other namespace connotes the SOAP envelope. The header element is optional, but when it is present, it should be the first immediate child of the envelope element. The body element must be present in all SOAP messages and must follow the header element if it is present. The body usually contains the specification of the actual message. In this example, the message contains the name (GetLastTradePrice) of the method as well as an input parameter value (COMPANY).

Another sample SOAP message is shown in Listing 8.4. The request carries a username and hashed password in the header to let the service know who is making the request. The header holds information about the message payload, possibly including elements such as security tokens and transaction contexts. The body holds the actual message content being passed between applications. The SOAP standard does not mandate what can go in a message header, giving SOAP its extensibility as new standards, such as WS-Security, can be specified just by defining new header elements and without requiring changes to the SOAP standard itself.

Listing 8.4 SOAP message sample

```
<?xml version = "1.0" encoding = "utf-8" ?>
<soap:Envelopexmlns:soap =
        "http://www.w3.org/2003/05/soap-envelope"
xmlns:xsi = "http://www.w3.org/2001/XMLSchema-instance"
xmlns:xsd = "http://www.w3.org/2001/XMLSchema"
xmlns:wsa = "http://schemas.xmlsoap.org/ws/2004/03/addressing"
xmlns:wsse = "http://docs.oasis-open.org/wss/2004/01/oasis-
    200401-wss-wssecurity-secext-1.0.xsd"
xmlns:wsu = "http://docs.oasis-open.org/wss/2004/01/oasis-
    200401-wss-wssecurity-utility-1.0.xsd">

<soap:Header>
<wsa:Action>
http://myCompany.com/getLastTradePrice</wsa:Action>
        <wsa:MessageID>uuid:4ec3a973-a86d-4fc9-bbc4-ade31d0370dc
        </wsa:MessageID>
        <wsse:Security soap:mustUnderstand = "1"
```

```
        <wsse:UsernameToken>
            <wsse:Username>NNK</wsse:Username>
            <wsse:PasswordType = "http://docs.oasisopen.
            org/wss/2004/01/oasis-200401-wss-username
                -token-profile-1.0#PasswordDigest">
            weYI3nXd8LjMNVksCKFV8t3rgHh3Rw ==
              </wsse:Password>
            <wsse:Nonce>WScqanjCEAC4mQoBE07sAQ ==
              </wsse:Nonce>
            <wsu:Created>2003-07-16T01:24:32Z</
                wsu:Created>
          </wsse:UsernameToken>
      </wsse:Security>
</soap:Header>

<soap:Body>
      <m:GetLastTradePrice
      xmlns:m = "http://myCompany.com/stockServices">
      <symbol>DIS</symbol>
      </m:GetLastTradePrice>
</soap:Body>
</soap:Envelope>
```

SOAP was the original Web Service standard and is still the most important and most widely used. Web Services do not depend solely on HTTP as a transport layer. There are a number of other standards included in the Web Service messaging category, including WS-Addressing and WS-Eventing. SOAP messages can be sent over any transport protocol, including TCP/IP, UDP, e-mail (SMTP), and message queues, and WS-Addressing provides transport-neutral mechanisms to address services and identify messages. WS-Eventing provides support for a publish–subscribe model by defining the format of the subscription request messages that clients send to publishers. Published messages that meet the provided filtering expression are sent to callers using normal SOAP messages.

8.5 UDDI

In addition to service interface declaration (WSDL) and the SOAP messaging standard, a large enterprise also needs a central place where the service provider can publish their services using WSDL and the service consumer can discover existing services. This is mainly due to the fact that in a large enterprise, developer resources may be dispersed geographically. In particular, service providers and service consumers may be located far apart. Such a central place is given the name registry. A registry is like a library card catalog used for recording the arrival of new books and other media as well as looking up books and other media. Another common analogy is the telephone system's Yellow Pages, used by service providers to publish their services and by service consumers to find services.

The Universal Description, Discovery, and Integration (UDDI) specification defines a standard way of registering, deregistering, and looking up services. Figure 9.n shows how UDDI enables the dynamic description, discovery, and integration of services. A service provider first registers a service with the UDDI registry. A service consumer looks up the required service in the UDDI registry. Then, when it finds the required service, the consumer directly binds with the provider to use the service.

The role of the UDDI registry in Web Services is similar to the role played by a search engine on the Internet. The power of the search engine comes from the keywords used to classify content. In a similar manner, a fine-grained search for a Web Service is possible only if a service is classified properly. The classification and identification taxonomies present in the UDDI registry provide a starting point for describing Web Services. Equally important is the classification of the businesses and organizations that offer Web Services.

SOAP services that are normally described using WSDL (Web Services Description Language) can be located by searching a UDDI (Universal Description, Discovery, and Integration) directory. Services can describe their requirements for things like security and reliability using policy statements, defined using the WS-Policy framework, and specialized policy standards such as WS-SecurityPolicy. These policies can be attached to a WSDL service definition or kept in separate policy stores and retrieved using WS MetadataExchange.

UDDI has proven to be the least used so far of the original three Web Service standards. Organizations are developing large complex Web Service systems today without the use of global UDDI directories, using other methods of finding services such as personal contact or published lists of services on websites. This could all change in the future, especially when industry associations start releasing common service definitions and need to publish directories of qualified service providers.

8.6 Security, Transactions, and Reliability

One of the problems faced by most middleware protocols is that they do not work well on the open Internet because of the connectivity barriers imposed by firewalls. Most organizations do not want outsiders to have access to the protocols and technologies they use internally for application integration and so block the necessary TCP/IP ports at their perimeter firewalls. The common technology response to this problem, and the one adopted by

Web Services, has been to coopt the Web protocol, HTTP, as a transport layer because of its ability to pass through most firewalls. This use of HTTP is convenient but also creates potential security problems as HTTP traffic is no longer just innocuously fetching Web pages. Instead, it may be making direct calls on internal applications.

WS-Security and its associated standards address these problems by providing strong cryptographic mechanisms to identify callers (authentication), protect content from eavesdroppers (encryption), and ensure information integrity (digital signatures). These standards are designed to be extensible, letting them be adapted easily to new security technologies and algorithms, and also supporting integration with legacy security technologies. WS-Security supports intermediary-based application architectures by allowing multiple security header elements, each labeled with the role of their intended recipient along the processing chain, and by supporting partial encryption and partial signatures. For instance, the sensitive credit card details can be hidden by encrypting them while leaving the rest of the message unencrypted so that it can be read by the routing application.

The final set of Web Service standards supports transactions and reliable messaging. There are two types of Web Service transactions supported by standards. WS-AtomicTransactions supports conventional distributed ACID transactions and assumes levels of trust and fast response times that make this standard suitable only for internal application integration tasks and unusable for Internet-scale application integration purposes. WS-BusinessActivity is a framework and a set of protocol elements for coordinating the termination of loosely coupled integrated applications. It provides some support for atomicity by invoking compensators when a distributed application finishes in failure.

The support for reliable messaging in Web Services simply ensures that all messages sent between two applications actually arrive at their destination in the order they were sent. WS-ReliableMessaging does not guarantee delivery in the case of failure, unlike queued messaging middleware using persistent queues. However, it is still a useful standard as it provides at most once in-order message delivery over any transport layer, even unreliable ones such as UDP or SMTP.

8.7 Semantic Web Services

Semantic Web Services (SWS) were proposed in order to pursue the vision of the Semantic Web presented in, whereby intelligent agents would be able to exploit semantic descriptions in order to carry out complex tasks on behalf of humans. Semantic Web Services were first proposed as an extension of Web Services with semantic descriptions in order to provide formal declarative definitions of their interfaces as well as to capture declaratively what the services do. The SWS approach is about describing services with metadata on the basis of domain ontologies as a means to enable their automatic location, execution, combination, and usage. If services are described and annotated using machine-understandable semantics, a service requestor may specify the service needed in terms of the problem domain, or in other words, using business terminology. This is the essential progress over what can be achieved with Web Services. While Web services could act as proxies for business services, which was the basis of their appeal for corporate companies earlier, a number of problems emerged related to how online Web Services could be found, invoked, and composed.

There are four main types of semantics that corresponding semantic descriptions can capture:

1. Data semantics: The semantics pertaining to the data used and exposed by the service
2. Functional semantics: Semantics pertaining to the functionality of the service
3. Nonfunctional semantics: Semantics related to the nonfunctional aspects of the service, for example, quality of service (QoS), security, or reliability
4. Execution semantics: Semantics related to exceptional behaviors such as runtime errors

The essential characteristic of SWS is therefore the use of languages with well-defined semantics covering the subset of the mentioned categories that are amenable to automated reasoning. Several languages have been used so far including those from the Semantic Web, for example, Resource Description Framework (RDF) and Web Ontology Language (OWL), SWS-specific languages such as the Web Service Modelling Language (WSML), or others originating from research on Knowledge-Based Systems such as F-Logic and Operational Conceptual Modelling Language (OCML).

SWS technologies seek to automate the tasks involved in the life cycle of service-oriented applications, which include the discovery and selection of services, their composition, their execution, and their monitoring among others.

8.8 Summary

This chapter discussed standards like XML, SOAP, WSDL, and UDDI. XML provides a middleware-independent format for the exchange of data and documents. SOAP provides a common message format for application interaction. WSDL provides a language- and platform-independent way to specify the interface offered by a service. A WSDL document consists of two parts. The first part describes in an abstract manner the operations, input and output parameters, and data types. The second part, which consists of a binding and implementation interface, specifies the transport protocol, message format, and service endpoint network address. The Universal Description, Discovery, and Integration (UDDI) specification defines a standard way of registering, deregistering, and looking up services. The last standard, WS-I Basic Profile, promotes the interoperability of services operating on different platforms by specifying additional constraints and clarifications on the aforementioned standards.

9

Enterprise Service Bus (ESB)

An ESB provides an implementation backbone for a SOA that treats applications as services. The ESB is about configuring applications rather than coding and hardwiring applications together. It is a lightweight infrastructure that provides plug-and-play enterprise functionality. It is ultimately responsible for the proper control, flow, and even translations of all messages between services, using any number of possible messaging protocols. An ESB pulls together applications and discrete integration components to create assemblies of services to form composite business processes, which in turn automate business functions in an enterprise. It establishes proper control of messaging as well as applying the needs of security, policy, reliability, and accounting, in an SOA architecture. With an ESB SOA implementation, previously isolated ERP, CRM, supply chain management, and financial and other legacy systems can become SOA enabled and integrated more effectively than when relying on custom, point-to-point coding or proprietary EAI technology. The end result is that with an ESB, it is then easier to create new composite applications that use pieces of application logic and/or data that reside in existing systems.

9.1 Defining Enterprise Service Bus (ESB)

The Enterprise Service Bus (ESB) is an open standard-based message backbone designed to enable the implementation, deployment, and management of SOA-based solutions with a focus on assembling, deploying, and managing distributed service-oriented architecture (SOAs). An ESB is a set of infrastructure capabilities implemented by middleware technology that enable an SOA and alleviate disparity problems between applications running on heterogeneous platforms and using diverse data formats. The ESB supports service invocations, message, and event-based interactions with appropriate service levels and manageability. The ESB is designed to provide interoperability between larger-grained applications and other components via standard-based adapters and interfaces. The bus functions as both transport and transformation facilitator to allow distribution of these services over disparate systems and computing environments.

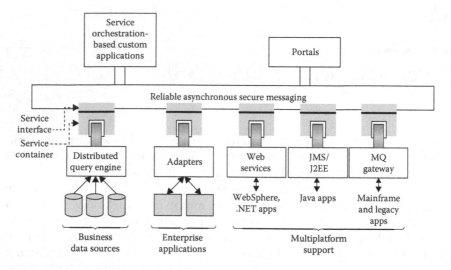

FIGURE 9.1
Enterprise Service Bus (ESB) linking disparate systems and computing environments.

The ESB distributed processing infrastructure is aware of applications and services and uses content-based routing facilities to make informed decisions about how to communicate with them. In essence, the ESB provides *docking stations* for hosting services that can be assembled and orchestrated and are available for use to any other service on the bus. Once a service is deployed into a service container, it becomes an integral part of the ESB and can be used by any application or service connected to it. The service container hosts, manages, and dynamically deploys services and binds them to external resources, for example, data sources, enterprise, and multiplatform applications, such as shown in Figure 9.1.

The distributed nature of the ESB container model allows individual event-driven services to be plugged into the ESB backbone on an as-needed basis. It allows them to be highly decentralized and work together in a highly distributed fashion, while they are scaled independently from one another. Applications running on different platforms are abstractly decoupled from each other and can be connected together through the bus as logical endpoints that are exposed as event-driven services. The WS-Notification family of specifications will bring the publish/subscribe functionality to ESB-focused current incarnations of Web Service standards (see Chapter 8, Section 8.1, "Web Service Standards").

To successfully build and deploy a distributed SOA, there are five design/deployment and management aspects that need to be addressed first:

1. *Service analysis and design*: A service development methodology should be used to enable service-oriented development and the reuse of existing applications and resources.

2. *Service enablement*: The service development methodology should determine which discrete application elements need to be exposed as services.

3. *Service orchestration*: Distributed services need to be configured and orchestrated in a unified and clearly defined distributed process.

4. *Service deployment*: Emphasis should also be placed on the production environment that addresses security, reliability, and scalability concerns.

5. *Service management*: Services must be audited, maintained, and reconfigured, and corresponding changes in processes must be made without rewriting the services or underlying application.

9.1.1 Evolution of ESB

Conceptually, the ESB has evolved from the store and forward mechanism found in middleware products and now is a combination of EAI, Web Services, XSLT, and orchestration technologies, such as BPEL. To achieve its operational objectives, the ESB draws from traditional EAI broker functionality in that it provides integration services such as connectivity and routing of messages based on business rules, data transformation, and adapters to applications. These capabilities are themselves SOA based in that they are spread out across the bus in a highly distributed fashion and hosted in separately deployable service containers. This is a crucial difference from traditional integration brokers, which are usually heavyweight, highly centralized, and monolithic in nature. The ESB approach allows for the selective deployment of integration broker functionality exactly where it is needed with no additional overbloating where it is not required.

To surmount problems of system heterogeneity and information model mismatches in an SOA implementation, an EAI middleware supporting hub-and-spoke integration patterns could be used. The hub-and-spoke approach introduces an integration layer between the client and server modules that must support interoperability among and coexist with deployed infrastructure and applications, and not attempt to replace them. However, this approach has its own drawbacks as a hub can be a central point of failure and can quickly become a bottleneck.

A scalable distributed architecture such as an SOA needs to employ a constellation of hubs. The requirements to provide an appropriately capable and manageable integration infrastructure for Web Services and SOA are coalescing into the concept of the Enterprise Service Bus (ESB), which will be the subject of this section. The two key ideas behind this approach are to loosely couple the systems taking part in the integration and break up the integration logic into distinct, easily manageable pieces.

Figure 9.1 above shows a simplified view of an ESB that integrates a J2EE application using JMS, a .NET application using a C# client, an MQ application that interfaces with legacy applications, and external applications and data sources using Web Services. In an ESB application, development tools allow new or existing distributed applications to be exposed as Web Services and be accessed via a portal. In general, resources in the ESB are modeled as services that offer one or more business operations. Technologies like J2EE Connector Architecture (JCA) may also be used to create services by integrating packaged applications (like ERP systems), which would then be exposed as Web Services.

An ESB enables the more efficient value-added integration of a number of different application components, by positioning them behind a service-oriented facade and by applying Web Service technology to the problem. For instance, in the figure above, a distributed query engine, which is normally based on XQuery or SQL, enables the creation of business data services, for example, sales order data or available product sets, by providing uniform access to a variety of disparate business data sources or organization repositories.

Endpoints in the ESB depicted in the Figure 9.1 above provide abstraction of physical destination and connection information (like TCP/IP host name and port number). In addition, they facilitate asynchronous and highly reliable communication between service containers using reliable messaging conventions. Endpoints allow services to communicate using logical connection names, which an ESB will map to actual physical network destinations at runtime. This destination independence gives the services that are part of the ESB the ability to be upgraded, moved, or replaced without having to modify code and disrupt existing ESB applications. For instance, an existing ESB invoicing service could be easily upgraded or replaced with a new service without disrupting other applications. Additionally, duplicate processes can be set up to handle failover if a service is not available. The endpoints can be configured to use several levels of QoS, which guarantee communication despite network failures and outages.

9.2 Elements of an ESB Solution

There are alternative ways to implement an ESB. The ESB itself can be a single centralized service or even a distributed system consisting of peer and sub-peer ESBs—in the form of an ESB federation—all working in tandem to keep the SOA system operational. In small-scale implementations of integration solutions, the physical ESB infrastructure is likely to be a centralized ESB topology. A centralized ESB topology is concentrated on a single cluster, or hub, of servers. This solution is reminiscent of hub-and-spoke middleware topologies, which use a central node that manages all interactions between applications and prevent an application having to integrate multiple times with several other applications (see Chapter 5, Section 5.12.1, "Replacing a Point-to-Point Integration Architecture with a Broker"). The hub-and-spoke approach simply carries out one integration process on the central node, which is a central point of control responsible for integration/translation activities, maintaining routing information, service naming, and so forth. The most popular hub-and-spoke EAI solution for the interenterprise arena is integration brokering.

Even though a hub-and-spoke solution is capable of being stretched out across organizational boundaries, it still does not allow the local autonomy that individual business units require to operate semi-independently of each other. This is usually caused by the integration broker's inability to easily span firewalls and network domains. However, as explained earlier in this chapter, the most serious drawback of this approach is that hub-and-spoke solutions can quickly become a point of contention for large-scale implementations. In an environment of loosely coupled units, it does not make sense for business process flow between localized applications or security domains to be managed by a single centralized authority like an integration broker.

In circumstances where organizational or geographically dispersed units need to act independently from one another, the infrastructure may become more physically distributed while retaining at least logically the central control over configuration. This calls for a federated hub solution. A federated ESB allows different enterprises such as manufacturers, suppliers, and customers to plug together their integration domains into a larger federated integration network. This topology allows for local message traffic, integration components, and adapters to be locally installed, configured, secured, and managed while allowing for a single integrated transaction and security model. In this figure, a federated ESB solution is used to form a virtual network of trading partners across industries and services able to take advantage of the wider range of options and partnering models.

The physical deployment of the ESB depends on candidate ESB technologies such as specialized MOM, integration brokers, and application servers. The use and combination of different candidate ESB technologies result in a variety of ESB patterns, each having its own requirements and constraints in connection with its physical deployment. Some ESB configurations might be suited to very widespread distribution to support integration over large geographical areas, while others might be more suited to deployment in localized clusters to support high availability and scalability. Matching the requirements for physical distribution to the capabilities of candidate technologies is an important aspect of ESB design. Also important is the ability to incrementally extend the initial deployment to reflect evolving requirements, to integrate additional systems, or to extend the geographical reach of the ESB infrastructure.

Irrespective of its implementation topology, the main aim of the ESB is to provide virtualization of the enterprise resources, allowing the business logic of the enterprise to be developed and managed independently of the infrastructure, network, and provision of those business services. Implementing an ESB requires an integrated set of middleware facilities that support the following interrelated architectural styles:

- Service-oriented architectures (SOAs), where distributed applications are composed of granular reusable services with well-defined, published, and standard-compliant interfaces
- Message-driven architectures, where applications send messages through the ESB to receiving applications
- Event-driven architectures, where applications generate and consume messages independently of one another

The ESB supports these architectural styles and service interaction capabilities and provides the integrated communication, messaging, and event infrastructure to enable them, as explained in the previous section. To achieve its stated objectives, the ESB amalgamates functional capabilities of application servers, integration brokers, and business process management technologies and product sets into a single integrated infrastructure. These middleware solutions are discussed in turn in the following sections.

9.2.1 Integration Brokers

To integrate disparate business applications, one must concentrate on the characteristics and functions of integration brokers, which we covered as part of the introduction to the distributing infrastructure in Chapter 3, Section 3.1 "Distributed Applications".

The integration broker is the system centerpiece. It facilitates information movement between two or more resources (source and target applications) and accounts for differences in application semantics and heterogeneous platforms. The various existing (or component) ESs, such as CRM, ERP systems, transaction processing monitors, and legacy systems, in this configuration, are connected to the integration broker by means of resource adapters.

The integration broker architecture presents several advantages given that integration brokers try to reduce the application integration effort by providing prebuilt functionality common to many integration scenarios. The value proposition rests on reuse (in terms of middleware infrastructure and the application integration logic) across multiple applications and initiatives. Modern integration brokers incorporate integration functionality such as transformation facilities, process integration, business process management and trading partner management functionality, packaged adapters, and user-driven applications through front-end capabilities such as Java Server Pages (JSPs).

In an ESB, the functionality of an integration broker, such as messaging and connectivity, application adapters, a data transformation engine, and routing of messages based on business rules, is spread out across a highly distributed architecture that allows selective deployment and independent scalability of each of those pieces. This is an important difference from the classic integration broker model where these capabilities are localized to a central monolithic server. In many situations, it is essential that newly developed ESB solutions be bridged to existing integration broker installations. In this scenario, the integration broker installation now becomes the asset, which the ESB utilizes to support new application development in future.

9.2.2 Application Servers

Another critical middleware infrastructure used in connection with ESBs is application servers. Application servers offer an integrated development environment for developing and deploying distributed Web- and non-web-based applications and services. Application servers typically provide Web connectivity for extending existing solutions and bring transaction processing mechanisms to the Web. An application server is a natural point for application integration as it provides a platform for development, deployment, and management of web-based, transactional, secure, distributed, and scalable enterprise applications. The application server middleware enables the functions of handling business processes and transactions as well as extending back-end business data and applications to the Web to which it exposes them through a single interface, typically a Web browser. This makes application servers ideal for portal-based ESB development. Unlike integration brokers, application servers do not integrate back-end systems directly but rather act as an integrated development and support framework for integrating business processes between enterprises. An application server expects the integration broker to function as a service provider providing data access, transformations, and content-based routing.

The adapter/component wrapper modules are responsible for providing a layer of abstraction between the application server and the component ES. This layer allows for ES component communications, as if the component ES were executed within the application server environment itself. Execution in this type of architecture occurs among component wrappers within the application server. The component wrappers in this facilitate point integration of component systems by wrapping legacy systems and applications and other back-end resources such as databases, ERP, CRM, and SRM, so that they can express data and messages in the standard internal format expected by the application server. The application server is oblivious to the fact that these components are only the external facade of existing ESs that do the real processing activities.

It is useful to clarify their meaning and intended purpose of wrappers. A (component) wrapper is nothing but an abstract component that provides a service implemented by legacy software that can be used to hide existing system dependencies. In a nutshell, a wrapper performs specific business functions, has a clearly defined API, and can appear as *standard* components for clients and be accessed by using modern industry protocols. Wrappers are normally deployed in modern, distributed environments, such as J2EE or .NET. A wrapper can provide a standard interface on one side, while on the other, it interfaces with existing application code in a way that is particular to the existing application code. The wrapper combines the existing application functionality that it wraps with other necessary service functionalities and represents it in the form of a virtual component that is accessed via a standard service interface by any other service in an ESB. When a legacy business process is wrapped, its realization comprises code to access an adapter that invokes the legacy system. An adapter, unlike a wrapper, does not contain presentation or business logic functions. It is rather a software module interposed between two software systems in order to convert between their different technical and programmatic representations and perceptions of their interfaces. Resource adapters translate the applications' messages to and from a common set of standards—standard data formats and standard communication protocols.

Application servers are principally J2EE based and include support for JMS, the Java 2 Connector Architecture, and Web Services; these technologies help implement application servers in the context of the ESB. Recall from Chapter 5, Section 5.5.2 "Java Messaging Service" that JMS is a transport-level vendor-agnostic API for enterprise messaging that can be used with many different MOM vendors. JMS frameworks not only function in asynchronous mode but also offer the capability of simulating a synchronous request/response mode. For application server implementations, JMS provides access to business logic distributed among heterogeneous systems. Having a message-based interface enables point-to-point and publish/subscribe mechanisms, guaranteed information delivery, and interoperability between heterogeneous platforms.

JCA is a technology that can be used to address the hardships of integrating applications in an ESB environment. It provides a standardized method for integrating disparate applications in J2EE application architectures. JCA defines a set of functionality that application server vendors can use to connect to back-end EISs, such as ERP, CRM, and legacy systems and applications. It provides support for resource adaptation, which maps the J2EE security, transaction, and communication pooling to the corresponding ES technology. When JCA is used in an ESB implementation, the ESB could provide a JCA container that allows packaged or legacy applications to be plugged into the ESB through JCA resource adapters. For instance, a process order service uses JCA to talk to a J2EE application that internally fulfills incoming orders.

The centralized nature of the application-server-centric model of integrating applications, just like the case of integration broker solutions, can quickly become a point of contention and introduce severe performance problems for large-scale integration projects. The application server model of integrating applications is generally based on developing integration code. The crucial difference between this model and the ESB integration model is that an ESB solution is more about configuration than coding. However, application servers have an important place in an enterprise architecture and are best when used for what they were originally intended for—to provide a component model for hosting business logic in the form of EJB and to serve up Web pages in an enterprise portal environment. Application servers can plug into an ESB using established conventions such as JMS and Message-Driven Beans.

9.2.3 Business Process Management

Today, enterprises are striving to become electronically connected to their customers, suppliers, and partners. To achieve this, they are integrating a wide range of discrete business processes across application boundaries of all kinds. Application boundaries may range from simple inquiries about a customer's order involving two applications to complex, long-lived transactions for processing an insurance claim involving many applications and human interactions and to parallel business events for advanced planning, production, and shipping of goods along the supply chain involving many applications, human interactions, and business-to-business interactions. When integrating on such a scale, enterprises need a greater latitude of functionality to overcome multiple challenges arising from the existence of proprietary interfaces, diverse standards, and approaches targeting the technical, data, automated business process, process analysis, and visualization levels. Such challenges are addressed by business process management technology. In this section, we shall only provide a short overview of BPM functionality in the context of ESB implementations.

BPM is the term used to describe the new technology that provides end-to-end visibility and control over all parts of a long-lived, multistep information request or transaction/process that spans multiple applications and human actors in one or more enterprises. BPM also provides the ability to monitor both the state of any single process instance and all process instances in an aggregate, using real-time metrics that translate actual process activity into key performance indicators. BPM is driven primarily by the common desire to integrate supply chains, as well as internal enterprise functions, without the need for even more custom software development. This means that the tools must be suitable for business analysts, requiring less (or no) software development. They reduce maintenance requirements because internally and externally integrated environments routinely require additions and changes to business processes.

Specialized capabilities such as BPM software solutions provide workflow-related business processes, process analysis, and visualization techniques in an ESB setting. In particular, BPM allows the separation of business processes from the underlying integration code. When sophisticated process definitions are called for in an ESB, a process orchestration engine—which supports BPEL or some other process definition language such as ebXML Business Process Specification Schema (BPSS)—may be layered onto the ESB. The process orchestration may support long-running stateful processes. It may also support parallel execution paths, with branching and merging of message flow execution paths based on join conditions or transition conditions being met. Sophisticated process orchestration can be combined with stateless itinerary-based routing to create an SOA that solves complex integration problems. An ESB uses the concept of itinerary-based routing to provide a message with a list of routing instructions. In an ESB, routing instructions, which represent a business process definition, are carried with the message as it travels through the bus across service invocations. The remote ESB service containers determine where to send the message next.

9.2.4 ESB Transport-Level Choices

Finally, before we close this section, it is important to understand the transport-level protocol choices that can be used in conjunction with an ESB. Web Services in the ESB can communicate using SOAP messages over a variety of protocols. Each protocol effectively provides a service bus connecting multiple endpoints. Currently, the most common service bus transport layer implementations include SOAP/HTTP(S) and SOAP/JMS.

The SOAP over HTTP service bus is the most familiar way to send requests and responses between service requestors and providers. As already explained in Chapter 2, Section 2.1.2 "TCP/IP Protocol", HTTP is a client–server model in which an HTTP client opens a connection and sends a request message to an HTTP server. The client request message is to invoke a Web Service. The HTTP server dispatches a response message containing the invocation and closes the connection. The use of an ESB enables the service requestor to communicate using HTTP and permits the service provider to receive the request using a different transport mechanism. Many ESB implementation providers have an HTTP service bus in addition to at least one other protocol. Any of these protocols can be used for ESB interactions and often are chosen based on service-level requirements.

JMS, part of the J2EE standard, provides a conventional way to create, send, and receive enterprise messages. While it does not quite provide the level of interoperability based on the wide adoption that the HTTP ESB can boast, the SOAP/JMS ESB brings advantages in terms of QoS. A SOAP/JMS

ESB can provide asynchronous and reliable messaging to a Web Service invocation. This means that the requestor can receive acknowledgment of assured delivery and communicate with enterprises that may not be available. A SOAP/JMS Web Service is a Web Service that implements a JMS queue–based transport. As in the case of the SOAP/HTTP, the SOA/JMS service bus enables service requestors and providers to communicate using different protocols.

9.2.5 Connectivity and Translation Infrastructure

For the most part, business applications in an enterprise are not designed to communicate with other applications. There is often an impedance mismatch between the technologies used within internal systems and with external trading partner systems. In order to seamlessly integrate these disparate applications, there must be a way in which a request for information in one format can easily be transformed into a format expected by the called service. For instance, in Figure 9.1, the functionality of a J2EE application needs to be exposed to non-J2EE clients such as .NET applications and other clients. In doing so, a Web Service may have to integrate with other instances of ESs in an organization, or the J2EE application itself may have to integrate with other ESs. In such scenarios, how the application exchanges information with the ESB depends on the application accessibility options. There are three alternative ways an application can exchange information with the ESB:

1. *Application-provided Web Service interface*: Some applications and legacy application servers have adopted the open standard philosophy and have included a Web Service interface. WSDL defines the interface to communicate directly with the application business logic. Where possible, taking a direct approach is always preferred.

2. *Non–Web Service interface*: The application does not expose business logic via Web Services. An application-specific adapter can be supplied to provide a basic intermediary between the application API and the ESB.

3. *Service wrapper as interface to adapter*: In some cases, the adapter may not supply the correct protocol (e.g., JMS) that the ESB expects. In this case, the adapter would be Web Service enabled.

As complementary technologies in an ESB implementation (resource), adapters and Web Services can work together to implement complex integration scenarios. Data synchronization (in addition to translation services) is one of the primary objectives of resource adapters. Adapters can thus take on the role of data synchronization and translation services, whereas Web Services will enable application functions to interact with each other. Web Services are an ideal mechanism for implementing a universally accessible application function (service) that may need to integrate with other applications to fulfill its service contract. The drivers of data synchronization and Web Services are also different. Web Services will generally be initiated by a user request/event, whereas data synchronization is generally initiated by state changes in data objects (e.g., customer, item, and order).

An event to which a Web Service reacts could be a user-initiated request such as a purchase order or an online bill payment. User events can naturally be generated by applications such as an order management application requiring a customer status check from an accounting system. On the other hand, a state change in a data object can be an activity like the addition of a new customer record in the customer service application or an update to the customer's billing address. These state changes trigger an adapter to add the new customer record or update the customer record in all other applications that keep their own copies of customer data.

For the J2EE to .NET application connectivity scenario, a connectivity service in the form of a resource adapter is required. In this implementation strategy, Web Services can become the interface between the company and its customers, partners, and suppliers, whereas the resource adapters become integration components tying up different ESs inside the company. This is just one potential implementation pattern in which Web Services and resource adapters can coexist. Another potential integration pattern in which Web Services and resource adapters are required to collaborate is in business process integration. Applications that use business processes will have to expose required functionality. Obviously, Web Services are ideal for this purpose. When the applications need to integrate with other EISs to fulfill their part in the business process, they will use resource adapters.

9.2.6 ESB Scalability

Scalability is a particularly important issue for any automated business integration solution. Scalability concerns in the case of ESB translate to how well the particular ESB implementation has been designed. The use of asynchronous communications, message itineraries, and message and process definitions allows different parts of the ESB to operate independently of one another. This results in a decentralized model providing complete flexibility in scaling any aspect of the integration network. Such a decentralized architecture enables independent scalability of individual services as well as the communications infrastructure itself. For instance, parallel execution of business operations and itinerary-based routing significantly contribute to the highly distributed nature of the ESB, as there is no centralized rule engine to refer back to for each step in the process.

Typical integration broker technologies handle scalability using a centralized hub-and-spoke model; that is, they handle changes in load and configuration by increasing broker capacity or by adding brokers in a centralized location. A centralized rule engine for the routing of messages can quickly become a bottleneck and also a single point of failure. In contrast, ESB allows capacity to be added where it is most needed—at the service itself.

When the capacity of a single broker is reached, brokers can be combined into clusters. These may act as a single virtual broker to handle increased demand from users and applications. The ESB's use of integration brokers and broker clusters increases scalability by allowing brokers to communicate and dynamically distribute load on the bus. For example, in the event that an increase in the use of the inventory services has overloaded the capacity of their host machine(s), new machines and new brokers can be added to handle the load without the need to change any of the services themselves and without requiring any additional development or administration changes to the messaging system. The notion of a separately deployable, separately scalable messaging topology combined with a separately deployable, separately scalable ESB service container model is what uniquely distinguishes this architectural configuration. The distributed functional pieces are able to work together as one logical piece with a single, globally accessible namespace for locating and invoking services.

9.3 Event-Driven Nature of ESB

In an ESB-enabled event-driven SOA, applications and services are treated as abstract service endpoints, which can readily respond to asynchronous events. Applications and event-driven services are tied together in an ESB-enabled event-driven SOA in a loosely coupled fashion, which allows them to operate independently from each other while still providing value to a broader business function. An event source typically sends messages through the ESB that publishes the messages to the objects that have subscribed to the events. The event itself encapsulates an activity and is a complete description of a specific action. To achieve its functionality, the ESB must support both the established Web Service technologies such as SOAP, WSDL, and BPEL, as well as emerging standards like WS-ReliableMessaging and WS-Notification.

An SOA requires an additional fundamental technology beyond the service aspect to realize its full potential: event-driven computing. Ultimately, the primary objective of most SOA implementations is to automate as much processing as necessary and to provide critical and actionable information to human users when they are required to interact with a business process. This requires the ESB infrastructure itself to recognize meaningful events and respond to them appropriately. The response could be either by automatically initiating new services and business processes or by notifying users of business events of interest, putting the events into topical context and, often, suggesting the best courses of action. In the enterprise context business events, such as a customer order, the arrival of a shipment at a loading dock, the payment of a bill, and so forth affect the normal course of a business process and can occur in any order at any point in time. Consequently, applications that use orchestrated processes that exchange messages need to communicate with each other using a broad capability known as an event-driven SOA.

An event-driven SOA is an architectural approach to distributed computing where events trigger asynchronous messages that are then sent between independent software components that need not have any information about each other by abstracting away from the details of underlying service connectivity and protocols. An event-driven SOA provides a more lightweight, straightforward set of technologies to build and maintain the service abstraction for client applications.

To achieve a more lightweight arrangement, an event-driven SOA requires that two participants in an event (server and client) be decoupled. With fully decoupled exchanges, the two participants in an event need not have any knowledge about each other before engaging in a business transaction. This means that there is no need for a service contract in WSDL that explicates the behavior of a server to the client. The only relationship is indirect, through the ESB, to which clients and servers are subscribed as subscribers and publishers of events. Despite the notion of decoupling in an event-driven SOA, recipients of events require metadata about those events. In such situations, recipients of events still have some information about those events. For instance, the publishers of the events often organize them on the basis of some (topical) taxonomy or, alternatively, provide details about the event, including its size and format, which is a form of metadata. In contrast to service interfaces, however, metadata that is associated to events is generated on an ad hoc basis as metadata tends to come along with the event rather than being contained in a separate service contract. In particular, ad hoc metadata describes published events that consumers can subscribe to, the interfaces that service clients and providers exhibit as well as the messages they exchange, and even the agreed format and context of this metadata, without falling into the formal service contracts themselves.

9.4 Key Capabilities of an ESB

In order to implement an SOA, both applications and infrastructure must support SOA principles. Enabling an application for SOA involves the creation of service interfaces to existing or new functions, either directly or through the use of adapters. Enabling the infrastructure, at the most basic level, involves the provision of the capabilities to route and deliver secure service requests to the correct service provider. However, it is also vital that the infrastructure supports the substitution of one service implementation by another with no effect on the clients of that service. This requires not only that the service interfaces be specified according to SOA principles but also that the infrastructure allow client code to invoke services irrespective of the service location and the communication protocol involved. Such service routing and substitution are among the many capabilities of the ESB. Additional capabilities can be found in the following list that describes detailed functional requirements of an ESB. It should be noted that not all of the capabilities described in the following are offered by current commercial ESB systems:

1. *Dynamic connectivity capabilities*: Dynamic connectivity is the ability to connect to Web Services dynamically without using a separate static API or proxy for each service. Most enterprise applications today operate on a static connectivity mode, requiring some static piece of code for each service. Dynamic service connectivity is a key capability for a successful ESB implementation. The dynamic connectivity API is the same regardless of the service implementation protocol (Web Services, JMS, EJB/RMI, etc.).

2. *Reliable messaging capabilities*: Reliable messaging can be primarily used to ensure guaranteed delivery of these messages to their destination and for handling events. This capability is crucial for responding to clients in an asynchronous manner and for a successful ESB implementation.

3. *Topic- and content-based routing capabilities*: The ESB should be equipped with routing mechanisms to facilitate not only topic-based routing but also more sophisticated content-based routing. Topic-based routing assumes that messages can be grouped into fixed, topical classes so that subscribers can explicate interest in a topic and as a consequence receive messages associated to that topic. Content-based routing, on the other hand, allows subscriptions on constraints of actual properties (attributes) of business events. Content-based routing forwards messages to their destination based on the context or content of the service. Content-based routing is usually implemented using techniques that can examine the content of a message and apply a set of rules to its content to determine which endpoints in the ESB infrastructure it may need to be routed to next. Content-based routing logic (rules) is usually expressed in XPath or a scripting language, such as JavaScript. For example, if a manufacturer provides a wide variety of products to its customers, only some of which are made in-house, depending on the product ordered, it might be necessary to route the message directly to an external supplier or route it internally to be processed by a warehouse fulfillment service. Content-based ESB

capabilities could be supported by emerging standard efforts such as WS-Notification.

4. *Transformation capabilities*: A critical ability of the ESB is the ability to route service interactions through a variety of transport protocols and to transform from one protocol to another where necessary. Another important aspect of an ESB implementation is the ability to support service messaging models and data formats consistent with the SOA interfaces. A major source of value in an ESB is that it shields any individual component from any knowledge of the implementation details of any other component. The ESB transformation services make it possible to ensure that messages and data received by any component are in the format it expects, thereby removing the need to make changes. The ESB plays a major role in transforming between differing data formats and messaging models, whether between basic XML formats and Web Service messages or between different XML formats (e.g., transforming an industry-standard XML message to a proprietary or custom XML format). The ESB connectivity and translation infrastructure is discussed in Section 9.4.5 below.

5. *Service enablement capabilities*: Service enablement includes the ability to access already existing resources such as legacy systems—technically obsolete mission-critical elements of an organization's infrastructure—and includes them in an SOA implementation. Tactically, legacy assets must be leveraged, service enabled, and integrated with modern service technologies and applications.

6. *Endpoint discovery with multiple QoS capabilities*: The ESB should support the basic SOA need to discover, locate, and bind to services. As many network endpoints can implement the same service contract, the ESB should make it possible for the client to select the best endpoint at runtime, rather than hard-coding endpoints at build time. The ESB should therefore be capable of supporting various QoSs and allow clients to discover the best service instance with which to interact based on QoS properties. Such capabilities should be controlled by declarative policies associated with the services involved using a policy standard such as the WS-Policy framework.

7. *Long-running process and transaction capabilities*: Service orientation, as opposed to distributed object architectures such as .NET or J2EE, more closely reflects real-world processes and relationships. Hence, SOA represents a much more natural way to model and build software that solves real-world business processing needs. Accordingly, the ESB should provide the ability to support business processes and long-running services—services that tend to run for long duration, exchanging message (conversation) as they progress.

Typical examples are an online reservation system, which interacts with the user as well as various service providers (airline ticketing, insurance claims, mortgage and credit product applications, etc.). In addition, in order to be successful in business environments, it is extremely important that the ESB provides certain transactional guarantees. More specifically, the ESB needs to be able to ensure that complex transactions are handled in a highly reliable manner, and if failure should occur, transactions should be capable of rolling back processing to the original, prerequest state. Long-duration transactional conversations could be made possible if implemented on the basis of messaging patterns using asynchrony, store and forward, and itinerary-based routing techniques. It should be noted that the base definition of an ESB as currently used by the ESB analyst and vendor community does not mandate a long-duration transaction manager.

8. *Security capabilities*: Generically handling and enforcing security is a key success factor for ESB implementations. The ESB needs both to provide a security model to service consumers and to integrate with the (potentially varied) security models of service providers. Both point-to-point (e.g., SSL encryption) and end-to-end security capabilities will be required. These end-to-end security capabilities include federated authentication, which intercepts service requests and adds the appropriate user name and credentials, validation of each service request and authorization to make sure that the sender has the appropriate privilege to access the service, and, lastly, encryption/decryption of XML content at the element level for both message requests and responses. To address these intricate security requirements, the ESB must rely on WS-Security and other security-related standards for Web Services that have been developed.

9. *Integration capabilities*: To support SOA in a heterogeneous environment, the ESB needs to integrate with a variety of systems that do not directly support service-style interactions. These may include legacy systems, packaged applications, or other EAI technologies. When assessing the integration requirements for ESB, several types or *styles* of integration must be considered, for example, process versus data integration.

10. *Management and monitoring capabilities*: In an SOA environment, applications cross system (and even organizational) boundaries, they overlap, and they can change over time. Managing these applications is a serious challenge. Examples include dynamic load balancing, failover when primary systems go down, and achieving topological or geographic affinity between the client and the service instance. Effective systems and application management in an ESB requires a management framework that is consistent across an increasingly

heterogeneous set of participating component systems, while supporting complex aggregate (cross component) management use cases, like dynamic resource provisioning and demand-based routing, and SLA enforcement in conjunction with policy-based behavior (e.g., the ability to select service providers dynamically based on the quality of service they offer compared to the business value of individual transactions).

An additional requirement for a successful ESB implementation is the ability to monitor the health, capacity, and performance of services. Monitoring is the ability to track service activities that take place via the bus and provide visibility into various metrics and statistics. Of particular significance is the ability to spot problems and exceptions in the business processes and move toward resolving them as soon as they occur. Chapter 18 examines the management and monitoring of distributed Web services based Cloudware platforms and applications.

11. *Scalability capabilities*: With a widely distributed SOA, there will be the need to scale some of the services or the entire infrastructure to meet integration demands. For example, transformation services are typically very resource intensive and may require multiple instances across two or more computing nodes. At the same time, it is necessary to create an infrastructure that can support the large nodes present in a global service network. The loose-coupled nature of an SOA requires that the ESB use a decentralized model to provide a cost-effective solution that promotes flexibility in scaling any aspect of the integration network. A decentralized architecture enables independent scalability of individual services as well as the communications infrastructure itself.

9.5 Leveraging Legacy Assets

There is a fundamental requirement in ESB settings to utilize functionality in existing applications and repurpose it for use in new applications. Enterprises are still burdened with older-generation operational applications that were constructed to run on various obsolescent hardware types, programmed in obsolete languages. Such applications are known as legacy applications.

Legacy applications are critical assets of any modern enterprise as they provide access to mission-critical business information and functionality and thus control the majority of an organization's business processes. Legacy applications could implement core business tasks such as taking and processing orders, initiating production and delivery, generating invoices, and crediting payments, distribution, inventory management, and related revenue-generating, cost-saving, and accounting tasks. Being able to leverage this value in new ESB-based solutions would provide an extremely attractive return on existing investments. Therefore, a best-of-breed ESB characteristic is to offer connectivity for legacy applications.

It is not possible to properly integrate legacy systems into Web Service solutions without extensive, intrusive modifications to these systems. Modifications are needed to reshape legacy systems to provide a natural fit with the Web Service architectural requirements and carefully retrofit business logic so that it can be used with new applications. Therefore, legacy applications need to be reengineered in order to reuse the core business processes entrenched in legacy applications. The legacy system reengineering process involves the disciplined evolution of an existing legacy system to a new *improved* environment by reusing as much of it (implementation, design, specification, requirements) as possible and by adding new capabilities. Through reengineering, business processes become more modular and granular exposing submodules that can be reused and are represented as services.

The primary focus of legacy reengineering and transformation is enterprises, business processes, the EAI, and how a legacy system can contribute to implementing the architecture without propagating the weaknesses of past designs and development methods. In its most fundamental form, the process of re-engineering involves three basic phases.

1. Understanding of an existing application, resulting in one or more logical descriptions of the application
2. Restructuring or transformation of those logical descriptions into new, improved logical descriptions
3. Development of the new application based on these improved logical descriptions

The re-engineering and transformation steps in the following have been considerably simplified. The purpose of these steps is to facilitate the process of legacy application modernization by modularizing legacy processes and business logic separately from presentation logic and data management activities, and representing them as components. These components can then be used to create interfaces for new services, thereby service enabling legacy applications.

These three broad phases comprise a series of six steps briefly described in the following:

1. *Understanding existing applications*: Before beginning the modernization process, the first task is to understand the structure and architecture of the existing application's architecture. This task includes gathering statistics about size, complexity, the amount of dead or unused code, and the amount of bad programming for each application. In addition, in selecting which programs to improve together, selecting the ones that affect common data is a critical step when planning to move through all of the modernization stages, including reengineering for reuse and/or migration.

2. *Rationalizing business logic*: A typical legacy system is composed of a large number of independent programs. These programs work together in a hardwired net of business process flows. Once an application's program code is clean, any programming anomalies have been removed, and nonbusiness logic has been filtered, it is possible to apply pattern-matching techniques across all of the application's programs to identify and segregate candidate common business logic.

3. *Identifying business rules*: When candidate reusable business logic has been rationalized to a subset of distinct, single occurrences of each service, it is then possible to determine whether each should become part of a process or express a business rule. For a definition and examples of use of business rules, refer to Section 9.2 "Event-Driven Nature of ESB". To achieve this, sophisticated algorithms are used to extract business rules from monolithic legacy systems within which business rules exist in many different guises. The extraction of business rules from legacy code is generally termed business rule recovery. The accuracy with which this task is performed is key to legacy application modernization.

4. *Extracting components*: Extracted business rules can be grouped together based on their contribution to achieve the intended business functionality. A group of rules is normally processing some common set of data to achieve intended business functions. Candidate business rules and associated business data are then extracted and appropriately represented as a cohesive legacy component. The number of rules that need to be grouped together is a matter of choice, depending on the desired granularity of the legacy component. A callable interface needs to be provided for these components.

5. *Wrapping component implementations*: Legacy systems were, for the most part, not implemented in a fashion that lends itself to componentization. Presentation logic is often intertwined with business logic, which is intertwined with systems and data access logic. During

this step, system-level and presentation-level legacy components are identified and separated from business-level legacy components. In this way, candidate components are identified for wrapping. Wrapping provides legacy functionality for new service-based solutions in much shorter time than it would take to build a replacement from scratch and recreate those dependencies. Wrapping requires that the appropriate level of abstraction for components be determined. When it comes to wrapping and legacy componentization, one should concentrate on identifying coarse-grained components. Reusing a larger component saves more effort, and thus larger components have greater value. Smaller components are more likely to be frequently used, but their use saves less effort. This implies that fine-grained components are less cost effective.

6. *Creating service interfaces*: Component wrappers result in well-defined boundaries of functionality and data. However, the modernized legacy system as a whole is still tightly coupled with components hardwired to each other via program-to-program calls. The SOA approach to large-scale system coupling requires removing from the individual component wrappers any direct knowledge of any other such components. This can be accomplished by breaking up program-to-program connectivity and replacing it with service-enabled APIs that can be used in conjunction with event-driven and business process orchestration mechanisms.

9.6 Summary

Reliable messaging protocols are at the heart of service-oriented computing architectural approaches that serve as the enabling facilitator for addressing the requirements of loosely coupled, standard-based, and protocol-independent distributed computing. This chapter explained how service-oriented architectures, techniques, and technologies when combined with those of event-based programming can offer the means to achieve the desired levels of business integration effectively, mapping IT implementations more closely to the business processes of the enterprises. Combining Web Service standards with an ESB infrastructure can potentially deliver the broadest connectivity between systems. An ESB supporting Web Services with more established application integration techniques enables an enterprise-wide solution that combines the best of both of these worlds.

10

Service Composition

Every enterprise has unique characteristics that are embedded in its business processes. Most enterprises perform a similar set of repeatable routine activities that may include the development of manufacturing products and services, bringing these products and services to market and satisfying the customers who purchase them. Automated business processes can perform such activities. We may view an automated business process as a precisely choreographed sequence of activities systematically directed toward performing a certain business task and bringing it to completion. Examples of typical processes in manufacturing firms include among other things new product development (which cuts across research and development, marketing, and manufacturing), customer order fulfillment (which combines sales, manufacturing, warehousing, transportation, and billing), and financial asset management. The possibility to design, structure, measure processes, and determine their contribution to customer value makes them an important starting point for business improvement and innovation initiatives.

The largest possible process in an organization is the value chain. The value chain is decomposed into a set of core business processes and support processes necessary to produce a product or product line. These core business processes are subdivided into activities. An activity is an element that performs a specific function within a process. Activities can be as simple as sending or receiving a message or as complex as coordinating the execution of other processes and activities. A business process may encompass complex activities, some of which run on back-end systems, such as a credit check, automated billing, a purchase order, stock updates and shipping, or even such frivolous activities as sending a document, and filling a form. A business process activity may invoke another business process in the same or a different business system domain. Activities will inevitably vary greatly from one company to another and from one business analysis effort to another.

At runtime, a business process definition may have multiple instantiations, each operating independently of the other, and each instantiation may have multiple activities that are concurrently active. A process instance is a defined thread of activity that is being enacted (managed) by a workflow engine. In general, instances of a process, its current state, and the history of its actions will be visible at runtime and expressed in terms of the business process definition so that

- users can determine the status of business activities and business
- specialists can monitor the activity and identify potential improvements to the business process definition.

10.1 Process

A process is an ordering of activities with a beginning and an end; it has inputs (in terms of resources, materials, and information) and a specified output (the results it produces). We may thus define a process as any sequence of steps that is initiated by an event; transforms information, materials, or commitments; and produces an output. A business process is typically associated with operational objectives and business relationships, for example, an insurance claims process or an engineering development process. A process may be wholly contained within a single organizational unit or may span different organizations, such as in a customer–supplier relationship. Typical examples of processes that cross organizational boundaries are purchasing and sales processes jointly set up by buying and selling organizations, supported by EDI and value-added networks. The Internet is now a trigger for the design of new business processes and the redesign of existing ones.

A business process is a set of logically related tasks performed to achieve a well-defined business outcome. A (business) process view implies a horizontal view of a business organization and looks at processes as sets of interdependent activities designed and structured to produce a specific output for a customer or a market. A business process defines the results to be achieved, the context of the activities, the relationships between the activities, and the interactions with other processes and resources. A business process may receive events that alter the state of the process and the sequence of activities. A business process may produce events for input to other applications or processes. It may also invoke applications to perform computational functions, and it may post assignments to human work lists to request actions by human actors. Business processes can be measured, and different performance measures apply, like cost, quality, time, and customer satisfaction.

A business process has the following behavior:

- It may contain defined conditions triggering its initiation in each new instance (e.g., the arrival of a claim) and defined outputs at its completion.
- It may involve formal or relatively informal interactions between participants.
- It has a duration that may vary widely.

- It may contain a series of automated activities and/or manual activities. Activities may be large and complex, involving the flow of materials, information, and business commitments.
- It exhibits a very dynamic nature, so it can respond to demands from customers and to changing market conditions.
- It is widely distributed and customized across boundaries within and between organizations, often spanning multiple applications with very different technology platforms.
- It is usually long running—a single instance of a process such as order to cash may run for months or even years.

Every business process implies processing: a series of activities (processing steps) leading to some form of transformation of data or products for which the process exists. Transformations may be executed manually or in an automated way. A transformation will encompass multiple processing steps. Finally, every process delivers a product, like a mortgage or an authorized invoice. The extent to which the end product of a process can be specified in advance and can be standardized impacts the way that processes and their workflows can be structured and automated.

Processes have decision points. Decisions have to be made with regard to routing and allocation of processing capacity. In a highly predictable and standardized environment, the trajectory in the process of a customer order will be established in advance in a standard way. Only if the process is complex and if the conditions of the process are not predictable will routing decisions have to be made on the spot. In general, the customer orders will be split into a category that is highly proceduralized (and thus automated) and a category that is complex and uncertain. Here, human experts will be needed, and manual processing is a key element of the process.

10.2 Workflow

A workflow system automates a business process, in whole or in part, during which documents, information, or tasks are passed from one participant to another for action, according to a set of procedural rules. Workflows are based on document life cycles and form-based information processing, so generally they support well-defined, static, *clerical* processes. They provide transparency, since business processes are clearly articulated in the software, and they are agile because they produce definitions that are fast to deploy and change.

A workflow can be defined as the sequence of processing steps (execution of business operations, tasks, and transactions), during which information and physical objects are passed from one processing step to another. Workflow is a concept that links together technologies and tools able to automatically route events and tasks with programs or users.

Process-oriented workflows are used to automate processes whose structure is well defined and stable over time, which often coordinate subprocesses executed by machines and which only require minor user involvement (often only in specific cases). An order management process or a loan request is an example of a well-defined process. Certain process-oriented workflows may have transactional properties. The process-oriented workflow is made up of tasks that follow routes, with checkpoints represented by business rules, for example, *pause for a credit approval*. Such business process rules govern the overall processing of activities, including the routing of requests, the assignment or distribution of requests to designated roles, the passing of workflow data from activity to activity, and the dependencies and relationships between business process activities.

A workflow involves activities, decision points, rules, routes, and roles. These are briefly described later. Just like a process, a workflow normally comprises a number of logical steps, each of which is known as an *activity*. An activity is a set of actions that are guided by the workflow. An activity may involve manual interaction with a user or workflow participant or might be executed using diverse resources such as application programs or databases. A work item or data set is created and is processed and changed in stages at a number of processing or decision points to meet specific business goals. Most workflow engines can handle very complex series of processes.

A workflow can depict various aspects of a business process including automated and manual activities, decision points and business rules, parallel and sequential work routes, and how to manage exceptions to the normal business process. A workflow can have logical *decision points* that determine which branch of the flow a work item may take in the event of alternative paths. Every alternate path within the flow is identified and controlled through a bounded set of logical decision points. An instantiation of a workflow to support a work item includes all possible paths from beginning to end.

Within a workflow, business rules in each decision point determine how workflow-related data are to be processed, routed, tracked, and controlled. Business *rules* are core business policies that capture the nature of an enterprise's business model and define the conditions that must be met in order to move to the next stage of the workflow. Business rules are represented as compact statements about an aspect of the business that can be expressed within an application, and as such, they determine the route to be followed. For instance, for a health-care application, business rules may include policies on how new claim validation, referral requirements, or special procedure approvals are implemented. Business rules can represent among other things typical business situations such as escalation ("send this document to a supervisor for approval") and managing exceptions ("this loan is more than $50,000; send it to the MD").

Workflow technology enables developers to describe full intra- or inter-organizational business processes with dependencies, sequencing selections, and iteration. It effectively enables the developers to describe the complex rules for processing in a business process, such as merging, selection based on field content, and time-based delivery of messages. To achieve these objectives, workflows are predicated upon the notion of pre-specified routing paths. *Routes* define the path taken by the set of objects making up the workflow. The routes of a workflow may be sequential, circular, or parallel work routes.

Routing paths can be sequential, parallel, or cyclic:

1. Sequential routing: A segment of a process instance under enactment by a workflow management system in which several activities are executed in sequence under a single thread of execution is called sequential routing.

2. Parallel routing: A segment of a process instance under enactment by a workflow management system where two or more activity instances are executing in parallel within the workflow, giving rise to multiple threads of control, is called parallel routing.

3. Condition routing: A point within the workflow where a single thread of control makes a decision as to which branch to take when having to select between multiple alternative workflow branches is known as condition routing.

A split point is a synchronization point within the workflow where a single thread of control splits into two or more threads that are executed in parallel within the workflow, allowing multiple activities to be executed simultaneously. A join point in the workflow is a synchronization point where two or more parallel executing activities converge into a single common thread of control. No split or join points occur during sequential routing. Parallel routing normally commences with an AND-Split (or split) and concludes with an AND-Join (or join or rendezvous) point.

Workflow routing includes two more synchronization points: OR-Split (or conditional routing) and OR-Join (or asynchronous join), which can be employed by both sequential and parallel routing constructs. A point within the workflow where two or more alternative activities workflow branches reconverge to form a single common activity as the next step within the workflow is known as asynchronous join. It must be noted that as no parallel activity execution has occurred at the join point, no synchronization is required.

Roles in a workflow define the function of the people or programs involved in the workflow. A role is a mechanism within a workflow that associates participants to a collection of workflow activity(ies). The role defines the context in which the user participates in a particular process or activity. The role often embraces organizational concepts such as structure and relationships, responsibility, or authority but may also refer to other attributes such as skill, location, value data, time, or date.

Workflow technology tends to relegate integration functions, such as synchronizing data between disparate packaged and legacy applications, to custom code within its activities—and thus outside the scope of the process model. Moreover, it uses a tightly coupled integration style that employs low-level APIs and that has confined workflow to local, homogeneous system environments, such as within a department or division. Therefore, traditional workflow implementations are closely tied to the enterprise in which they are deployed and cannot be reliably extended outside organizational borders to customers, suppliers, and other partners. As a consequence, one of the major limitations of WMSs is integration: they are not good at connecting cross-enterprise systems together. Modern workflow technology tries to address this deficiency by extending this functionality to cross-enterprise process integration by employing business process management functionality. They achieve this by integrating middleware, process sequencing, and orchestration mechanisms as well as transaction processing capabilities (see "Service Composition" later).

The definition, creation, and management of the execution of workflow are achieved by a workflow management system running on one or more workflow engines. A workflow management system is capable of interpreting the process and activity definitions, interacting with workflow participants, and, where required, invoking the use of software-enabled tools and applications. Most WMSs integrate with other systems used by an enterprise, such as document management systems, databases, e-mail systems, office automation products, geographic information systems, and production applications.

10.3 Business Process Management (BPM)

BPM is a commitment to expressing, understanding, representing, and managing a business (or the portion of business to which it is applied) in terms of a collection of business processes that are responsive to a business environment of internal or external events. The term *management of business processes* includes process analysis, process definition and redefinition, resource allocation, scheduling, measurement of process quality and efficiency, and process optimization. Process optimization includes collection and analysis of both real-time measures (monitoring) and strategic measures (performance management) and their correlation as the basis for process improvement and innovation. A BPM solution is a graphical productivity tool for modeling, integrating, monitoring, and optimizing process flows of all sizes, crossing any application, company boundary, or human interaction. BPM codifies value-driven processes and institutionalizes their execution within the enterprise. This implies that BPM tools can help analyze, define, and enforce process standardization. BPM provides a modeling tool to visually construct, analyze, and execute cross-functional business processes.

BPM is more than process automation or traditional workflow. BPM within the context of EAI and e-business integration provides the flexibility necessary to automate cross-functional processes. It adds conceptual innovations and technology from EAI and e-business integration and reimplements it on an e-business infrastructure based on Web and XML standards. Conventional applications provide traditional workflow features that work well only within their local environment. However, integrated process management is then required for processes spanning organizations. Automating cross-functional activities, such as checking or confirming inventory between an enterprise and its distribution partners, enables corporations to manage processes by exception based on real-time events driven from the integrated environment. Process execution then becomes automated, requiring human intervention only in situations where exceptions occur; for example, inventory level has fallen below a critical threshold or manual tasks and approvals are required.

The distinction between BPM and workflow is mainly based on the management aspect of BPM systems: BPM tools place considerable emphasis on management and business functions. Although BPM technology covers the same space as workflow, its focus is on the business user and provides more sophisticated management and analysis capabilities. With a BPM tool, the business user is able to manage all the process of a certain type, for example, claim processes, and should be able to study them from historical or current data and produce costs or other business measurements. In addition, the business user should also be able to analyze and compare the data or business measurements based on the different types of claims. This type of functionality is typically not provided by modern workflow systems.

10.4 Business Processes via Web Services

Business processes management and workflow systems today support the definition, execution, and monitoring of long-running processes that coordinate the activities of multiple business applications. However, because these systems are activity oriented and not communication (message) oriented, they do not separate internal implementation from external protocol description. When processes span business boundaries, loose coupling based on precise external protocols is required because the parties involved do not share application and workflow implementation technologies and will not allow external control over the use of their back-end applications. Such business interaction protocols are by necessity message centric; they specify the flow of messages representing business actions among trading partners, without requiring any specific implementation mechanism. With such applications, the loosely coupled, distributed nature of the Web enables exhaustive and full orchestration, choreography, and monitoring of the enterprise applications that expose the Web Services participating in the message exchanges.

Web Services provide standard and interoperable means of integrating loosely coupled web-based components that expose well-defined interfaces, while abstracting the implementation- and platform-specific details. Core Web Service standards such as SOAP, WSDL, and UDDI provide a solid foundation to accomplish this. However, these specifications primarily enable the development of simple Web Service applications that can conduct simple interactions. However, the ultimate goal of Web Services is to facilitate and automate business process collaborations both inside and outside enterprise boundaries. Useful business applications of Web Services in EAI and business-to-business environments require the ability to compose complex and distributed Web Service integrations and the ability to describe the relationships between the constituent low-level services. In this way, collaborative business processes can be realized as Web Service integrations.

A business process specifies the potential execution order of operations originating from a logically interrelated collection of Web Services, each of which performs a well-defined activity within the process. A business process also specifies the shared data passed between these services, the external partners' roles with respect to the process, joint exception handling conditions for the collection of Web Services, and other factors that may influence how Web Services or organizations participate in a process. This would enable long-running transactions between Web Services in order to increase the consistency and reliability of business processes that are composed out of these Web Services.

The orchestration and choreography of Web Services is enabled under three specification standards, namely, the Business Process Execution Language for Web Services (BPEL4WS or BPEL for short), WS-Coordination (WS-C), and WS-Transaction (WS-T). These three specifications work together to form the bedrock for reliably choreographing Web Service-based applications, providing BPM, transactional integrity, and generic coordination facilities. BPEL is a workflow-like definition language that describes sophisticated business processes that can orchestrate Web Services. WS-Coordination and WS-Transaction complement BPEL to provide mechanisms for defining specific standard protocols for use by transaction processing systems, workflow systems, or other applications that wish to coordinate multiple Web Services.

The next section describes briefly the problem of service composition before moving on to an overview of BPEL.

10.4.1 Service Composition

The platform-neutral nature of services creates the opportunity for building composite services by combining existing elementary or complex services (the component services) from different enterprises and in turn offering them as high-level services or processes. Composite services (and, thus, processes) integrate multiple services—and put together new business functions—by combining new and existing application assets in a logical flow.

The definition of composite services requires coordinating the flow of control and data between the constituent services. Business logic can be seen as the ingredient that sequences, coordinates, and manages interactions among Web Services. By programming a complex cross-enterprise workflow task or business transaction, it is possible to logically chain discrete Web Services activities into cross-enterprise business processes. This is enabled through orchestration and choreography (because Web Services technologies support coordination and offer an asynchronous and message-oriented way to communicate and interact with application logic).

10.4.1.1 Orchestration

Orchestration describes how Web Services can interact with each other at the message level, including the business logic and execution order of the interactions from the perspective and under control of a single endpoint. This is, for instance, the case of the process flow where the business process flow is seen from the vantage point of a single supplier. Orchestration refers to an executable business process that may result in a long-lived, transactional, multistep process model. With orchestration, business process interactions are always controlled from the (private) perspective of one of the business parties involved in the process.

10.4.1.2 Choreography

Choreography is typically associated with the public (globally visible) message exchanges, rules of interaction, and agreements that occur between multiple business process endpoints, rather than a specific business process that is executed by a single party. Choreography tracks the sequence of messages that may involve multiple parties and multiple sources, including customers, suppliers, and partners, where each party involved in the process describes the part it plays in the interaction and no party *owns* the conversation. Choreography is more collaborative in nature than orchestration. It is described from the perspectives of all parties (common view) and, in essence, defines the shared state of the interactions between business entities. This common view can be used to determine specific deployment implementations for each individual entity. Choreography offers a means by which the rules of participation for collaboration can be clearly defined and agreed to, jointly. Each entity may then implement its portion of the choreography as determined by their common view.

10.5 Business Process Execution Language (BPEL)

The development of the BPEL language was guided by the requirement to support service composition models that provide flexible integration, recursive composition, separation of composability of concerns, stateful conversation and life-cycle management, and recoverability properties. BPEL has now emerged as the standard to define and manage business process activities and business interaction protocols comprising collaborating Web Services. This is an XML-based flow language for the formal specification of business processes and business interaction protocols. By doing so, it extends the Web Service interaction model and enables it to support complex business processes and transactions. Enterprises can describe complex processes that include multiple organizations—such as order processing, lead management, and claims handling—and execute the same business processes in systems from other vendors.

BPEL as a service composition (orchestration) language provides several features to facilitate the modeling and execution of business processes based on Web Services. These features include

1. Modeling business process collaboration (through <partnerLink>s)
2. Modeling the execution control of business processes (through the use of a self-contained block and transition-structured language that support the representation of directed graphs)
3. Separation of abstract definition from concrete binding (static and dynamic selection of partner services via endpoint references)
4. Representation of participants' roles and role relationships (through <partnerLinkType>s)
5. Compensation support (through fault handlers and compensation)
6. Service composability (structured activities can be nested and combined arbitrarily)
7. Context support (through the <scope>mechanism)
8. Spawning off and synchronizing processes (through <pick> and <receive> activities)
9. Event handling (through the use of event handlers)

BPEL can also be extended to provide other important composition language properties such as support for Web Service policies and security and reliable messaging requirements. In this section, we summarize the most salient BPEL features and constructs.

10.5.1 Background of WSDL

BPEL's composition model makes extensive use of Web Services Description Language, WSDL. It is therefore necessary to provide an overview of WSDL before going into the details of BPEL itself. A WSDL description consists of two parts: an abstract part defining the offered functionality and a concrete part defining how and where this functionality may be accessed. By separating the abstract from the concrete, WSDL enables an abstract component to be implemented by multiple code artifacts and deployed using different communication protocols and programming models.

The abstract part of a WSDL definition consists of one or more interfaces, called portTypes in WSDL. PortTypes specify the operations provided by the service and their input and/or output message structures. Each message consists of a set of parts; the types of these parts are usually defined using XML schema. The concrete part of a WSDL definition consists of three parts. It binds the portType to available transport protocol and data encoding formats in a set of one or more bindings. It provides the location of endpoints that offer the functionality specified in a portType over an available binding in one or more ports. Finally, it provides a collection of ports as services.

10.5.2 BPEL4WS

BPEL4WS is a workflow-based composition language geared toward service-oriented computing and layered as part of the Web Service technology stack. BPEL composes services by defining control semantics around a set of interactions with the services being composed. The composition is recursive; a BPEL process itself is naturally exposed as a Web Service, with incoming messages and their optional replies mapped to calls to WSDL operations offered by the process. Offering processes as services enables interwork flow interaction, higher levels of reuse, and additional scalability.

Processes in BPEL are defined using only the abstract definitions of the composed services, that is, the abstract part (portType/operations/messages) of their WSDL definitions. The binding to actual physical endpoints and the mapping of data to the representation required by these endpoints is intentionally left out of the process definition, allowing the choice to be made at deployment time, at design time, or during execution. Added to the use of open XML specifications and standards, this enables two main goals: flexibility of integration and portability of processes.

The BPEL language is designed to specify both business protocols and executable processes. A business protocol, called an *abstract process* in BPEL, specifies the flow of interactions that a service may have with other services. For example, one may accompany a WSDL description with an abstract BPEL process to inform parties using it in what order and in what situations operations in the WSDL should be called (e.g., a call to a "request for quote" operation must precede a call to a "place order" operation). An *executable process* is similar to an abstract process, except that it has a slightly expanded BPEL vocabulary and includes information that enables the process to be interpreted, such as fully specifying the handling of data values, and including interactions with private services that one does not want to expose in the business protocol. For example, when an order is placed, the executable BPEL process might have to invoke a number of internal applications wrapped as services (e.g., applications related to invoicing, customer relationship management, stock control, and logistics), but these calls should not be visible to the customer and would be omitted from the abstract process the customer sees. In the executable variant, the process can be seen as the implementation of a Web Service. Most work in BPEL has been focused on the executable variant of the language.

10.5.3 BPEL Process Model

BPEL has its roots in both graph- and calculus-based process models, giving designers the flexibility to use either or both graph primitives (nodes and links) and complex control constructs creating implicit control flow. The two process modeling approaches are integrated through BPEL's exception handling mechanism. The composition of services results from the use of predefined interaction activities that can invoke operations on these services and handle invocations to operations exposed by the process itself. The unit of composition in BPEL is the activity. Activities are combined through nesting in complex activities with control semantics and/or through the use of conditional links. In contrast to traditional workflow systems in which data-flow is explicitly defined using data links, BPEL gives activities the read/write access to shared, scoped variables. In addition to the main forward flow, BPEL contains fault handling and rollback capabilities, event handling, and life-cycle management.

The role of BPEL is to define a new Web Service by composing a set of existing services through a process-integration-type mechanism with control language constructs. The entry points correspond to external WSDL clients invoking either input-only (request) or input/output (request–response) operations on the interface of the composite BPEL service. BPEL provides a mechanism for creating implementation- and platform-independent compositions of services woven strictly from the abstract interfaces provided in the WSDL definitions. The definition of a BPEL business process also follows the WSDL convention of strict separation between the abstract service interface and service implementation. In particular, a BPEL process represents parties and interactions between these parties in terms of abstract WSDL interfaces (by means of <portType>s and <operation>s), while no references are made to the actual services (binding and address information) used by a process instance. Both the interacting process and its counterparts are modeled in the form of WSDL services. Actual implementations of the services themselves may be dynamically bound to the partners of a BPEL composition, without affecting the composition's definition. Business processes specified in BPEL are fully executable portable scripts that can be interpreted by business process engines in BPEL conformant environments.

BPEL distinguishes five main sections:

1. The message flow section of BPEL is handled by basic activities that include invoking an operation on some Web Service, waiting for a process operation to be invoked by some external client, and generating the response of an input/output operation.

2. The control flow section of BPEL is a hybrid model principally based on block-structured definitions with the ability to define selective state transition control flow definitions for synchronization purposes.

3. The dataflow section of BPEL comprises variables that provide the means for holding messages that constitute the state of a business process. The messages held are often those that have been received from partners or are to be sent to partners. Variables can also hold data that are needed for holding state related to the process and never exchanged with partners. Variables are scoped, and the name of a variable should be unique within its own scope.

4. The process orchestration section of BPEL uses partner links to establish peer-to-peer partner relationships.

5. The fault and exception handling section of BPEL deals with errors that might occur when services are being invoked with handling compensations of units of work and dealing with exceptions during the course of a BPEL computation.

BPEL consists of the following basic activities:

receive: The receive activity initiates a new process when used at its start or does a blocking wait for a matching message to arrive when used during a process.

reply: The reply activity sends a message in reply.

invoke: The invoke activity calls a Web Service operation of a partner service. This can either be a one-way or a request–response call. One way means that the called service will not send a response, whereas request–response blocks the process until a response is received.

assign: The assign activity updates the values of variables or partner links with new data.

validate: The validate activity checks the correctness of XML data stored in variables.

wait: The wait activity pauses the process, either for a given time period or until a certain point in time has passed.

empty: The empty activity is a no-op instruction for a business process.

Another element of the WS-BPEL language is a *variable*. BPEL supports both global (i.e., process level) and local (i.e., scope level) variables. BPEL variables may be typed using an XML schema (XSD) type or element or a WSDL message. For initializing or assigning variables, BPEL provides the assign activity. Each assign consists of one or more copy statements. In each copy, the *from* element specifies the assignment source for data elements or partner links, and the *to* element specifies the assignment target.

Additionally, there are basic activities that deal with fault situations:

throw: The throw activity generates a fault from inside the business process.

rethrow: The rethrow activity propagates a fault from inside a fault handler to an enclosing scope, where the process itself is the outermost scope.

compensate: The compensate activity invokes compensation on all completed child scopes in default order.

compensateScope: The compensateScope activity invokes compensation on one particular (completed) child scope.

exit: The exit activity immediately terminates the execution of a business process instance.

Furthermore, WS-BPEL offers structured activities. Structured activities can have other activities as children; that is, they represent container activities. WS-BPEL consists of the following structured activities:

flow: The activities contained in a flow are executed in parallel, partially ordered through control links. A flow activity represents a directed graph. Note that cyclic control links are not allowed.

sequence: The activities contained in a sequence are performed sequentially in lexical order.

if: The if activity represents a choice between multiple branches. However, exactly one branch is selected.

while: The contained activity of a while loop is executed as long as a specified predicate evaluates to true.

repeatUntil: The contained activity of a repeatUntil loop is executed until a specified predicate evaluates to true.

forEach: The activity contained in a forEach loop is performed sequentially or in parallel, controlled by a specified counter variable. This loop can be terminated prematurely by means of a completion condition.

pick: The pick activity blocks and waits either for a suitable message to arrive or for a time out, whichever occurs first.

scope: A container that associates its contained activity with its own local elements, such as variables, partner links, correlation sets, and handlers (please see the following).

To handle exceptional situations, WS-BPEL offers four different handlers:

1. *catch and catchAll*: Fault handlers for dealing with fault situations in a process. A fault handler can be compared to the catchpart of a try{}... catch{}-block in programming languages like Java.

2. *onEvent and onAlarm*: Event handlers for processing unsolicited inbound messages or timer alarms concurrently to the regular control flow.

3. *compensationHandler*: A compensation handler undoes the persisted effects of a successfully completed scope.

4. *terminationHandler*: A termination handler can be used for customizing a forced scope termination, for example, caused by an external fault.

In addition to concepts introduced already, there are three more concepts for communication: partner links, correlation sets, and (variable) properties:

1. *PartnerLinks* describe the relationship between a process and its services. A partner link points to a Web Service interface the process provides via a myRole attribute. Consequently, a partnerRole attribute points to the Web Service interface that is required from the partner. A partner link can only have one myRole attribute (inbound partner), only one partnerRole attribute (outbound partner), and both attributes (bidirectional partner).

2. *CorrelationSets* are of help in identifying (stateful) process instances. Each process instance will get one or more unique keys based on business data, which are used to correlate a process instance with an incoming message. A correlation set consists of one or more properties.

3. A *property* is business data that creates a name that has a semantic meaning beyond an associated XML type, for example, a social security number versus a plain XML schema integer type. Therefore, properties help to isolate the process logic from the details of a variable definition. Such typed properties are then mapped (aliased) to the parts of a WSDL message or an XSD element.

Business Process Modeling Notation (BPMN): BPMN is a notation used to graphically depict business processes. The language provides users the capability to capture their internal business procedures in a graphical notation. In other words, BPMN is a graph-oriented visual language that allows to model business processes in a flowchart-like fashion. Such a standardized graphical notation for business processes allows to explain and exchange processes in a standard manner and to better understand collaborations and business transactions between organizations. Basically, the BPMN language consists of four core elements:

1. *Flow objects* are the nodes of the BPMN graph. There are three kinds of flow objects: activities, events, and gateways.
2. *Connecting objects* are the edges of a BPMN graph. BPMN allows three different kinds of connecting objects: sequence flow, message flow, and association.
3. *Swimlanes* are used to group other modeling elements in two distinct ways: a pool represents a process. It can be divided up into multiple lanes, where each lane is a subpartition of that process and is used to organize and categorize activities (e.g., activities that are performed by the same department are grouped in the same lane).
4. *Artifacts.* As an example, a data object is an artifact that represents the data that an activity requires before it can be performed or that an activity produced after it has been performed. For the sake of completeness, there are two more artifacts mentioned in the BPMN standard, text annotation and group.

10.6 Summary

This chapter described the characteristics and features of process, workflow, and business process management systems. It explains how business processes can be realized and composed using Web Services. It gives an overview of the Business Process Execution (BPEL) and how processes can be described using BPEL.

11

Application Service Providers (ASPs)

Applications are evolving from those that facilitated a single business function (e.g., accounting and payroll) to integrated application environments that facilitate business processes spanning entire enterprises or the extended enterprises. This evolution entails more people, planning, and ongoing management, especially for mission-critical applications that must maintain a high level of availability. With such increasing complexity and the increased amount of time and skill needed to keep up with the rapidly changing technology cycles, organizations are increasingly seeking outside assistance to deploy, manage, and enhance their applications.

Application Outsourcing (AO) is a service wherein responsibility for the deployment, management, and enhancement of a packaged or customized software application is handed out contractually to an external service provider. AO entails specific activities and expertise aimed at managing the software application or set of applications. Contractual service-level agreements (SLAs) are set at the application level and include responsibilities for

- Application availability
- Application performance
- Application enhancement

Figure 11.1 presents the positioning of the various types of application outsourcers based on the type of application being outsourced (proprietary vs. packaged) and the configuration/location of the service provision (one to one or one to many).

The salient characteristics of the various categories are as follows:

1. Top-left quadrant—represents *data center service providers* that provide application-hosting services and support to other divisions and are typically managed internally by the enterprises.
2. Bottom-left quadrant—represents the *application maintenance service providers* that provide hosting, maintenance, and support of custom-developed application at the enterprises' site(s).
3. Top-right quadrant—represents the *application service providers* that provide operations and maintenance of both packaged and custom-developed application to disparate enterprises using the same outsourcing infrastructure (hardware, software, networks,

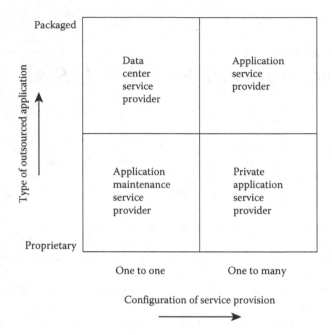

FIGURE 11.1
Types of application outsourcers.

operational staff, etc.). We will describe this category in Section 11.6 "Oracle Siebel On Demand."

4. Bottom-right quadrant—represents the *private application service providers* that provide operations and maintenance especially of custom applications to an ecosystem of enterprises involved in a particular area of business. We will describe this category in Section 11.7 "Private ASPs".

The top reasons cited by enterprises for resorting to outsourcing are as follows:

- Free internal resources for more strategic projects
- Offload functions that are difficult to manage
- Receiving cash infusion from transfer of assets to outsourcers
- Replacing capital expenditure with pay-as-you-use operational expenses
- Gaining access to world-class IT expertise
- Reducing costs for research, development, and successful deployment
- Reducing investment risk in a rapidly changing environment
- Obtaining additional manpower on an as-and-when-needed basis

In the simplest case, the application is merely hosted for the customer. AO typically does not include responsibility for the underlying business processes or functions; a more comprehensive engagement like Business Processing Outsourcing (BPO) may include operations assessment, process improvement, and change management services. When standard applications are being outsourced, the level of staff and asset transfer to this facility may be minimal and the application itself may be provided via a license or lease agreement included in the overall service agreement. On the other hand, for more customized applications, the contract may include the transfer of people and assets associated with the outsourced application.

11.1 Enterprise Application Service Providers (ASPs)

The basic idea behind the Application Service Provider (ASP) computing model is that a provider hosts and manages applications that users can access over networks like the Internet. This is the final step toward software manifesting as a service. The market is evolving toward a state where companies will pay for the software as services on a usage basis, as they do traditionally with utility services like electricity and gas. This market was pioneered in 1998 by start-ups such as Breakaway Solutions, USinternetworking, and Corio. The ASP model typically involves lease-to-own options on software. It entails renting access at a low rate to the functionality of ERPs, CRMs, SCMs, etc., over the Internet. In contrast, traditional outsourcing requires upfront purchase of software licenses and often charges large fees for contracted services. Recently, however, many ASPs are also experimenting with other pricing models that involve charges for initial customization, migration, and integration, plus a flat monthly fee depending on the use of functionality and services, or even a percentage of the customer's revenue. The ASP services could range from infrastructure, colocation, cohosting, dedicated hosting right to even hosted businesses such as hosted buying services or hosted customer relationship management services.

The level of the packaged application could vary from discrete applications (like web-based transaction application for conducting a commerce-based transaction) to environment applications (like ERP and CRM). The level of service provided by an ASP can range from simple hosting of application(s) to managing the application environment. Hosting is a standardized service that is based on a high-volume and low-cost business model in which applications are hosted at a site and accessed remotely. Managing an application environment goes beyond simple hosting and may include upfront consulting, customization, and extensions of the application and ongoing application technical support.

11.2 Fundamentals of ASP

International Data Corp. (IDC) describes an ASP scenario as "an end user accesses an application resident on a server, just as he or she would on a LAN or in the enterprise data center. However, the server resides at the ASP's third-party data center and is reached via a dedicated line or the Internet (or extranet). The applications can range from low-end, productivity programs (e.g., word processing) to high-end ERP modules. The service is provided on a subscription basis and can bundle a full range of hosted application services."

Compare the process of leasing a car. Not much upfront money is required, consumers get something they might not be able to buy outright, they pay for it monthly, and, at the end of the lease, they decide what to do with the car. By just licensing a few seats from an ASP, organizations get a full-functioning application that might be something more powerful and sophisticated than they could buy outright. They have access to the application without having to pay for hardware, software, or installation. Organizations can realize financial cost savings, reduce capital investments, and lower IT management requirements. Such an option also allows organizations to focus on their core businesses and react quickly to changes in the marketplace—both opportunities and threats.

In traditional outsourcing arrangements, the entire business process is handed off to the outsourcing company—operations, the legacy application itself, the infrastructure it was built on, and some of the internal IT staff to support it. Today, every level of the IT infrastructure (network, data, messaging and system management) can be selectively outsourced. With the ASP model, the software and its required infrastructure (including support) are provided by the application service provider, but the actual business process operations are still handled by the organization. If an insurance company outsources its claims processing, the outsourcer receives the claims and processes the claims on its hardware using its software and its staff. With the ASP model, the insurance company's staff receives the claims and processes the claims on the ASP's hardware using the ASP's software and infrastructure.

An ASP service is not the time-sharing of the 1960s or the outsourcing of the 1980s. The ASP model is much more than the rental of a slice of time. The model allows an organization to decide the location of the computing capability based on economic and financial grounds. It provides an option for sharing information and conducting transactions. The ASP model uses client–server architecture and relies on secure, cost-effective data communications. The IT staff does not need to have expertise in the application or the infrastructure that is being handled by the ASP. ASPs can be used to fill gaps in an application portfolio. So the focus is on saving time as well as cost, and time and cost are two major variables of any IT project.

Packaged software developers can use the ASP model to convert the infrequent buyer into a steady revenue stream customer. Usually, a customer buys the latest version of a software product and then elects not to upgrade for two or three generations. Under the ASP model, an ASP customer is provided with the latest version of a software package and pays a monthly fee to use that software, thereby generating a steady stream of revenue for the software package developer. The software vendor upgrades only the master copies on licensed ASP servers. The software vendor is not required to maintain old code or support multiple versions of a product. If customers do not like the upgrade, they cannot go back to a prior version.

Customers of ASPs usually pay a flat fee to sign up and from then on a monthly fee. For that monthly fee, the customer gets all upgrades automatically as soon as they are released—all the new drivers, the new features and everything. However, because the ASP is monitoring the customer's payments, if the customer stops paying, it no longer gets the use of the software.

11.3 ASP Business Model

ASP provides access to and management of an application. An ASP owns the software or has a contractual agreement with the software vendor to license it. Customers gain access to the environment without making investments in application license fees, hardware, and staff. The application is managed from a central location (the ASP site) rather than the customer's sites. Customers access the application via the Internet or leased lines. The ASP is responsible for delivering on the customer's contract regardless of its structure—sole provider or partnered. If a problem arises, the ASP is responsible for resolving the issue. Service guarantees usually address availability, security, networked storage, and management and are spelled out in service-level agreements (SLAs). ASPs enforce these guarantees by closely monitoring the server environments and often add proprietary modifications to ensure performance uptime and security.

An ASP provides the application service as its primary business. The service may be delivered from beginning to end by a single vendor or via partnerships among several vendors. A single-source vendor controls everything from implementation to ongoing operations and maintenance of the application. The customer deals with only one vendor, and that vendor has complete control over the process. Under this model, the vendor must have expertise in a variety of areas, maintain a data center infrastructure, and have high capital requirements.

11.3.1 Service Level Agreements (SLAs)

SLAs spell out the customer's expectations for service, which might range from expected response times to minimum bandwidth. Some ASPs include guarantees such as 99.9% uptime and disaster recovery. ASPs will add security to an already secure platform (e.g., Windows NT/2000 or .Net) to guarantee security levels.

An SLA details the day-to-day expected service. There should be means to award exceeded minimum requirements that can be offset against days that failed to meet expectations. The SLA might also include provisions for days when the ASP's servers are offline for maintenance. An SLA should also include a clause that allows the customer to terminate the contract without penalty if it receives poor service. A customer should also make sure that it can get out of the deal with whatever it needs to bring a new ASP on board—data, customized software, and the like. Customers should keep the contract term as short as possible—no more than 3 years. It is difficult to know what hosting will look like in 5 years. Make sure that the performance penalties truly motivate the ASP to address the organization's issues (remember, the ASP has other customers) and that penalties escalate each time the problem occurs. Establish metrics that truly measure growth. Choose two simple ones and agree on a firm price for the service as usage grows. Furthermore, customers should not try to trade reduced service for lower monthly fees. The only way for an ASP to lower the organization's costs is to cut service levels. The quality of the service is key to the customer's successful use of, and, therefore the benefit derived from, the ASP.

11.4 ASP Value Drivers

Factors driving the ASP solutions are as follows:

1. Enabling technologies
   a. Pervasiveness of the Internet
   b. Access and declining cost of bandwidth capacity
   c. Shared applications in a client–server environment
   d. Browsers as an accepted GUI application
   e. Potential of e-commerce and e-business solutions
2. Technical drivers
   a. Utilization of emerging technologies and best-of-breed applications
   b. Accelerated application development

    c. Rapidly changing and increasing complexity of technology

    d. Obtaining business domain, functional and vertical industry expertise

    e. Shortage skilled IT labor

    f. Transfer of risk regarding application ownership by the super users

3. Business drivers

    a. Reduction in total cost of ownership (TCO) by at least 30–50%

    b. Predictability of cash flows by eliminating the uncertainties of post-implementation software-related expenditures

    c. Focus on core competencies and strategic objectives

    d. Improvement in the efficiency of internal IT staff by freeing them to focus on processes and systems to leverage core competencies

    e. Improvement in coordination efforts on a global basis

However, merely hosting and managing an application is not adequate and needs to be augmented for the creation of proprietary and sustainable customer relationships with value-added variations like

- Domain expertise emphasis
- Vertical industry emphasis
- Vertical exchange emphasis
- Infrastructure emphasis
- Security infrastructure emphasis
- Full-service provider emphasis
- Aggregator emphasis

11.5 ASP Benefits, Risks, and Challenges

Application Service Providers (ASPs) present a real opportunity to replace the in-house IT department. Smaller companies also need the full functionality of high-end applications such as ERP, SCM, or CRM that cannot be afforded by them. ASPs meet the needs of MME companies that cannot afford the substantial upfront investments in establishing the infrastructure, knowledgeable and experienced staff, ongoing administration and monitoring services, better backup and recovery methods, and so forth. The top reasons that are cited by companies for renting applications are guaranteed performance levels, high availability, and responsive service and support. The cost of ownership of hosted applications can be 25% cheaper than managing the applications internally. Other criteria for companies to contract ASPs are lower upfront costs, faster implementations, higher redundancy, larger scalability of hardware and bandwidth, automatic upgrades, quicker distribution and deployment, and data storage, backup, and recovery capabilities.

ASP services that include application-hosting services enable customers to evaluate, implement, and operate enterprise applications online. This enables companies to perform detailed evaluations and prototype company-specific solutions via the Internet prior to deciding to purchase or actually purchasing the ERP, CRM, or SCM product. Customers also have the facility to implement and continue using applications functionality via the Web browser.

Using application service providers (ASPs) has many advantages, such as

- Enabling a company to concentrate on enhancing the competitiveness of its core functions
- Enabling a company to outsource the enhancing competitiveness of its noncore but still-critical functions like implementing call center operations
- Reengineering critical but noncore functions and processes quickly
- Comprehensive industry-specific and company-specific evaluations
- Rapid prototyping of company-specific implementations
- Lower upfront investments in hardware and software, technical manpower resources, and training
- Reduced risks of initial erroneous decisions in pricing, reliability, scalability, bandwidth, and security
- Flexibility of options like hosting, cohosting, and colocation
- Time to implement at a lower cost
- Time to go live at a lower cost
- Time to benefit at a lower cost
- Time to full ownership at a lower cost
- Time to deploy at other sites at a lower cost
- Higher service-level guarantees
- Better customer service
- Reduced administration, management, maintenance, and support costs

ASPs need to gain customer acceptance as well as IT acceptance. The selling focus is usually on business management—pitching a business solution or business service. The focus is on the value added. The IT organization often comes into the discussions to address security and network issues. IT needs to view the ASP as an alternative, not an interloper. IT needs to become advisors, not turf protectors. Potential customers must be convinced that their application and its data will be available to them 24×7 but yet secure from outsiders. As Internet traffic continues to grow, potential customers must also be convinced that their access will not slow down. As mobile use escalates, ASPs need to deal with the needs of employees out of the office and the security requirements mobile access demands.

Some of the challenges faced by the ASP concept are

- Security of information
- Scope and flexibility of services
- Overall quality of service and support
- Adaptability of software

11.6 Oracle SAP CRM On Demand

Oracle SAP CRM On Demand provides organizations with all the comprehensive SAP CRM capabilities along with the power of SAP's software development, infrastructure management, and operational services to maximize the business value of their software investment. Delivered to users over the Internet, SAP CRM On Demand frees IT organizations from day-to-day maintenance, upgrades, and software management while providing unparalleled levels of availability, reliability, and security. This results in a far more efficient and cost-effective approach to managing CRM and SAP CRM technology over the long term.

SAP CRM On Demand achieves operational excellence and low total cost of ownership (TCO) using standardization and automation as key enablers to improve customers' experience across the entire software ownership life cycle, including in the core service areas like

- Infrastructure management
- Service-level management
- Security management
- Software management
- IT governance

11.7 Private ASPs

It has been well established that the largest cost component in an application's life cycle is the effort expended on maintenance and upgrades, and this is valid even in cases of packaged software implementations by reason of the customizations that are indispensable. The key idea underlying the private ASP is that *the high cost of maintenance and upgrades (and operations) is spread across the boundary of the enterprise to include the associated ecosystem of at least the exclusive suppliers.* The viability of this business model can be illustrated from the fact that the customization requirements for implementing enterprise systems (ERP, SCM, CRM, etc.) at General Motors's partner/subsidiary Delphi Electronics are going to be similar to or will be heavily influenced by the customization requirements of General Motors itself. From there, it is not very difficult to see that there will be major advantage in combining the implementation projects and teams, training programs, testing centers, rollout projects, IT centers, disaster recovery centers, and so on.

We will describe this concept by taking Ford as the hypothetical example. A private ASP is an extended ASP service, managed and operated by a major player like Ford for itself and its partners—both customers and vendors. All customization and upgrade issues are dictated mainly by Ford, unlike in the case of a traditional ASP where they may have to customize for each customer independently, *which basically makes the traditional ASP model unviable.*

$$\text{Private ASP} = (\text{SCM} + \text{CRM like SAP} + \text{Net Markets}) \text{on ASP basis for}$$
$$\text{Ford and its participating partners}$$

None of the partners need to be compelled to join the private ASP. Ford, however, must mandate that such partners need to integrate and be compatible with its private ASP at their own cost. Licensing issues are simplified in that all software licenses are owned via the private ASP instituted jointly by Ford and/or its partners.

The composite value proposition for Ford and its partners involves the following:

1. Easy to add/integrate new vendor partner.
2. Value proposition of a NetMarket—Ford can service an enlarged base of vendors for more competitive bidding but within the class of these preferred partners/vendors.

3. Value proposition of an ASP for reduced application management, operations, and maintenance costs.
4. Value proposition of implementation, management, and maintenance of (say) a CRM for individual partners who do not have a CRM that is also the same CRM as used by Ford (at much reduced cost than outside of the private ASP).
5. Value proposition of an EAI provider, especially for integrating internal and proprietary applications of Ford and its individual partners (at much reduced cost than outside of the private ASP).
6. Value proposition of profitable IT services via the private ASP for Ford and/or its partners.
7. Value proposition of an outsourcing service in terms of assets, manpower, support, training, etc.
8. Value proposition of SCM implementation, operations, and management for itself and its partners—this would be of great benefit in the long term.

11.7.1 What Does a Private ASP Offer?

A private ASP offers the following;

- It provides application services to a restricted business ecosystem of Ford and its participating partners.
- It provides individual enterprise-oriented services for Ford and each of the partners through separate instances of each of these applications.
- It also provides individual company-specific services for internal applications like HRMS, CRM, and e-mail to Ford and its partners separately. Thus, if there are 100 partners who have joined Ford's private ASP program, there may be 100 different implementations of HRMS, CRM, etc., (from different vendors) systems for the specific requirements for each of these partners but handled by the same entity—the private ASP.

Ford can own the private ASP facility fully and run it as a profit center. It can charge its partners on subscription (and/or even on transaction basis) along with a possible lump sum enrolment fee. Or Ford can float a separate company along with its participating partners to establish and run this private ASP; the private ASP can then charge via a uniform model for all its delivered services, to all members of the private ASP.

11.8 Summary

This chapter introduced the concept of Application Service Provider (ASP). It followed this discussion on the value drivers, benefits, risks, and challenges of the ASP model. In the latter part of the chapter, we discussed the concept of private ASP and its benefits.

12

Grid Computing

The grid computing paradigm based on resource sharing was brought to broader public by the popular project (http://setiathome.berkeley.edu) SETI@home. The goal of the Search for Extraterrestrial Intelligence (SETI) project is the detection of intelligent life outside earth. The project uses radio telescopes to listen for narrow-bandwidth radio signals from space. As such signals are not known to occur naturally, it is expected that a detection of them would provide evidence of extraterrestrial technology. The analysis of radio telescope signals involves huge amounts of data and is very computing intensive. No single research lab could provide the computing power needed for it. Given the tremendous number of household PCs, the involved scientists came up with the idea to invite owners of PCs to participate in the research by providing the computing power of their computers when they are idle. Users download a small program on their desktop. When the desktop is idle, the downloaded program would detect it and use the idle machine cycles. When the PC is connected back to the Internet, it would send the results back to the central site. The SETI initiative recently celebrated the 10th anniversary (it was launched on May 17, 1999) and has at present more than 3 million users that participate with their PCs.

12.1 Background to Grid Computing

Grid computing means that computing power and resources can be obtained as utility similar to electricity—the user can simply request information and computations and have them delivered to him or her without necessity to care where the data he or she requires reside or which computer is processing his or her request. From the technical perspective, grid computing means the virtualization and sharing of available computing and data resources among different organizational and physical domains. By means of virtualization and support for sharing of resources, scattered computing resources are abstracted from the physical location and their specific features and provided to the users as a single resource that is automatically allocated to their computing needs and processes. Almost every organization has significant unused computing capacity, widely distributed among a tribal arrangement of PCs, midrange platforms, mainframes, and supercomputers. For example, if a company has 5000 PCs, at an average computing power of 333 MIPS, this equates to an aggregate 1.5 tera (1012) floating-point operations per second (TFLOPS) of potential computing power. As another example, in the United States, there are an estimated 300 million computers. At an average computing power of 333 MIPS, this equates to a raw computing power of 100,000 TFLOPS. Mainframes are generally idle 40% of the time; Unix servers are actually *serving* something less than 10% of the time; most PCs do nothing for 95% of a typical day. This is an inefficient situation for customers. TFLOPS speeds that are possible with grid computing enable scientists to address some of the most computationally intensive scientific tasks, from problems in protein analysis that will form the basis for new drug designs to climate modeling and deducing the content and behavior of the cosmos from astronomical data.

Prior to the deployment of grid computing, a typical business application had a dedicated platform of servers and an anchored storage device assigned to each individual server. Applications developed for such platforms were not able to share resources, and, from an individual server's perspective, it was not possible, in general, to predict, even statistically, what the processing load would be at different times. Consequently, each instance of an application needed to have its own excess capacity to handle peak usage loads. This predicament typically resulted in higher overall costs than would otherwise need to be the case. To address these lacunae, grid computing aims at exploiting the opportunities afforded by the synergies, the economies of scale, and the load smoothing that result from the ability to share and aggregate distributed computational capabilities and deliver these hardware-based capabilities as a transparent service to the end user.

At the core of grid computing, therefore, are virtualization and virtual centralization as well as availability of heterogeneous and distributed resources based on collaboration among and sharing of existing infrastructures from different organizational domains that together build the computing grid. The key concept is the ability to negotiate resource-sharing arrangements among a set of participating parties (providers and consumers) and then to use the resulting resource pool for some purpose. The sharing that we are concerned with is not primarily file exchange but rather direct access to computers, software, data, and other resources, as is required by a range of collaborative problem-solving and resource-brokering strategies emerging in industry, science, and engineering. This sharing is, necessarily, highly controlled, with resource providers and consumers defining clearly and carefully just what is shared, who is allowed to share, and the conditions under which sharing occurs. A set of individuals and/or institutions defined by such sharing rules form what we call a virtual organization (VO).

The benefits gained from grid computing can translate into competitive advantages in the marketplace. Grids enable the following:

- Enable resource sharing
- Provide transparent access to remote resources
- Make effective use of computing resources, including platforms and data sets
- Reduce significantly the number of servers needed by (25–75%)
- Allow on-demand aggregation of resources at multiple sites
- Reduce execution time for large-scale data processing applications
- Provide access to remote databases and software
- Provide load smoothing across a set of platforms
- Provide fault tolerance
- Take advantage of time zone and random diversity (in peak hours, users can access resources in off-peak zones)
- Provide the flexibility to meet unforeseen emergency demands by renting external resources for a required period instead of owning them
- Enable the realization of a virtual data center

Grid computing emerged in the early 1990s, when high-performance computers were connected by fast data communication with the aim to support calculation- and data-intensive scientific applications. Since the mid-1990s, the concept of grid has evolved. Similar to other infrastructure innovations— for example, the Internet—the grid was first introduced and adopted in science for the support of research in various scientific disciplines that require high-performance computing (HPC) together with huge amounts of data stored in dedicated databases. Examples of such sciences are earth science, astroparticle physics, and computational chemistry. They are summarized under the term *e-science*. To support e-science, many national and international initiatives have been started by governments in many countries in order to leverage existing investments in research infrastructure and to enable sharing and efficient use of available computational resources, data, and specialized equipment.

Cloud computing is frequently compared to grid computing. Grid computing also has the same intent of abstracting out computing resources to enable utility models and was proposed at least a decade earlier than cloud computing, and there are many aspects of grid computing that have formed the basis of the requirements placed on a cloud. However, there are also very specific differences between a grid computing infrastructure and the features one should expect from a cloud computing infrastructure. All resources constituting a grid computing infrastructure are predefined and predetermined corresponding to the spare capacity that is available to promise with the constituents for participating in the grid. In contrast, cloud computing releases or augments pre-identified and dedicated resources dynamically, depending on the demand. Grid computing is somewhat akin to an application service provider (ASP) environment, but with a much higher level of performance and assurance.

12.2 Introduction to Grid Computing

The vision of grid computing is to enable computing to be delivered as a utility. This vision is most often presented with an analogy to electrical power grids, from which it derives the name *grid*. So grid computing was meant to be used by individual users who gain access to computing devices without knowing where the resource is located or what hardware it is running and so on. In this sense, it is pretty similar to cloud computing. However, just as electrical power grids can derive power from multiple power generators and deliver the power as needed by the consumer, the key emphasis of grid computing was to enable sharing of computing resources or forming a pool of shared resources that can then be delivered to users. So most of the initial technological focus of grid computing was limited to enabling shared use of resources with common protocols for access. Also, since the key takers of this fascinating vision were educational institutions, a particular emphasis was given to handle heterogeneous infrastructure, which was typical of a university data center. From a technical perspective, a software-only solution was proposed (Globus) and implemented on this heterogeneous infrastructure to enable use of these resources for higher computing needs. Once reasonably successful within universities, grid computing faced a serious issue when it came to sharing resources across commercial institutions. Establishing trust and security models between infrastructure resources pooled from two different administrative domains became even more important.

The first most cited definition of grid computing reflected these origins and was suggested by Foster and Kesselman (1998): "A computational grid is a hardware and software infrastructure that provides dependable, consistent, pervasive, and inexpensive access to high-end computational capabilities."

The main resources that can be shared in a grid are

- Computing/processing power
- Data storage/networked file systems
- Communications and bandwidth
- Application software
- Scientific instruments

Grid middleware is specific software, which provides the necessary functionality required to enable sharing of heterogeneous resources and establishing of virtual organizations. From a market perspective, Grid middleware provides a special virtualization and sharing layer that is placed among the heterogeneous infrastructure and the specific user applications using it. Grid computing is basically the deployed grid middleware or the computing enabled by grid middleware based on flexible, secure, coordinated resource sharing among a dynamic collection of individuals, institutions, and resources. Grid computing means on the one hand that heterogeneous pools of servers, storage systems, and networks are pooled together in a virtualized system that is exposed to the user as a single computing entity. On the other hand, it means programming that considers grid infrastructure and applications that are adjusted to it. Grid infrastructure refers to the combination of hardware and grid middleware that transforms single pieces of hardware and data resources into an integrated virtualized infrastructure that is exposed to the user as a single computer despite of heterogeneity of the underlying infrastructure. Utility computing is the provision of grid computing and applications as a service either as an open grid utility or as a hosting solution for one organization or VO. Utility computing is based on pay-per-use business models.

Thus, grid computing is a new computing paradigm based on IT resource sharing and on provisioning of IT resources and computing in a way similar to how electricity is consumed today. It is enabled by specific grid middleware provided on the market either as packaged or open-source software or in the form of utility computing. The major potential advantages of grid computing for an improved management of IT in companies can be summarized as follows:

- Grids harness heterogeneous systems together into a single large computer and, hence, can apply greater computational power to a task and enable greater utilization of available infrastructure. In particular, with grid computing existing, underutilized resources can be exploited better.
- Grid computing enables greater scalability of infrastructure by removing limitation inherent in the artificial IT boundaries existing between separate groups or departments.

- Grid computing results in improved efficiency of computing, data, and storage resources due to parallel CPU capacity, load balancing, and access to additional resources. As computing and resources can be balanced on demand, grid computing results also in increased robustness and reliability—failing resources can be replaced easier and faster with other resources available in the grid.

- Grid computing furthermore enables a more efficient management of distributed IT resources of companies. With the help of virtualization, physically distributed and heterogeneous resources can be better and uniformly managed. This makes possible to centrally set priorities and assign distributed resources to tasks.

- Grid computing, in combination with utility computing, enables the transformation of capital expenditure for IT infrastructure into operational expenditure and provides the opportunity for increased scalability and flexibility. However, the usage of utility computing results in comparatively higher security and privacy risks.

- Grid computing enables cost savings in the IT departments of companies due to reduced total cost of ownership (TCO). Instead of investing in new resources, greater demand can be met by higher utilization of existing resources or by taking advantage of utility computing.

12.2.1 Virtualization

Grid computing also differs from virtualization. Resource virtualization is the abstraction of server, storage, and network resources in order to make them available dynamically for sharing, both inside and outside an organization. The universal problem that virtualization is solving in a data center is that of dedicated resources. While this approach does address performance, this method lacks fine granularity. Typically, IT managers take an educated guess as to how many dedicated servers they will need to handle peaks, purchase extra servers, and then later find out that a significant number of these servers are significantly underutilized. A typical data center has a large amount of idle infrastructure, bought and set up online to handle peak traffic for different applications; virtualization offers a way of moving resources from one application to another dynamically. Three representative products are HP's Utility Data Center, EMC's VMware, and Platform Computing's Platform LFS. With virtualization, the logical functions of the server, storage, and network elements are separated from their physical functions (e.g., processor, memory, I/O, controllers, disks and switches). In other words, all servers, storage, and network devices can be aggregated into independent pools of resources. Some elements may even be further subdivided (server partitions, storage LUNs) to provide an even more granular level of control. Elements from these pools can then be allocated, provisioned, and managed, manually or automatically, to meet the changing needs and priorities of one's business.

Virtualization has somewhat more of an emphasis on local resources, whereas grid computing has more of an emphasis on geographically distributed interorganizational resources. Virtualization is a step along the way on the road to utility computing (grid computing).

12.2.2 Cluster

The distinction between clusters and grids relates to the way resources are managed. In the case of clusters (aggregations of processors in parallel-based configurations), the resource allocation is performed by a centralized resource manager and scheduling system. Also, nodes cooperatively work together as a single unified resource. In the case of grids, each node has its own resource manager and does not aim at providing a single system view. A cluster is comprised of multiple interconnected independent nodes that cooperatively work together as a single unified resource. This means all users of clusters have to go through a centralized system that manages the allocation of resources to application jobs. Unlike grids, cluster resources are almost always owned by a single organization. These cluster management systems have centralized control, complete knowledge of system state and user requests, and complete control over individual components.

Actually, many grids are constructed using clusters or traditional parallel systems as their nodes, although this is not a requirement.

12.2.3 Web Services

Grid computing also differs from basic Web Services, although it now makes use of these services. Whereas the Web is mainly focused on communication, grid computing enables resource sharing and collaborative resource interplay toward common business goals. Web Services provide standard infrastructure for data exchange between two different distributed applications, whereas grids provide an infrastructure for aggregation of high-end resources for solving large-scale problems in science, engineering, and commerce. While most Web Services involve static processing and moveable data, many grid computing mechanisms involve static data (on large databases) and moveable processing. These Web Services will play a key constituent role in the standardized definition of grid computing, since Web Services have emerged in the past few years as a standard-based approach for accessing network applications. The recent trend is to implement grid solutions using Web Services technologies, for example, the Globus Toolkit 3.0 middleware. In this context, low-level grid services are instances of Web Services (a grid service is a Web Service that conforms to a set of conventions that provide for controlled, fault-resilient, and secure management of services).

12.2.4 P2P Network

Both peer-to-peer computing and grid computing are concerned with the same general problem, namely, the organization of resource sharing within VOs. As is the case with Peer-to-Peer (P2P) environments, grid computing allows users to share files, but unlike P2P, grid computing allows many-to-many sharing. Furthermore, with grid computing, the sharing is not only in reference to files but other resources as well. The grid community generally focuses on aggregating distributed high-end machines such as clusters, whereas the P2P community concentrates on sharing low-end systems such as PCs connected to the Internet.

12.3 Comparison with Other Approaches

There is no single or unique solution to a given computing problem; grid computing is one of a number of available solutions in support of optimized distributed computing. Corporate IT professionals will have to perform appropriate functional, economic, business-case, and strategic analyses to determine which computing approach ultimately is best for their respective organizations. Furthermore, it should be noted that grid computing is an evolving field and, so, there is not always one canonical, normative, universally accepted, or axiomatically derivable view of *doing it with the grid*.

Like virtualization technologies, grid computing enables the virtualization of IT resources. But, unlike virtualization technologies, which virtualize a single system, grid computing enables the virtualization of broad-scale and disparate IT resources. Similarly, like clusters and distributed computing, grids bring computing resources together. But, unlike clusters and distributed computing, which need physical proximity and operational homogeneity, grids can be geographically distributed and heterogeneous.

12.4 Characteristics of a Grid

In 2002, Ian Foster from Argonne National Laboratories proposed a three-point checklist for determining whether a system is a grid or not. Ian Foster along with Steve Tucker in the popular article "Anatomy of Grid" defined grid computing as "coordinated resource sharing and problem solving in dynamic, multi-institutional virtual organizations."

So the key concept emphasized was the ability to negotiate resource sharing agreements among a set of participating parties—where sharing did not really mean *exchange* but direct access to computing resources either in a collaborative resource sharing or negotiated resource-brokering strategies. Further, this sharing was highly controlled with resource providers and consumers grouped into virtual organizations primarily based on sharing conditions.

The following is the precise simple checklist that was proposed: a grid is a system that functions as follows:

1. *Coordinates resources that are not subject to centralized control*: The first criterion states that a grid should integrate computing resources from different control domains (say servers from computer centers of different universities, each center having a different system administrator in each university). Technologically, this requirement addresses the issues of cross-domain security, policy management, and membership.

2. *Uses standard, open, general-purpose protocols and interfaces*: The use of a common standard for authentication, authorization, resource discovery, and resource access becomes a necessity in such cases and hence the second criterion.

3. *Delivers nontrivial quality of service*: Finally, in an effort toward commercializing the usage of shared resources, it is important to support various quality-of-service parameters such as response time, throughput, availability, or even co-allocation of resources to meet user demands.

12.5 Types of Grids

Broadly speaking, there are three types of grid: computational grids, data grids, and service grids. A computational grid is a distributed set of resources that are dedicated to aggregate computational capacity. Computational grids are highly suitable for task farming or high-throughput computing applications where there is typically one data set and a huge parameter space through which the scientist wishes to search. A data grid is a collection of distributed resources that are specifically set up for processing and transferring large amounts of data. A service grid is a collection of distributed resources that provides a service that cannot possibly be achieved through one single computer. In this example, therefore, the grid will typically consist of several different resources, each providing a specific function that needs to be aggregated in order to collectively perform the desired services. For example, you could have a service that obtained its functionality by integrating and connecting databases from two separate VOs (representing two data streams from physics sensors/detectors) in order to output their correlation. Such a service could not be provided by one organization or the other since the output relies on the combination of both.

The term *computational grid* comes from an analogy with the electric power utility grid. A computational grid is focused on setting aside resources specifically for computing power and uses networks of computers as a single, unified computing resource. It is possible to cluster or couple a wide variety of resources including supercomputers, storage systems, data sources, and special classes of devices distributed geographically and use them as a single unified resource. Such pooling requires significant hardware infrastructure to achieve the necessary interconnections and software infrastructure to monitor and control the resulting ensemble. The majority of the computational grids are centered on major scientific experiments and collaborative environments. Computational grid applications exhibit several functional computational requirements. These include the ability to manage a variety of computing resources, select computing resources capable of running a user's job, predict loads on grid resources, and decide about resource availability, dynamic resource configuration, and provisioning. Other useful mechanisms for the management of resources include failure detection, failover, and security mechanisms.

A data grid is responsible for housing and providing access to data across multiple organizations and makes them available for sharing and collaboration purposes. These data sources can be databases, file systems, and storage devices. The requirement for managing large data sets is a core underpinning of any grid computing environment. Data grids can also be used to create a single, virtual view of a collection of data sources for large-scale collaboration. This process is called *data federation*. In data grids, the focus is on the management of data that are being held in a variety of data storage facilities in geographically dispersed locations. For example, medical data grids are designed to make large data sets, such as patient records containing clinical information and associated digital x-rays, medication history, doctor reports, symptoms history, genetic information, and so on, available to many processing sites. By coupling the availability of these massive data sets with the large processing capability of grid computing, scientists can create applications to analyze the aggregated information. Searching the information for patterns or signatures enables scientists to potentially reach new insights regarding the environmental or genetic causes of diseases. Data grid systems must be capable of providing data virtualization services to provide the ability to discover data, transparency for data access, integration, and processing as well as the ability to support flexible data access and data filtering mechanisms. In addition, the provision of security and privacy mechanisms for all respective data in a grid system is an essential requirement for data grids.

12.6 Grid Technologies

First of all, grid computing defines a notion of a virtual organizations to enable flexible, coordinated, secure resource sharing among participating entities. A virtual organizations (VO) is basically a dynamic collection of individuals or institutions from multiple administrative domains. A VO forms a basic unit for enabling access to shared resources with specific resource-sharing policies applicable for users from a particular VO. The key technical problem addressed by grid technologies is to enable resource sharing among mutually distrustful participants of a VO who may have varying degrees of prior relationship (perhaps none at all) and enable them to solve a common task.

The five layers of grid computing are interrelated and depend on each other. Each subsequent layer uses the interfaces of the underlying layer. Together, they create the grid middleware and provide a comprehensive set of functionalities necessary for enabling secure, reliable, and efficient sharing of resources (computers, data) among independent entities. This functionality includes low-level services such as security, information, directory, resource management (resource trading, resource allocation, quality of service), and high-level services/tools for application development, resource management, and scheduling. In addition, there is a need to provide the functionality for brokerage of resources and accounting and billing purposes.

The main functionalities of a grid middleware are

- Virtualization and integration of heterogeneous autonomous resources
- Provision of information about resources and their availability
- Flexible and dynamic resource allocation and management
- Brokerage of resources either based on company policies
- Security and trust (Security includes authentication [assertion and confirmation of the identity of a user] and authorization [check of rights to access certain services or data] of users as well as accountability.)
- Management of licenses
- Billing and payment
- Delivery of nontrivial Quality of Service (QoS)

An extensible and open grid architecture shown in Figure 12.1 was defined by Ian Forster in The Anatomy of the Grid in which protocols, services, application programming interfaces (APIs), and system development kits (SDKs) are categorized according to their roles in enabling resource sharing.

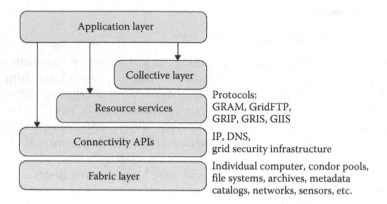

FIGURE 12.1
Layered grid architecture.

a. The grid fabric layer provides the resources to which shared access is mediated by grid protocols. These can be computational resources, storage systems, catalogs, network resources, or even a logical entity, such as a distributed file system, computer cluster, or distributed computer pool. A well-known toolkit for the fabric layer is the Globus Toolkit that provides local resource-specific operations on existing computing elements.

b. The connectivity layer includes the core protocols for communication and authentication for internode communication. The key aspects of these protocols include single sign-on, delegation, user-based trust relationships, and integration with local security solutions. One important protocol whose reference implementation is available in Globus is the public key–based GSI protocol (Grid Security Infrastructure), which extends TLS (Transport Layer Security) to address these issues.

c. The resource layer includes APIs and SDKs for secure negotiation, monitoring, control, accounting, and payment for operations on a single shared resource. An example protocol at this layer is the GRAM (Grid Resource Access and Management) protocol used for allocation, monitoring, and control of computational resources and the GRIP (Grid Resource Information Protocol) and Grid FTP (File Transfer Protocol), which are extensions of LDAP and FTP protocols.

d. The collective layer implements a variety of sharing behaviors with directory services, brokering services, programming systems community accounting and authorization services, and even collaborative services. One such service is the GIIS (Grid Information Index Servers) that supports arbitrary views on resource subsets, which can be used with LDAP and the DUROC library that supports resource co-allocation.

e. The application layer involves the user applications that are deployed on the grid. It is important to note that not any user application can be deployed on a grid. Only a grid-enabled or gridified application, that is, an application that is designed or adjusted to run in parallel and use multiple processors of a grid setting or that can be executed on different heterogeneous machines, can take advantage of a grid infrastructure.

12.7 Grid Computing Standards

One of the challenges of any computing technology is getting the various components to communicate with each other. Nowhere is this more critical than when trying to get different platforms and environments to interoperate. It should, therefore, be immediately evident that the grid computing paradigm requires standard, open, general-purpose protocols and interfaces. Standards for grid computing are now being defined and are beginning to be implemented by the vendors. To make the most effective use of the computing resources available, these environments need to utilize common protocols. We are now indeed entering a new phase of grid computing in which standards will define grids in a consistent way by enabling grid systems to become easily built *off-the-shelf* systems. Standard-based grid systems have been called by some *third-generation grids* or 3G grids.

First-generation or *1G grids* involved local *metacomputers* with basic services such as distributed file systems and site-wide single sign-on, upon which early-adopter developers created distributed applications with custom communication protocols. Test beds extended 1G grids across distances and attempts to create *metacenters* explored issues of interorganizational integration. First-generation grids were totally custom-made proofs of concept. Second-generation or *2G* grid systems began with projects such as Condor, I-WAY (the origin of Globus), and Legion (origin of Avaki), in which underlying software services and communication protocols could be used as a basis for developing distributed applications and services. Second-generation grids offered basic building blocks, but deployment involved significant customization and filling in many gaps. Independent deployments of 2G grid technology today involve enough customized extensions that interoperability is problematic and interoperability among 2G grid systems is rather difficult. This is why the industry needs 3G grids.

By introducing standard technical specifications, 3G grid technology will have the potential to allow both competition and interoperability not only among applications and toolkits but also among implementations of key services. The goal is to mix and match components, but this potential will only be realized if the grid community continues to work at defining standards. The Global Grid Forum community is applying lessons learned from 1G and 2G grids and from Web Service technologies and concepts to create 3G architectures.

12.8 Globus

Current implementations of open grid architecture follow a Web Service–based interface enabling interoperability between different implementations of the protocols. Since Web Services by definition are stateless, the grid community (Globus alliance) introduced a set of enhanced specifications called Web Services Resource Framework (WSRF) that Web Services could implement to become stateful. Open Grid Services Architecture now defines a service-oriented grid computing environment, which not only provides standardized interfaces but also removes the need for layering in the architecture and defines a concept of virtual domains, allowing dynamic grouping of resources as well.

The standard bodies involved in evolving the grid protocols were

a. The Global Grid Forum (GGF)

b. Organization for the Advancement of Structured Information Standards (OASIS)

c. World Wide Web Consortium (W3C)

d. Distributed Management Task Force (DMTF)

e. Web Services Interoperability Organization (WS-I)

A reference implementation of these protocols is available in a popular open-source software toolkit called Globus Toolkit (GT), which was developed by the Globus alliance, a community of organizations and individuals developing fundamental technologies behind the grid. The nice thing about this software is that it enables existing resources to easily join a grid pool by enabling the required protocols locally. To get started on setting up a grid, one just needs to download and install GT on any of the supported platforms. To create a resource pool, it is a good idea to install a resource scheduler such as the Condor cluster scheduler and configure that as a grid gateway for resource allocation. After some initial security configurations (obtaining signed certificates and setting up access rights), the grid can be up and running!

12.9 Summary

This chapter presented an overview of grid computing. It described the grid's fundamental characteristics, types, and constituent technologies. It presents the grid computing standards as well as an overview of the Globus grid computing environment.

Section III

Cloudware

Since cloud services are resultant of a combination of factors like hardware, software, networks, services, etc., there is need to coin a new word to identify this newer category of offerings; hence, the new name *cloudware*.

Chapter 13 introduces the basics of cloud computing, cloud delivery models such as IaaS, PaaS, and SaaS, as well as deployment models such as public, private, and hybrid clouds. Chapter 14 presents the advantages of cloud computing solutions. While Chapter 15 describes the various cloud computing technologies, Chapter 16 deals with the commercially available cloud computing environments. Chapter 17 describes cloudware development paradigms and environments including Google MapReduce and the Hadoop ecosystem. Chapter 18 presents operations and management issues related to cloud computing solutions, while Chapter 19 provides an overview of governance, risks, and compliance issues involved. This sections ends with Chapter 20 that presents details on how companies can successfully prepare and transition to cloudware environments.

13

Cloudware Basics

Many motivating factors have led to the emergence of cloud computing. Businesses require services that include both infrastructure and application workload requests, while meeting defined service levels for capacity, resource tiering, and availability. IT delivery often necessitates costs and efficiencies that create a perception of IT as a hindrance, not a strategic partner. Issues include underutilized resources, overprovisioning or underprovisioning of resources, lengthy deployment times, and lack of cost visibility. Virtualization is the first step toward addressing some of these challenges by enabling improved utilization through server consolidation, workload mobility through hardware independence, and efficient management of hardware resources.

The virtualization system is a key foundation for the cloud computing system. We stitch together compute resources so as to appear as one large computer behind which the complexity is hidden. By coordinating, managing, and scheduling resources such as CPUs, network, storage, and firewalls in a consistent way across internal and external premises, we create a flexible cloud infrastructure platform. This platform includes security, automation and management, interoperability and openness, self-service, pooling, and dynamic resource allocation. In the view of cloud computing we are advocating, applications can run within an external provider, in internal IT premises, or in combination as a hybrid system—it matters how they are run, not where they are run.

Cloud computing builds on virtualization to create a service-oriented computing model. This is done through the addition of resource abstractions and controls to create dynamic pools of resources that can be consumed through the network. Benefits include economies of scale, elastic resources, self-service provisioning, and cost transparency. Consumption of cloud resources is enforced through resource metering and pricing models that shape user behavior. Consumers benefit through leveraging allocation models such as pay-as-you-go to gain greater cost efficiency, lower barrier to entry, and immediate access to infrastructure resources.

13.1 Cloud Definition

Here is National Institute of Standards and Technology (NIST) working definition:

Cloud computing is a model for enabling convenient, on-demand network access to a shared pool of configurable computing resources (e.g., networks, servers, storage, applications, and services) that can be rapidly provisioned and released with minimal management effort or service provider interaction. This cloud model promotes availability and is composed of five essential characteristics, three delivery models, and four deployment models.

The five essential characteristics are as follows:

1. On-demand self-service
2. Broad network access
3. Resource pooling
4. Rapid elasticity
5. Measured service

The three delivery models are as follows:

1. Infrastructure as a Service (IaaS)
2. Platform as a Service (PaaS)
3. Software as a Service (SaaS)

The four deployment models are as follows:

1. Public cloud
2. Private cloud
3. Hybrid cloud
4. Community cloud

Cloud computing is the IT foundation for cloud services and it consists of technologies that enable cloud services. The key attributes of cloud computing are shown in Table 13.1. Key attributes of cloud services are described in Table 13.2.

13.2 Cloud Characteristics

Large organizations such as IBM, Dell, Microsoft, Google, Amazon, and Sun have already started to take strong positions with respect to cloud

TABLE 13.1

Key Attributes of Cloud Computing

Attributes	Description
Offsite, third-party provider	In the cloud execution, it is assumed that third party provides services. There is also a possibility of in-house cloud service delivery.
Accessed via the Internet	Services are accessed via standard-based, universal network access. It can also include security and quality-of-service options.
Minimal or no IT skill required	There is a simplified specification of requirements.
Provisioning	It includes self-service requesting, near real-time deployment, and dynamic and fine-grained scaling.
Pricing	Pricing is based on usage-based capability and it is fine grained.
User interface	User interface includes browsers for a variety of devices and with rich capabilities.
System interface	System interfaces are based on Web Services APIs providing a standard framework for accessing and integrating among cloud services.
Shared resources	Resources are shared among cloud services users; however, via configuration options with the service, there is the ability to customize.

TABLE 13.2

Key Attributes of Cloud Services

Attributes	Description
Infrastructure systems	It includes servers, storage, and networks that can scale as per user demand.
Application software	It provides web-based user interface, Web Services APIs, and a rich variety of configurations.
Application development and deployment software	It supports the development and integration of cloud application software.
System and application management software	It supports rapid self-service provisioning and configuration and usage monitoring.
IP networks	They connect end users to the cloud and the infrastructure components.

computing provision. They are so much behind this latest paradigm that the success is virtually guaranteed. The essential characteristics of cloud environment include the following:

- On-demand self-service that enables users to consume computing capabilities (e.g., applications, server time and network storage) as and when required.
- *Broad network access*: Capabilities are available over the network and accessed through standard mechanisms that promote use by heterogeneous thin or thick client platforms (e.g., mobile phones, tablets, laptops, and workstations).

- Multitenancy and resource pooling that allows combining hetero-geneous computing resources (e.g., hardware, software, processing servers and network bandwidth) to serve multiple consumers—such resources being dynamically assigned.

- Rapid elasticity and scalability that allows functionalities and resources to be rapidly, elastically, and automatically scaled out or in, as demand rises or drops.

- Measured provision to automatically control and optimize resource allocation and to provide a metering capability to determine the usage for billing purpose, allowing easy monitoring, controlling, and reporting.

13.3 Cloud Delivery Models

Cloud computing is not a completely new concept for the development and operation of Web applications. It allows for the most cost-effective develop-ment of scalable Web portals on highly available and fail-safe infrastructures. In the cloud computing system, we have to address different fundamentals like virtualization, scalability, interoperability, quality of service, failover mechanism, and the cloud deployment models (private, public, hybrid) within the context of the taxonomy. The taxonomy of cloud includes the dif-ferent participants involved in the cloud along with the attributes and tech-nologies that are coupled to address their needs and the different types of services like *XaaS* offerings where X is software, hardware, platform, infra-structure, data, and business (Figures 13.1 and 13.2).

13.3.1 Infrastructure as a Service (IaaS)

The IaaS model is about providing compute and storage resources as a ser-vice. According to NIST, IaaS is defined as follows:

The capability provided to the consumer is to provision processing, storage, net-works, and other fundamental computing resources where the consumer is able to deploy and run arbitrary software, which can include operating systems and applica-tions. The consumer does not manage or control the underlying cloud infrastructure but has control over operating systems, storage, deployed applications, and possibly limited control of select networking components (e.g., host firewalls).

The user of IaaS has single ownership of the alloted hardware infrastruc-ture and can use it as if it is his or her own machine on a remote network and has control over the operating system and software on it. IaaS is illustrated in Figure 13.1. The IaaS provider has control over the actual hardware, and the

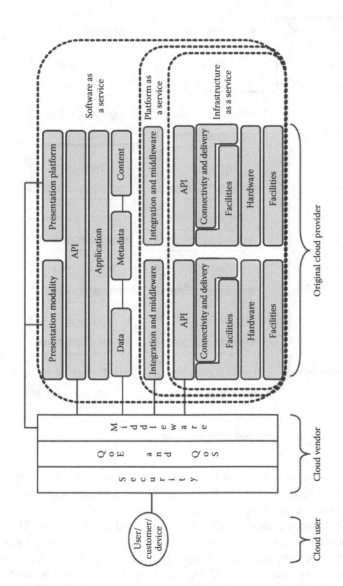

FIGURE 13.1
The cloud reference model.

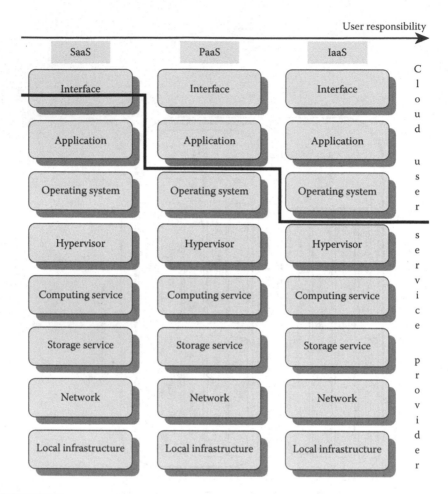

FIGURE 13.2
Portfolio of services for the three cloud delivery models.

cloud user can request allocation of virtual resources, which are then allocated by the IaaS provider on the hardware (generally without any manual intervention). The cloud user can manage the virtual resources as desired, including installing any desired OS, software, and applications. Therefore, IaaS is well suited for users who want complete control over the software stack that they run; for example, the user may be using heterogeneous software platforms from different vendors, and they may not like to switch to a PaaS platform where only selected middleware is available. Well-known IaaS platforms include Amazon EC2, Rackspace, and RightScale. Additionally, traditional vendors such as HP, IBM, and Microsoft offer solutions that can be used to build private IaaS.

13.3.2 Platform as a Service (PaaS)

The PaaS model is to provide a system stack or platform for application deployment as a service. NIST defines PaaS as follows:

The capability provided to the consumer is to deploy onto the cloud infrastructure consumer-created or acquired applications created using programming languages and tools supported by the provider. The consumer does not manage or control the underlying cloud infrastructure including network, servers, operating systems, or storage but has control over the deployed applications and possibly application hosting environment configurations.

Figure 13.1 shows a PaaS model diagrammatically. The hardware, as well as any mapping of hardware to virtual resources, such as virtual servers, is controlled by the PaaS provider. Additionally, the PaaS provider supports selected middleware, such as a database and Web application server shown in the figure. The cloud user can configure and build on top of this middleware, such as define a new database table in a database. The PaaS provider maps this new table onto their cloud infrastructure. Subsequently, the cloud user can manage the database as needed and develop applications on top of this database. PaaS platforms are well suited to those cloud users who find that the middleware they are using matches the middleware provided by one of the PaaS vendors. This enables them to focus on the application. Windows Azure, Google App Engine, and Hadoop are some well-known PaaS platforms. As in the case of IaaS, traditional vendors such as HP, IBM, and Microsoft offer solutions that can be used to build private PaaS.

13.3.3 Software as a Service (SaaS)

SaaS is about providing the complete application as a service. SaaS has been defined by NIST as follows:

The capability provided to the consumer is to use the provider's applications running on a cloud infrastructure. The applications are accessible from various client devices through a thin client interface such as a Web browser (e.g., web-based e-mail). The consumer does not manage or control the underlying cloud infrastructure including network, servers, operating systems, storage, or even individual application capabilities, with the possible exception of limited user-specific application configuration settings.

Any application that can be accessed using a Web browser can be considered as SaaS. These points are illustrated in Figure 13.1. The SaaS provider controls all the layers apart from the application. Users who log in to the SaaS service can both use the application and configure the application for their use. For example, users can use Salesforce.com to store their customer data. They can also configure the application, for example, requesting additional space for storage or adding additional fields to the customer data that is already being used. When configuration settings are changed, the SaaS infrastructure performs any management tasks needed (such as allocation of additional storage) to support the changed configuration. SaaS platforms are targeted toward users who want to use the application without any software installation (in fact, the motto of Salesforce.com, one of the prominent SaaS vendors, is "No Software"). However, for advanced usage, some small amount of programming or scripting may be necessary to customize the application for usage by the business (e.g., adding additional fields to customer data). In fact, SaaS platforms like Salesforce.com allow many of these customizations to be performed without programming but by specifying business rules that are simple enough for nonprogrammers to implement. Prominent SaaS applications include Salesforce.com for CRM, Google Docs for document sharing, and Web e-mail systems like Gmail, Hotmail, and Yahoo! Mail. IT vendors such as HP and IBM also sell systems that can be configured to set up SaaS in a private cloud; SAP, for example, can be used as an SaaS offering inside an enterprise.

Table 13.3 presents a comparison of the three cloud delivery models.

13.4 Cloud Deployment Models

13.4.1 Private Clouds

A private cloud has an exclusive purpose for a particular organization. The cloud resources may be located on or off premise and could be owned and managed by the consuming organization or a third party. This may be an example of an organization who has decided to adopt the infrastructure cost-saving potential of a virtualized architecture on top of existing hardware. The organization feels unable to remotely host their data, so they are looking to the cloud to improve their resource utilization and automate the management of such resources. Alternatively, an organization may wish to extend its current IT capability by using an exclusive, private cloud that is remotely accessible and provisioned by a third party. Such an organization may feel uncomfortable with their data being held alongside a potential competitor's data in the multitenancy model.

13.4.2 Public Clouds

A public cloud, as its name implies, is available to the general public and is managed by an organization. The organization may be a business (such as Google), academic, or a governmental department. The cloud computing provider owns and manages the cloud infrastructure. The existence of many different consumers within one cloud architecture is referred to as a multitenancy model.

TABLE 13.3

Comparison of Cloud Delivery Models

Service Type	IaaS	PaaS	SaaS
Service category	VM rental; online storage	Online operating environment, online database, online message queue	Application and software rental
Service customization	Server template	Logic resource template	Application template
Service provisioning	Automation	Automation	Automation
Service accessing and using	Remote console, Web 2.0	Online development and debugging, integration of offline development tools and cloud	Web 2.0
Service monitoring	Physical resource monitoring	Logic resource monitoring	Application monitoring
Service level management	Dynamic orchestration of physical resources	Dynamic orchestration of logic resources	Dynamic orchestration of application
Service resource optimization	Network virtualization, server visualization, storage visualization	Large-scale distributed file system. Database, middleware, etc.	Multitenancy
Service measurement	Physical resource metering	Logic resource usage metering	Business resource usage metering
Service integration and combination	Load balance	SOA	SOA, mashup
Service security	Storage encryption and isolation, VM isolation, VLAN; SSL/SSH	Data isolation, operating environment isolation, SSL	Data isolation, operating environment isolation, SSL; Web authentication and authorization

13.4.3 Hybrid Clouds

Hybrid clouds are formed when more than one type of cloud infrastructure is utilized for a particular situation. For instance, an organization may utilize a public cloud for some aspect of its business, yet also have a private cloud on premise for data that is sensitive. As organizations start to exploit cloud service models, it is increasingly likely that a hybrid model is adopted as the specific characteristics of each of the different service models are harnessed. The key enabler here is the open standards by which data and applications are implemented, since if portability does not exist, then vendor lock-in to a particular cloud computing provider becomes likely. Lack of data and application portability has been a major hindrance for the widespread uptake of grid computing, and this is one aspect of cloud computing that can facilitate much more flexible, abstract architectures.

13.4.4 Community Clouds

Community clouds are a model of cloud computing where the resources exist for a number of parties who have a shared interest or cause. This model is very similar to the single-purpose grids that collaborating research and academic organizations have created to conduct large-scale scientific experiments (e-science). The cloud is owned and managed by one or more of the collaborators in the community, and it may exist either on or off premise.

13.5 Cloud Benefits

Cloud computing is an attractive paradigm that promises numerous benefits, inherent in the characteristics, as mentioned earlier. These include

- Optimization of a company's capital investment by reducing costs of purchasing hardware and software, resulting in a much lower total cost of ownership and, ultimately, a whole new way of looking at the economics of scale and operational IT
- Simplicity and agility of operations and use, requiring minimal time and effort to provision additional resources
- Enabling an enterprise to tap into a talent pool, as and when needed, for a fraction of the cost of hiring staff or retaining the existing staff and, thus, enabling the key personnel in the organizations to focus more on producing value and innovation for the business
- Enabling small organizations to access the IT services and resources that would otherwise be out of their reach, thus placing large organizations and small businesses on a level playing field
- Providing novel and complex computing architectures and innovation potential
- Providing mechanism for disaster recovery and business continuity through a variety of fully outsourced ICT services and resources

Cloud computing can be massively scalable, and there are built-in benefits of efficiency, availability, and high utilization that, in turn, result in reduced capital expenditure and reduced operational costs. It permits seamless sharing and collaboration through virtualization. In general, cloud computing promises cost savings, agility, innovation, flexibility, and simplicity. The offerings from vendors, in terms of services of the application, platform, and infrastructure nature, are continuing to mature, and the cost savings are becoming particularly attractive in the current competitive economic climate. Another broader aim of cloud technology is to make supercomputing available to the enterprises, in particular, and the public, in general.

Table 13.4 presents a comparison of cloud benefits for Small and Medium enterprises (SMEs) and Large enterprises.

The major benefits of the cloud paradigm can be distilled to its inherent flexibility and resiliency, the potential for reducing costs, availability of very large amounts of centralized data storage, means to rapidly deploy computing resources, and scalability.

13.5.1 Flexibility and Resiliency

A major benefit of cloud computing is the flexibility, though cloud providers cannot provide infinite configuration and provisioning flexibility and will seek to offer structured alternatives. They might offer a choice among a number of computing and storage resource configurations at different capabilities and costs, and the cloud customer will have to adjust his or her requirements to fit one of those models.

The flexibility offered by cloud computing can be in terms of

- Automated provisioning of new services and technologies
- Acquiring increased resources on an as-needed basis
- Ability to focus on innovation instead of maintenance details
- Device independence
- Freedom from having to install software patches
- Freedom from concerns about updating servers

Resiliency is achieved through the availability of multiple redundant resources and locations. As autonomic computing becomes more mature, self-management and self-healing mechanisms can ensure the increased reliability and robustness of cloud resources. Also, disaster recovery and business continuity planning are inherent in using the provider's cloud computing platforms.

13.5.2 Reduced Costs

Cloud computing offers reductions in system administration, provisioning expenses, energy costs, software licensing fees, and hardware costs. The cloud paradigm, in general, is a basis for cost savings because capability and resources can be paid for incrementally without the need for large investments in computing infrastructure. This model is especially true for adding storage costs for large database applications. Therefore, capital costs are reduced and replaced by manageable, scalable operating expenses.

TABLE 13.4

Comparison of Cloud Benefits for Small and Medium Enterprises (SMEs) and Large Enterprises

Economic Benefits	Small and Medium Enterprises (SMEs)	Large Enterprises
Strategic flexibility	Critical in getting quickly to market. Cloud services allow start-ups to rapidly develop and deploy their products as long as they can use the open source or proprietary development platforms of the cloud providers. As the cloud market offerings mature, there will be many more platform options available.	Cloud services can provide large enterprises the same strategic benefits as start-ups for new initiatives as long as legacy software integration and data issues are not significant. With appropriate software development talent, operating units can rapidly develop and market test new innovations without putting additional strain on IT budgets, staff, or hardware. Long-standing internal IT management policies and standards may have to be reexamined and modified to allow this to happen.
Cost reduction	Pay-as-you-go pricing may be critical if operating capital or venture capital funding is not available. With cloud services, growth can more easily be funded through operating revenues and there may be tax advantages to converting what would have been longer-term depreciation expenses to fully loaded current expenses.	Cloud services provide the same cost–benefits for isolated and exploratory initiatives. Instant availability and low setup costs for new development and deployment environments allow operating units to explore new initiatives quickly at low cost without increasing internal IT hardware or staff overheads. For high data traffic volumes, it may become more economical to bring the operations in-house. Because maintaining legacy hardware and software absorbs the majority of IT costs, large corporations may see significant costs savings by selectively moving noncritical applications and processes to external clouds.
Software availability	Software as a service (SaaS) and platform as a service (PaaS) provide necessary software and infrastructure at low entry cost. Limited online version functionality may be more than offset by dramatic cost savings.	Existing volume licensing of legacy desktop and process-integrated enterprise software may make the status quo more attractive if end-user retraining, process modifications, and other change costs are high. Legacy desktop software may have more features and functionality than is currently available in SaaS versions. But the legacy software licensing costs may dramatically increase if it is hosted in a private cloud environment.

Scalability	One of the most dramatic benefits for SMEs and start-ups. If successful, applications designed to autoscale can scale endlessly in a cloud environment to meet the growing demand.	Large enterprises with significant hardware, legacy software, and staff resources can benefit from cloud scalability by identifying CPU-intensive processes such as image processing, PDF conversion, and video encoding that would benefit from the massively scalable parallel processing available in clouds. While this may require modifying legacy applications, the speed benefits and reduced local hardware requirements may far outweigh the software modification costs.
Skills and staffing	While the proper design of cloud applications requires high-level software development skills, their maintenance and support is vastly simplified in the cloud environment. Cloud providers handle all maintenance and support issues for both hardware and platform software at costs that are either bundled into the usage fees or available in various configurations as premium services. This allows significant cost savings through reduced staff overheads.	Because the majority of enterprise IT costs goes to support legacy applications and hardware, the greatest staffing benefits will be seen in new cloud initiatives that do not add to the staffing burden. Longer-term, as the enterprise begins to analyze cloud technology potential for its legacy operations, retraining of existing staff or bringing in new staff with cloud technology skills will be necessary to take advantage of the new paradigm. Thus, some investment will have to be made before large-scale or long-term benefits will be seen. The staffing investment may be significant if the enterprise is attempting to create a private cloud to handle dynamic resource allocation and scalability across its operating units. In this case, it may face significant staff investment as well as the required hardware, software, and network investment to implement and maintain their private cloud.
Energy efficiency	Because SMEs can dramatically reduce or eliminate local servers, cloud computing provides direct utility cost savings as well as environmental benefits.	Even very large enterprise IT data centers cannot achieve the energy efficiencies found in the massive facilities of public cloud providers even with aggressive high-density server and virtualization strategies. In periods of economic downturns, green initiatives typically cannot compete for scarce capital funds. By employing a mixed strategy that off-loads applications and processing to external clouds when feasible, IT managers are able to minimize their energy costs and carbon footprint.
System redundancy and data backup	This is a large benefit for SMEs, the majority of which are poorly prepared for hardware failures and disaster recovery. Cloud storage can reduce downside risks at low cost.	Because cloud technologies distribute both data storage and data processing across potentially large number of servers, the likelihood of data loss due to hardware failure is much lower than in most large private data centers. The cloud data storage can provide a cost-effective supplemental backup strategy.

> There might be some instances, particularly for long-term, stable computing configuration usage, where cloud computation might not have a cost advantage over using one's internal resources or directly leasing equipment. For example, if the volume of data storage and computational resources required are essentially constant and there is no need for rapid provisioning and flexibility, an organization's local computational capabilities might be more cost effective than using a cloud.

Resources are used more efficiently in cloud computing, resulting in substantial support and energy cost savings. The need for highly trained and expensive IT personnel is also reduced; client organizational support and maintenance costs are reduced dramatically because these expenses are transferred to the cloud provider, including 24/7 support that in turn is spread onto a much larger base of multiple tenants or clients.

Another reason for migrating to the cloud is the drastic reduction in the cost of power and energy consumption.

13.5.3 Centralized Data Storage

Many data centers are an ensemble of legacy applications, operating systems, hardware, and software and are a support and maintenance nightmare. This situation requires more specialized maintenance personnel, increased costs because of lack of standardization, and a higher risk of crashes. The cloud not only offers a larger amounts of data storage resources than are normally available in local, corporate computing systems, it also enables decrease or increase in the resources used per requirements—with the corresponding adjustments in operating cost. This centralization of storage infrastructure results in cost efficiencies in utilities, real estate, and trained personnel. Also, data protections mechanisms are much easier to implement and monitor in a centralized system than on large numbers of computing platforms that might be widely distributed geographically in different parts of an organization.

13.5.4 Reduced Time to Deployment

In a competitive environment where rapid evaluation, development, and deployment of new approaches, processes, solutions, or offerings are critical, the cloud offers the means to use powerful computational or large storage resources on short notice within a short period of time frame, without requiring sizeable initial investments of finances, efforts, or time (in hardware, software, and personnel). Thus, this rapid provisioning of latest technologically upgraded and enhanced resources can be accomplished at relatively small cost (with minimal cost of replacing discontinued resources) and offers the client access to advanced technologies that are constantly being acquired by the cloud provider. Improved delivery of services obtained by rapid cloud provisioning improves time to market and, hence, market growth.

13.5.5 Scalability

Cloud computing provides the means, within limits, for a client to rapidly provision computational resources to meet increases or decreases in demand. Cloud scalability provides for optimal resources so that computing resources are provisioned per requirements seamlessly ensuring maximum cost-benefit to the clients. Since the cloud provider operates on a multitenancy utility model, the client organization has to pay only for the resources it is using at any particular time.

SCALABILITY OF COMPLEX SYSTEMS LIKE THE INTERNET

Complex systems exhibit different patterns of behavior than traditional systems and require new design principles based on a deeper understanding of the physical properties of their components and of the manner in which they interact with one another, and with the environment. The behavior of any system cannot be explained without precise knowledge of the interactions among its components, hence, such an effort is undertaken whenever we need to make progress in our understanding of the behavior of the physical world surrounding us or a system that we engineer. Predicting all possible interactions among the components of a system during the design process is an even more daunting task.

The topology of a network used to model the interactions in complex biological, social, economic, and computing systems is described by means of graphs in which vertices represent the entities and edges represent their interactions. The number of edges incident upon a vertex is called the degree of the vertex. Systems interconnected by scale-free networks have the property that we can limit the number of interaction paths among its components without limiting the number of components of this network system. Because the degrees of nodes obey a power-law distribution, while the vast majority of components have one or very few connections, the number of highly connected components is very small. This ensures and assures the scalability of scale-free networks like the Internet.

13.6 Cloud Challenges

There are multiple technical challenges that any cloud platform or application needs to address in order to truly provide a utility-like computing infrastructure. The three key challenges are described in the following:

1. *Scalability*: Ability to scale to millions of clients simultaneously accessing the cloud service
2. *Multitenancy*: Ability to provide the isolation as well as good performance to multiple tenants using the cloud infrastructure
3. *Availability*: Ability that ensures that the infrastructure and applications are highly available regardless of hardware and software faults

13.6.1 Scalability

On-demand scaling (and descaling) of computation is one of the critical needs of any cloud computing platform. Compute scaling can be either done at the infrastructure level or platform level. At the infrastructure level, it is about increasing the capacity of the compute power, while at the platform level, the techniques are mainly to intelligently manage the different client requests in a manner that best utilizes the compute infrastructure without requiring the clients to do anything special during peaks in demand:

1. Scale-up or vertical scaling is about adding more resources to a single node or a single system to improve performance—such as addition of CPUs, use of multicore systems instead of single-core, or adding additional memory. In order to support on-demand scaling of the cloud infrastructure, the system should be able to increase its compute power dynamically without impacting the platform or application executing over it. Unless a system is virtualized, it is generally not possible to increase the capacity of a compute system dynamically without bringing down the system. The more powerful compute resource can now be effectively used by a virtualization layer to support more processes or more virtual machines—enabling scaling to many more clients. The advantage of scale-up systems is that the programming paradigm is simpler, since it does not involve distributed programming, unlike scale-out systems.
2. Scale out or horizontal scaling, on the other hand, is about expanding the compute resources by adding a new computer system or node to a distributed application. A Web server (like Apache) is a typical example for such a system. In fact, given that most cloud applications

are service-enabled, they need to be developed to expand on demand using scaling-out techniques. The advantage of scale-out systems is that commodity hardware, such as disk and memory, can be used for delivering high performance. A scale-out system such as interconnected compute nodes forming a cluster can be more powerful than a traditional supercomputer, especially with faster interconnect technologies. Scale-out systems will essentially be distributed systems with a shared high-performance disk storage used for common data. Unlike scale-up systems, in order to leverage full power of scale-out systems, there should be an effort from the programmer to design applications differently. Many design patterns exist for applications designed for scale-out systems like MapReduce.

Scale-out solutions have much better performance and price/performance over scale-up systems. This is because a search application essentially consists of independent parallel searches, which can easily be deployed on multiple processors. Scale-out techniques can be employed at application level as well. For example, a typical Web search service is scalable where two client query requests can be processed completely as parallel threads. The challenge in scale-out systems, however, is the complex management of the infrastructure, especially when the infrastructure caters to dynamic scaling of resources. Additionally, as noted, applications that do not consist of independent computations are difficult to scale out.

13.6.2 Multitenancy

This deals with implementation of multitenancy with fine-grained resource sharing while ensuring security and isolation between customers and also allowing customers to customize the database:

1. *Ad hoc/custom instances*: In this lowest level, each customer has their own custom version of the software. This represents the situation currently in most enterprise data centers where there are multiple instances and versions of the software. It was also typical of the earlier ASP model that represented the first attempt to offer software for rent over the Internet (see Chapter 11, "Application Service Providers (ASPs))". The ASP model was similar to the SaaS model in that ASP customers (normally businesses), upon logging in to the ASP portal, would be able to rent the use of a software application like CRM. However, each customer would typically have their own instance of the software being supported. This would imply that each customer would have their own binaries, as well as their own dedicated processes for implementation of the application. This makes management extremely difficult, since each customer would need their own management support.

2. *Configurable instances*: In this level, all customers share the same version of the program. However, customization is possible through configuration and other options. Customization could include the ability to put the customer's logo on the screen and tailoring of workflows to the customer's processes. In this level, there are significant manageability savings over the previous level, since there is only one copy of the software that needs to be maintained and administered. For instance, upgrades are seamless and simple.

3. *Configurable, multitenant efficient instances*: Cloud systems at this level in addition to sharing the same version of the program also have only one instance of the program running that is shared among all the customers. This leads to additional efficiency since there is only one running instance of the program.

4. *Scalable, configurable, multitenant efficient instances*: In addition to the attributes of the previous level, the software is also hosted on a cluster of computers, allowing the capacity of the system to scale almost limitlessly. Thus, the number of customers can scale from a small number to a very large number, and the capacity used by each customer can range from being small to very large. Performance bottlenecks and capacity limitations that may have been present in the earlier level are eliminated. For instance, in a cloud e-mail service like Gmail or Yahoo Mail, multiple users share the same physical e-mail server as well as the same e-mail server processes. Additionally, the e-mails from different users are stored in the same set of storage devices and perhaps the same set of files. This results in management efficiencies; as a contrary example, if each user had to have a dedicated set of disks for storing e-mail, the space allocation for each user would have to be managed separately. However, the drawback of shared storage devices is that security requirements are greater; if the e-mail server has vulnerabilities and can be hacked, it is possible for one user to access the e-mails of another.

13.6.3 Availability

Cloud services also need special techniques to reach high levels of availability. Mission-critical enterprise services generally have availability in the 99.999% range. This corresponds to a downtime of 5 min in an entire year! Clearly, sophisticated techniques are needed to reach such high levels of reliability. Even for non-mission-critical applications, downtime implies loss of revenue. It is therefore extremely important to ensure high availability for both mission-critical as well as non-mission-critical cloud services. There are basically two approaches to ensuring availability. The first approach is to ensure high availability for the underlying application upon which the cloud service is built. This generally involves one of the three techniques:

infrastructure availability ensuring redundancy in infrastructure, such as servers, so that new servers are always ready to replace failed servers; middleware availability achieved with middleware redundancy; and application availability achieved via application redundancy.

> *To the 9s* (measures of application availability): Service-level agreements (SLAs) on availability are often measured in 9s. This describes the target percent of unplanned availability to be achieved, typically on a monthly or annual basis. Each 9 corresponds to a 10-fold decrease in the amount of downtime. For an important application, such as e-mail or a CRM system, three 9s might be a reasonable target, whereas critical services such as public utilities would tend to target five 9s.
> The following table describes the amount of acceptable downtime per year for the corresponding level of availability:
>
> | # of 9s | SLA Target (%) | Maximum Downtime per Year |
> | --- | --- | --- |
> | 2 | 99 | 3 days, 15 h, and 40 min |
> | 3 | 99.9 | 8 h and 46 min |
> | 4 | 99.99 | 52 min and 36 s |
> | 5 | 99.999 | 5 min and 16 s |
> | 6 | 99.9999 | 31.56 s |

The other approach is to build support for high availability into the cloud infrastructure, which is of two types:

1. *Failure detection*, where the cloud infrastructure detects failed application instances and avoids routing requests to such instances
2. *Application recovery*, where failed instances of application are restarted

13.6.3.1 Failure Detection

Many cloud providers, such as Amazon Web Services' Elastic Beanstalk, detect when an application instance fails and avoid sending new requests to the failed instance. In order to detect failures, one needs to monitor for failures.

13.6.3.1.1 Failure Monitoring

There are two techniques of failure monitoring. The first method is heartbeats, where each application instance periodically sends a signal (called a heartbeat) to a monitoring service in the cloud. If the monitoring service does not receive a specified number of consecutive heartbeats, it may declare the application instance as failed. The second is the method of probes. Here, the monitoring service periodically sends a probe, which is a lightweight service request, to the application instance. If the instance does not respond to a specified number of probes, it may be considered failed. There is a trade-off between speed and accuracy of detecting failures. To detect failures rapidly, it may be desirable to set a low value for the number of missed heartbeats or probes. However, this could lead to an increase in the number of false failures. An application instance may not respond due to a momentary overload or some other transient condition. Since the consequences of falsely declaring an application instance failed are severe, generally a high threshold is set for the number of missed heartbeats or probes to virtually eliminate the likelihood of falsely declaring an instance failed.

13.6.3.1.2 Redirection

After identifying failed instances, it is necessary to avoid routing new requests to these instances. A common mechanism used for this in HTTP-based protocols is HTTP redirection.

13.6.3.2 Application Recovery

In addition to directing new requests to a server that is up, it is necessary to recover old requests. An application-independent method of doing this is checkpoint/restart. Here, the cloud infrastructure periodically saves the state of the application. If the application is determined to have failed, the most recent checkpoint can be activated, and the application can resume from that state.

13.6.3.2.1 Checkpoint/Restart Paradigm

Checkpoint/restart can give rise to a number of complexities. First, the infrastructure should checkpoint all resources, such as system memory, otherwise the memory of the restarted application may not be consistent with the rest of the restarted application. Checkpointing storage will normally require support from the storage or file system, since any updates that were performed have to be rolled back. This could be complex in a distributed application, since updates by a failed instance could be intermingled with updates from running instances. Also, it is difficult to capture and reproduce activity on the network between distributed processes.

In a distributed checkpoint/restart, all processes of distributed application instances are checkpointed, and all instances are restarted from a common checkpoint if any instance fails. This has obvious scalability limitations and also suffers from correctness issues if any interprocess communication data is in-transit at the time of failure. For instance, Ubuntu Linux has support for checkpoint/restart of distributed programs. Even sequential applications can be transparently checkpointed if linked with the right libraries, using the Berkeley lab checkpoint/restart library. It can also invoke application-specific code (which may send out a message to the users or write something in a log file, etc.) during checkpointing and restart.

13.6.3.2.2 *Transactional Paradigm*

When an application instance fails, some signal (e.g., closing of network connections) is sent to the other instances, which abort all work in progress for the failed instance. When the failed instance restarts, it restarts all transactions in progress. Additionally, other instances restart any requests they made to the failed instance.

13.7 Summary

This chapter introduced the concept of cloud computing. It describes its definition, presents the cloud delivery and deployment models, and highlights its benefits for enterprises. In the last part of the chapter, we discussed the primary challenges faced while provisioning of cloud services, namely, scalability, multitenancy, and availability. This prepares the background for understanding the economics of cloud computing solutions, which we take up in the next chapter.

14

Cloudware Economics

14.1 Drivers for Cloud Computing in Enterprises

As the technology has been maturing, greater management emphasis is being placed on the need to realize business benefits from cloud investment. With the pace of technology maturity, there will be a shift in how vendors supply their cloud technology and solutions and how cloud consumers build on new business models through the technology enhancements offered by cloud. Investment in cloud computing should be more than a business enabler and cost reduction exercise; it should be an exercise in capability building leveraged to drive the business, with increased adaptability, agility, flexibility, scalability, and mobility across the enterprise system landscape.

Some of the key business drivers that are linked to creating value beyond cost efficiencies and business scalability are as follows:

- Cloud computing enables business agility by enabling the business to respond faster to the demanding needs of the market; by facilitating access, prototyping, and rapid provisioning, organizations can adjust processes and services to meet the changing needs of the markets. Faster, and easier, prototyping and experimentation can also serve as a platform for innovation. This allows shorter development cycles and faster time to products and value.

- Virtualization offers a tangible benefit of abstracting away the operational system complexity, resulting in better user experience and productivity. This, in turn, can significantly reduce maintenance and upgrade costs, while providing flexibility for innovative enhancements and developments in the background.

- Expanded computing power and capacity allows cloud computing to offer simple, yet context-driven, variability. It can improve user experience and increase product relevance by allowing a more enhanced, and subtle customization of products and services, and personalized experience.

Developing a robust business case that demonstrates the return on investment of cloud can benefit all parties in a cloud venture. The cloud customers and consumers can justify the investment in terms of costs and benefits of the key technology features and the new operating models. This exercise needs to identify any interdependencies and trade-offs. The output of this exercise will also serve as a key ingredient of the strategic planning process. It will also serve as a good performance benchmark tool and metric to monitor the investment effectiveness and determine if the cloud provision is delivering both the business and technology promises, while also identifying any potential scope for fine-tuning and improvement.

In this chapter, we discuss some of the pertinent financial and accounting techniques that form an important aspect of measuring or appraising investment decisions. It is important, however, to emphasize that no investment decision should be based purely on financial metrics. Organizations need to consider both financial and nonfinancial indicators to determine the value of cloud. Some contributing factors to this value will be qualitative and challenging to express in monetary value.

However, as technology and business models mature, IT will continue on its path to commoditization. Most approaches build a business case for cloud that is predominantly viewed through operational efficiencies, focusing on cost optimizations that are evaluated using cost-based calculations linked to resource utilization. The metrics often used (especially by SMEs) are linked to cost efficiencies achieved as a result of a perceived shift from CapEx (capital expenditure) to OpEx (operating expenditure), TCO (total cost of ownership), and at best, looking at ROI (return on investment) and NPV (net present value) of cloud investment.

For traditional IT, the organizations invest in infrastructure assets such as hardware and software code, which requires capital expenditure. Capital expenditure poses some risks:

- Capital is limited, especially in the case of public sector or SMEs.
- CapEx raises the barrier for entry by making it difficult to access the latest technology, especially in the case of SMEs.
- Precious capital will be tied down in physical assets that rapidly depreciate, and there is the associated cost of maintenance and upgrade. This poses an opportunity cost as part of this capital that could be invested elsewhere to drive innovation.
- Large investment in physical IT hardware and software, especially in the case of large enterprises, risks vendor lock-in that reduces business flexibility and agility.
- For growth or scaling, in addition to the need for modernizing old technology, substantial investment in the infrastructure, architecture, and integration is needed.

One value proposition offered by cloud providers is the opportunity to reduce the IT capital expenditure and address the issues earlier. Instead, such larger capital investment is made by the providers themselves who require cloud computing platforms or private cloud users. In their case, they benefit from economies of scale through shared service models. Some other implications of using OpEx are as follows:

- There will be much faster rate of cost reduction using cloud.
- Cost of ownership will be transformed.
- Removal of up-front capital and release funds.
- Shift from balance sheet to operating statement.
- Cash flow implications where revenue generation and expenditure will be based on service usage.
- There will be a fresh focus on productivity and revenue generation while keeping capital costs down through greater efficiencies of working capital.
- Minimizing up-front investment to drive improved asset usage ratios, average revenue per unit, average margin per user, and cost of asset recovery.
- Maximizing the use of capital by moving funding toward optimizing capital investment leverage and risk management of sources of funding.

When the cost of capital is high, shifting CapEx to OpEx may more easily be justified. For OpEx to be beneficial is that there should be a reliable mechanism to measure and predict usage and tie this to business performance metrics or opt for a monthly or annual baseline fixed rate. A business may still choose to invest in CapEx for differentiated business processes yet adopt a usage-based model to improve financial efficiency.

14.1.1 Total Cost of Ownership (TCO)

TCO is an accounting metric that takes all direct and indirect costs of technology acquisition and operation into account over the IT project life cycle. The costs include everything from initial investment in hardware and software acquisition to installation, administration, training, maintenance and upgrades, service and support, security and disaster recovery, power, and any other associated costs. The typical cost components are broadly categorized as acquisition costs versus operational costs, each incurring administrative and management costs. A simple allocation of these costs is illustrated in Table 14.1.

TABLE 14.1

IT Cost Components

Direct Costs	Indirect Costs	Overheads
Server	Network	Facilities
Storage	Storage	Power
Software (application)	Software (infrastructure) labor (operational)	Bandwidth
Implementation	Maintenance and upgrades	Labor (admin)
	Support	
	Training	

Moving from traditional on-premise IT to on-demand cloud service requires examination of the assumptions underlying TCO. The cloud environment tends to abstract asset virtualization, obfuscate labor, and deliver IT services at a contracted rate. In comparison, cloud services are supplied and metered on the resources consumed, and the cloud provider will typically have clear pricing models that cover the cost of the consumed resources. Hence, in cloud TCO calculations, there is an opportunity to consolidate and simplify some of the cost components, as the main infrastructure and up-front costs are displaced by service subscription and reassigned as operational costs.

While calculating TCO, it is common to evaluate multiple scenarios by carrying out a sensitivity analysis to understand how various patterns of usage influence cost derivers and overall TCO. It is then important to have a baseline cost advantage target (in %) to be able to benchmark the cloud deployment costs against it. Depending on their level of maturity, organizations engage differently with cloud technology. In the case of many SMEs, reactive response to incident management, undocumented or unrepeatable processes, and unplanned implementations tend to increase the complexity and cost of any IT service regardless of the delivery mechanism.

Organizations need to study the fully loaded costs in the light of the business benefits gained and the opportunity costs of not moving to the cloud. Often, it might be a case of paying a premium for much improved, optimized, or secure IT provision. Hence, it is imperative to benchmark costs beyond an equivalent amount of internal server capability.

Baseline TCO can be calculated as follows:

1. Identify all different cost streams both business and technical. Some common sources of cost include amount of compute capacity, network traffic, and storage. Certain services may be on a pay-for-use basis but some costs such as static IP address for certain applications; in the same, there are some service support and management

costs, as well as cost of skills upgrade (offset by perhaps a leaner IT team), that need also be accounted for. By definition, cloud implies a dynamic service that assures optimal utilization. This, on the other hand, means that fluctuations in service use could become challenging, unlike the static resources in a traditional IT environment that can be accounted for more easily.

2. Incorporate hidden cost. Whereas in-house provisioning incurs hidden costs such as additional administrative headcount, additional property and facilities requirements, inevitable overprovisioning costs, and additional costs for ensuring redundancy, the cloud provision's hidden costs could come from potential costs such as service interruptions, inappropriate service scaling, mismanagement or a denial of service attack, extra security, and contingency disaster preparedness and recovery plan costs, as well as the initial cost of cloud readiness including costs associated with setup, interfacing and integrating with discrete local infrastructure or resources, and administrating the whole new operating system.

3. Evaluate the application profiles and service mix. Applications utilize computing resources at varying rates. Some are more compute intensive, whereas others do a small amount of processing across an enormous amount of data. This exercise helps to create a clearer TCO picture by assigning costs to the different cloud services, according to application profile.

4. Calculate the TCO under a number of different application topologies to understand costs under different loads. Identify and cost the required compute instances according to application load variations. Technically speaking, study the horizontal and vertical scaling patterns of the applications. If the load on an application varies significantly, it will most likely require a larger deployment of multiple compute instances to reduce the application bottlenecks.

5. Evaluate the role of load variation. It is important to identify the periods and patterns of application requiring larger loads or experience load variation. A static pattern assumption is hardly useful to calculate cloud TCO. Carrying out a statistical (e.g., Monte Carlo) and scenario analysis to explicitly assess TCO under different load patterns can assist with more accurate estimation of TCO.

In practice to derive value from TCO analysis, it should be included in the calculation of other measures such as return on investment (ROI), net present value (NPV), internal rate of return (IRR), or Economic Value Added (EVA). That way, value planning for cloud is not one-dimensionally cost focused, but it will take into account the quantified business benefits as well.

14.1.1.1 Payback Method

The payback method calculates the number of years it will take before the initial investment of the project is paid back. The shorter the payback time, the more attractive a project is as it reduces the risk of longer-term payouts. The method is quite popular due to its simplicity; the weakness of the method is it ignores the time value of money:

$$\text{Number of years to pay back} = \frac{\text{Original investment}}{\text{Annual net cash flow}}$$

Although a popular investment appraisal method, payback period only qualifies as a first screening technique to initially appraise a project. Its scope is limited to the period the investment is recovered; hence, it ignores potential benefits as a result of investment gains or shortfalls thereafter.

14.1.1.2 Accounting Rate of Return on Investment (ROI)

This method calculates the return on investment (ROI) by calculating the resulting cash inflows (produced by the investment) for depreciation. The investment inflows are totaled and the investment costs are subtracted to derive the profit. The profit is divided by the number of years invested, then by the investment cost, to estimate the annual rate of return.

An ROI analysis calculates the difference between the stream of benefits and the stream of costs over the lifetime of the system discounted by the applicable interest rate. In order to find the ROI, the average net benefit has to be calculated:

$$\text{Net benefits} = \frac{\text{Total benefits} - \text{total costs} - \text{depreciation}}{\text{Useful life}}$$

leading to

$$\text{ROI} = \frac{\text{Net benefits}}{\text{Initial investment}}$$

14.1.1.3 Net Present Value (NPV)

The Net Present Value (NPV) approach calculates the amount of money that an investment is worth, taking into account its costs, earnings, and time value of money (inflation). Thus, it compares the economic value of a project today with the value of the same project in future, taking inflation and returns into account. If NPV of a prospective project is positive, it should be accepted. If the NPV is negative, the project should probably be rejected because the resulting cash flows will also be negative.

First, the present value is calculated as

$$\text{Payment} = \frac{1 - (1 + \text{interest})^{-n}}{\text{Interest}}$$

leading to

Net Present Value (NPV)

  = Present Value of Expected Cash Value – Initial Investment Costs

14.1.1.4 Cost–Benefit Ratio

This calculation method views the total benefits of an investment over the costs consumed to deliver these benefits:

$$\text{Cost–benefit ratio} = \frac{\text{Total benefits}}{\text{Total costs}}$$

14.1.1.5 Profitability Index

The profitability index attempts to identify the relationship between the costs and benefits of the project through the ration calculated as

$$\text{Profitability index} = \frac{\text{Present value of cash flows}}{\text{Investment}}$$

The lowest acceptable value of profitability index is 1.0; any value lower than 1.0 would indicate that the project's present value is less than the initial investment. As values of the profitability index increase, so does the financial attractiveness of the proposed project.

14.1.1.6 Internal Rate of Return (IRR)

The IRR calculates the rate of return that an investment is expected to earn, taking into consideration the time value of money. The higher is the project's IRR, the more desirable is it to carry out the project.

Internal rate of return (IRR) is a capital investment measure that indicates how efficient an investment is (yield), using a compounded return rate. If the cost of capital used to discount future cash flows is increased, the NPV of the project will fall. As the cost of capital continues to increase, the net present value will become zero before it becomes negative. The IRR is the cost of the capital (or a required rate of return) that produces an NPV of zero.

> For the NPV method, we assume that the generated cash flows over the life of the project can be invested elsewhere, at a rate equal to the cost of capital, as the cost of capital represents an opportunity cost. The IRR, on the other hand, assumes that generated cash flows can be reinvested elsewhere at the internal rate of return. The larger the IRR in relation to the cost of capital, the less likely that the alternative returns can be realized; hence, the underlying investment assumption in the IRR method is a doubtful one, whereas for NPV, the reinvestment assumption seems more realistic. In the same way, NPV can accommodate conventional cash flows, whereas in comparison, we may get multiple results through the IRR method.

If a company has several competing cloud computing projects, the IRR can be used in selecting which project to prioritize.

14.1.1.7 Economic Value Added (EVA)

Economic Value Added (EVA™), also known as economic profit, is a measure used to determine the company's financial performance based on the residual wealth created. It depicts the investor or shareholder value creation above the required return or the opportunity cost of the capital. It measures the economic profit created when the return on the capital employed exceeds the cost of the capital. Reducing costs increases profits and economic value added. Unlike ROI, EVA takes into account the residual values for an investment.

14.2 Capital Budgeting Models

The business case of the cloud computing project should contain the cost–benefit analysis. The evaluation point is to justify that the benefits have outweighed the costs. In this section, six capital budgeting models will be examined briefly. These models are

- The payback method
- The accounting rate of Return on Investment (ROI)
- The Net Present Value (NPV)
- The cost–benefit ratio
- The profitability index
- The Internal Rate of Return (IRR)
- The Economic Value Added (EVA)

14.3 Provisioning Configurations

14.3.1 Traditional Internal IT

In the traditional internal IT model, or zero-outsource model, all aspects that constitute an IT application or service are purchased and managed using internal resources. The most common form is office IT infrastructure. In many offices, an Internet connection is provisioned from an ISP and connected to the internal network via a router. This internal network is then provisioned with firewalls, switches, central file and print servers, desktop computers, and perhaps a wireless network and laptops. Internal IT purchases, installs, and operates all this equipment as well as general office software. IT for more specialized business applications can be handled in the same manner, with custom or packaged applications that are loaded onto hardware provisioned for that purpose.

You can also deploy applications for external audiences, such as a corporate website in a traditional IT model. Depending on the scale of such an application, it can either share the network connection (typically on a separate VLAN to isolate it from internal traffic for security reasons) or be provisioned with its own dedicated Internet connectivity and an isolated network.

14.3.2 Colocation

Another possible model for deploying an application is within a third-party data center, otherwise known as a colocation facility. In this model, the company is still responsible for purchasing the server hardware and developing or purchasing the required software for running the application. The colocation facility provides that third party with power, cooling, rack space, and network connectivity for their hardware. The colocation facility typically also provides redundant network connectivity, backup power, and physical security.

Colocation services are typically purchased as annual contracts with an initial service fee and monthly charges based on the amount of rack space (usually bundled with a specified allocation of power) and committed bandwidth. For hardware housed in facilities that are not in close proximity to a company's IT resources, you can purchase what some call *remote-hands* capability in case a manual intervention is required on your behalf.

14.3.3 Managed Service

In the managed-service model, in addition to outsourcing the core infrastructure, such as power and network connectivity, the company no longer purchases server and networking hardware. The managed-service provider rents these to the company and also takes on the responsibility of managing the hardware systems and base operating system software. In some cases, the provider also rents standard software such as databases and rudimentary database management services as part of their service offering.

Similar to the colocation scenario, contracting with a managed-service provider typically involves at minimum an annual commit, with an initial setup fee followed by a recurring monthly charge based on the configuration of hardware and software being rented. In this model, bandwidth is not typically charged for separately; instead, you get a standard allotment based on the number of servers for which you contracted. You can also contract for ancillary services, such as backups. Typically, the charge is based on the amount of storage required on a monthly basis (Tables 14.2 through 14.4).

14.3.4 IaaS Cloud Model

Finally, we get to the cloud model. In this model, as in the managed-service model, the company outsources the infrastructure and hardware, but in an entirely different way. Instead of dedicated hardware resources, *the company utilizes virtualized resources that are dynamically allocated only at the time of need.*

We can think of this as the analog of just-in-time manufacturing, which brought tremendous efficiencies to the production of goods. Instead of stockpiling large inventories, manufacturers can reduce their carrying costs by having inventory delivered just as it is needed in manufacturing. Similarly, the dynamic allocation of resources in a cloud service allows a customer to use computing resources only when necessary. Servers do not have to sit idle during slack periods.

The billing model for cloud services is aligned with this sort of usage profile, with service provisioning often requiring no up-front cost and monthly billing based on the actual amount of resources consumed that month. This may translate into significant cost advantages over traditional deployment models.

14.4 Quality of Service (QoS)

QoS refers to the ability of the cloud service to respond to expected invocations and to perform them at the level commensurate with the mutual expectations of both its provider and its customers. Several quality factors that reflect customer expectations, such as constant service availability, connectivity, and high responsiveness, become key to keeping a business competitive and viable as they can have a serious impact upon service provision. QoS thus becomes an important criterion that determines the service usability and utility, both of which influence the popularity of a particular cloud service, and an important selling and differentiating point between cloud services providers.

TABLE 14.2

Economic Costs of Cloud Adoption

Economic Costs	Small and Medium Enterprises (SMEs)	Large Enterprises
Data security	SMEs are better able to use third-party services such as payment processing to handle secure transactions.	Data are an enterprise's most important IT and operating asset. Current uncertainty regarding the security of the data assets stored in public clouds is one of the most significant barriers in cloud adoption. Large enterprises may not want their data stored in countries where intellectual property piracy is prevalent. Some companies may not want their data stored on equipment used by their competitors.
Data confidentiality	SMEs face the same data confidentiality issues as large enterprises.	One of the advantages of cloud computing and storage for confidentiality is that the data transfer and storage algorithms encrypt the data into units that are difficult to reconstruct without the specialized algorithms/keys if the data are intercepted in transfer or the cloud security is compromised.
Data regulations	SMEs face the same regulatory data location issues as large enterprises.	Depending on the company's industry, there may be significant regulatory issues regarding data location. Data that identify the individual in certain health and financial contexts are subject to LIS regulations. Similarly, the EU has laws that restrict the transfer of certain data outside of its borders.
Data integrity	The data integrity and reliability of cloud suppliers may be higher than that provided by the existing internal systems.	Cloud technologies are relatively new and storage and data transfer algorithms slice the data into small units, which are stored and transferred dynamically within the storage region. Estimating and factoring the risks of potential data corruption of mission-critical data at this early stage of cloud implementation may be difficult leading to nonadoption, especially if the existing internal systems, processes, and protocols are working.
Data transfer costs	For new initiatives that do not require the transfer of legacy data to the clouds, transfer costs are minimal. Getting locked into a particular cloud service provider is currently a market concern due to the lack of open standards among the providers.	Moving the existing data sets to clouds inquires data integrity check to ensure that all of the data have been transferred fully and that they have not been corrupted. For very large data sets, this may represent significant staff costs. Cloud vendors typically charge data transfer costs. If the data set is large and there is significant data chum due to transaction processing, it may be more cost effective to look at more traditional hosting options.

(Continued)

TABLE 14.2 (*Continued*)

Economic Costs of Cloud Adoption

Economic Costs	Small and Medium Enterprises (SMEs)	Large Enterprises
Integration costs and legacy application reengineering	In start-ups and small companies, potentially little or no integration is required between cloud applications and legacy applications.	Potentially significant costs to have new cloud applications interact with legacy applications or to modify legacy applications to offload processing to cloud-based components. Conversely, there may be advantages to reengineering legacy applications and hosting them in a public cloud when integrating Web 2.0 functionality with legacy applications.
Software licensing	Cloud services (SaaS and PaaS) provide significant software licensing cost savings for start-ups and small companies.	Migrating large enterprises to cloud-based SaaS may not be cost effective relative to the existing enterprise licensing agreements. Depending on the licensing agreements for third-party software, especially if licensing fees are based on the number of CPUs using the software, hosting legacy applications in a cloud environment may involve significantly increased licensing costs or noncompliance with the agreements if the software is installed on a machine image used for autoscaling as the user demand increases.
Cloud availability— *rolling brownouts*	Unavailability of the cloud services or slow performance due to heavy traffic is a serious concern when choosing a cloud vendor.	Same as with SMEs. Currently, even large vendors have experienced slow performance or suspended service due to overwhelming utilization.
Data security	SMEs are better able to use third-party services such as payment processing to handle secure transactions.	Data are an enterprise's most important IT and operating asset. Current uncertainty regarding the security of the data assets stored in public clouds is one of the most significant barriers in cloud adoption. Large enterprises may not want their data stored in countries where intellectual property piracy is prevalent. Some companies may not want their data stored on equipment used by their competitors.
Data confidentiality	SMEs face the same data confidentiality issues as large enterprises.	One of the advantages of cloud computing and storage for confidentiality is that the data transfer and storage algorithms encrypt the data into units that are difficult to reconstruct without the specialized algorithms/keys if the data are intercepted in transfer or the cloud security is compromised.

Data regulations	SMEs face the same regulatory data location issues as large enterprises.	Depending on the company's industry, there may be significant regulatory issues regarding data location. Data that identify the individual in certain health and financial contexts are subject to US regulations. Similarly, the EU has laws that restrict the transfer of certain data outside of its borders.
Data integrity	The data integrity and reliability of cloud suppliers may be higher than that provided by the existing internal systems.	Cloud technologies are relatively new and storage and data transfer algorithms slice the data into small units, which are stored and transferred dynamically within the storage region. Estimating and factoring the risks of potential data corruption of mission-critical data at this early stage of cloud implementation may be difficult leading to nonadoption, especially if the existing internal systems, processes, and protocols are working.
Data transfer costs	For new initiatives that do not require the transfer of legacy data to the clouds, transfer costs are minimal. Getting locked into a particular cloud service provider is currently a market concern due to the lack of open standards among the providers.	Moving the existing data sets to clouds requires data integrity check to ensure that all of the data have been transferred fully and that they have not been corrupted. For very large data sets, this may represent significant staff costs. Cloud vendors typically charge data transfer costs. If the data set is large and there is significant data churn due to transaction processing, it may be more cost effective to look at more traditional hosting options.
Integration costs and legacy application reengineering	In start-ups and small companies, potentially little or no integration is required between cloud applications and legacy applications.	Potentially significant costs to have new cloud applications interact with legacy applications or to modify legacy applications to offload processing to cloud-based components. Conversely, there may be advantages to reengineering legacy applications and hosting them in a public cloud when integrating Web 2.0 functionality with legacy applications.
Software licensing	Cloud services (SaaS and PaaS) provide significant software licensing cost savings for start-ups and small companies.	Migrating large enterprises to cloud-based SaaS may not be cost effective relative to the existing enterprise licensing agreements. Depending on the licensing agreements for third-party software, especially if licensing fees are based on the number of CPUs using the software, hosting legacy applications in a cloud environment may involve significantly increased licensing costs or noncompliance with the agreements if the software is installed on a machine image used for autoscaling as the user demand increases.
Cloud availability— *rolling brownouts*	Unavailability of the cloud services or slow performance due to heavy traffic is a serious concern when choosing a cloud vendor.	Same as with SMEs. Currently, even large vendors have experienced slow performance or suspended service due to overwhelming utilization.

TABLE 14.3

Value Comparison on Colocation, Managed Services, and IaaS for Providers

	Colocation	Managed Services	IaaS with Cloud Computing
Profit margin	Low; intense competition	Low; intense competition	High; cost saving by resource sharing
Value add service	Very few	Few	Rich, such as IT service management and software renting
Operation	Manual operation; complex	Manual operation; complex	Automatic and integrated operation; end-to-end request management
Response to customer request	Manual action; slow	Manual action; slow	Automatic process; fast
Power consumption	Normal	Normal	Reduce power by server consolidation and sharing; scheduled power off

TABLE 14.4

Value Comparison on Colocation, Managed Services, and IaaS for Users

	Colocation	Managed Services	IaaS Using Cloud
Performance	Depends on hardware	Depends on hardware	Guaranteed performance
Price	Server investment plus bandwidth and space fee	Bandwidth and server renting fee	CPU, memory, storage, bandwidth fee; pay per use
Availability	Depends on single hardware	Depends on single hardware	Highly available by hardware failover
Scalability	Manual scale out	Manual scale out	Automated scale out
System management	Manual hardware setup and configuration; complex	Manual hardware setup and configuration; complex	Automated OS and software installation; remote monitoring and control; simple
Staff	High labor cost and skill requirement	High labor cost and skill requirement	Low labor cost and skill requirement
Usability	Need on site operation	Need on site operation	All work done through Web UI; quick action

A significant requirement for a cloud-based application is to operate in such a way that it functions reliably and delivers a consistent service at a variety of levels. This requires not only focusing on the functional properties of services but also concentrating on describing the environment hosting the cloud service, that is, describing the nonfunctional capabilities of services. Each service hosting environment may offer various choices of QoS based on technical requirements regarding demands for around-the-clock levels of service availability, performance and scalability, security and privacy policies, and so on, all of which must be described. It is thus obvious that the QoS offered by a cloud service is becoming the highest priority for service providers and their customers.

Delivering QoS on the Internet is a critical and significant challenge because of its dynamic and unpredictable nature. Applications with very different characteristics and requirements compete for all kinds of network resources. Changes in traffic patterns, securing mission-critical business transactions, and the effects of infrastructure failures, low performance of Web protocols, and reliability issues over the Web create a need for Internet QoS standards. Often, unresolved QoS issues cause critical transactional applications to suffer from unacceptable levels of performance degradation.

Traditionally, QoS is measured by the degree to which applications, systems, networks, and all other elements of the IT infrastructure support availability of services at a required level of performance under all access and load conditions. While traditional QoS metrics apply, the characteristics of cloud services environments bring both greater availability of applications and increased complexity in terms of accessing and managing services and thus impose specific and intense demands on organizations, which QoS must address. In the cloud services' context, QoS can be viewed as providing assurance on a set of quantitative characteristics. These can be defined on the basis of important functional and nonfunctional service quality properties that include implementation and deployment issues as well as other important service characteristics such as service metering and cost, performance metrics (e.g., response time), security requirements, (transactional) integrity, reliability, scalability, and availability. These characteristics are necessary requirements to understand the overall behavior of a service so that other applications and services can bind to it and execute it as part of a business process.

The key elements for supporting QoS in a cloud services environment are summarized in the following:

1. *Availability*: Availability is the absence of service downtimes. Availability represents the probability that a service is available. Larger values mean that the service is always ready to use while smaller values indicate unpredictability over whether the service will be available at a particular time. Also associated with availability is time to repair (TTR). TTR represents the time it takes to repair a service that has failed. Ideally, smaller values of TTR are desirable.

2. *Accessibility*: Accessibility represents the degree with which a cloud service request is served. It may be expressed as a probability measure denoting the success rate or chance of a successful service instantiation at a point in time. A high degree of accessibility means that a service is available for a large number of clients and that clients can use the service relatively easily.

3. *Conformance to standards*: This describes the compliance of a cloud service with standards. Strict adherence to correct versions of standards by service providers is necessary for proper invocation of cloud services by service requestors. In addition, service providers must stick to the standards outlined in service-level agreements (SLAs) between service requestors and providers.

4. *Integrity*: This describes the degree with which a cloud service performs its tasks according to its Web Service's WSDL description as well as conformance with Service-Level Agreement (SLA). A higher degree of integrity means that the functionality of a service is closer to its Web Service's WSDL description or SLA.

5. *Performance*: Performance is measured in terms of two factors—throughput and latency. Throughput represents the number of cloud service requests served at a given time period. Latency represents the length of time between sending a request and receiving the response. Higher throughput and lower latency values represent good performance of a cloud service. When measuring the transaction/request volumes handled by a cloud service, it is important to consider whether these come in a steady flow or burst around particular events like the open or close of the business day or seasonal rushes.

6. *Reliability*: Reliability represents the ability of a service to function correctly and consistently and provides the same service quality despite system or network failures. The reliability of a cloud service is usually expressed in terms of the number of transactional failures per month or year.

7. *Scalability*: Scalability refers to the ability to consistently serve the requests despite variations in the volume of requests. High accessibility of cloud services can be achieved by building highly scalable systems.

8. *Security*: Security involves aspects such as authentication, authorization, message integrity, and confidentiality. Security has added importance because cloud service invocation occurs over the Internet. The amount of security that a particular cloud service requires is described in its accompanying SLA, and service providers must maintain this level of security.

9. *Transactionality*: There are several cases where cloud services require transactional behavior and context propagation. The fact that a particular cloud service requires transactional behavior is described in its accompanying SLA, and service providers must maintain this property.

14.4.1 Service-Level Agreement (SLA)

As organizations depend on business units, partners, and external service providers to furnish them with services, they rely on the use of SLAs to ensure that the chosen service provider delivers a guaranteed level of service quality. An SLA is a formal agreement (contract) between a provider and client, formalizing the details of a Web Service (contents, price, delivery process, acceptance and quality criteria, penalties, and so on, usually in measurable terms) in a way that meets the mutual understandings and expectations of both the service provider and the service requestor.

An SLA is basically a QoS guarantee typically backed up by chargeback and other mechanisms designed to compensate users of services and to influence organizations to fulfill SLA commitments. Understanding business requirements, expected usage patterns, and system capabilities can go a long way toward ensuring successful deployments. An SLA is an important and widely used instrument in the maintenance of service provision relationships as both service providers and clients alike utilize it.

An SLA may contain the following parts:

- *Purpose*: This field describes the reasons behind the creation of the SLA.
- *Parties*: This field describes the parties involved in the SLA and their respective roles, for example, service provider and service consumer (client).
- *Validity period*: This field defines the period of time that the SLA will cover. This is delimited by start time and end time of the agreement term.
- *Scope*: This field defines the services covered in the agreement.
- *Restrictions*: This field defines the necessary steps to be taken in order for the requested service levels to be provided.
- *Service-level objectives*: This field defines the levels of service that both the service customers and the service providers agree on and usually includes a set of service-level indicators, like availability, performance, and reliability. Each of these aspects of the service level will have a target level to achieve.
- *Penalties*: This field defines what sanctions should apply in case the service provider underperforms and is unable to meet the objectives specified in the SLA.

- *Optional services*: This field specifies any services that are not normally required by the user but might be required in case of an exception.
- *Exclusion terms*: These specify what is not covered in the SLA.
- *Administration*: This field describes the processes and the measurable objectives in an SLA and defines the organizational authority for overseeing them.

SLAs can be either static or dynamic in nature. A static SLA is an SLA that generally remains unchanged for multiple service time intervals. Service time intervals may be calendar months for a business process that is subject to an SLA or may be a transaction or any other measurable and relevant period of time for other processes. They are used for assessment of the QoS and are agreed between a service provider and service client. A dynamic SLA is an SLA that generally changes from service period to service period, to accommodate changes in provision of service.

14.5 Summary

The chapter started with the discussion of the drivers of cloud computing. This was followed with the presentation of the concept of total cost of ownership (TCO) and capital budgeting models. Costs for various models of cloud service provisioning, namely, traditional internal IT, colocation, managed services, and IaaS are considered for producing a comparison for assessment. In the last part, we discuss aspects related to the Quality of Service (QoS) and Service Level Agreement (SLA) for provisioning of cloud services.

15

Cloudware Technologies

Virtualization is widely used to deliver customizable computing environments on demand. Virtualization technology is one of the fundamental components of cloud computing. Virtualization allows the creation of a secure, customizable, and isolated execution environment for running applications without affecting other users' applications. The basis of this technology is the ability of a computer program—or a combination of software and hardware—to emulate an executing environment separate from the one that hosts such programs. For instance, we can run Windows OS on top of a virtual machine, which itself is running on Linux OS. Virtualization provides a great opportunity to build elastically scalable systems that can provision additional capability with minimum costs.

15.1 Virtualization

Resource virtualization is at the heart of most cloud architectures. The concept of virtualization allows an abstract, logical view on the physical resources and includes servers, data stores, networks, and software. The basic idea is to pool physical resources and manage them as a whole. Individual requests can then be served as required from these resource pools. For instance, it is possible to dynamically generate a certain platform for a specific application at the very moment when it is needed—instead of a real machine, a virtual machine is instituted.

Resource management grows increasingly complex as the scale of a system as well as the number of users and the diversity of applications using the system increase. Resource management for a community of users with a wide range of applications running under different operating systems is a very difficult problem. Resource management becomes even more complex when resources are oversubscribed and users are uncooperative. In addition to external factors, resource management is affected by internal factors, such as the heterogeneity of the hardware and software systems, the ability to approximate the global state of the system and to redistribute the load, and the failure rates of different components. The traditional solution for these in a data center is to install standard operating systems on individual systems and rely on conventional OS techniques to ensure resource sharing, application protection, and performance isolation. System administration, accounting, security, and resource management are very challenging for the providers of service in this setup; application development and performance optimization are equally challenging for the users.

The alternative is resource virtualization, a technique analyzed in this chapter. Virtualization is a basic tenet of cloud computing—which simplifies some of the resource management tasks. For instance, the state of a virtual machine (VM) running under a virtual machine monitor (VMM) can be saved and migrated to another server to balance the load. At the same time, virtualization allows users to operate in environments with which they are familiar rather than forcing them to work in idiosyncratic environments. Resource sharing in a virtual machine environment requires not only ample hardware support and, in particular, powerful processors but also architectural support for multilevel control. Indeed, resources such as CPU cycles, memory, secondary storage, and I/O and communication bandwidth are shared among several virtual machines; for each VM, resources must be shared among multiple instances of an application. There are two distinct approaches for virtualization, namely, the full virtualization and the paravirtualization. Full virtualization is feasible when the hardware abstraction provided by the VMM is an exact replica of the physical hardware. In this case, any operating system running on the hardware will run without modifications under the VMM. In contrast, paravirtualization requires some modifications of the guest operating systems because the hardware abstraction provided by the VMM does not support all the functions the hardware does.

One of the primary reasons that companies have implemented virtualization is to improve the performance and efficiency of processing of a diverse mix of workloads. Rather than assigning a dedicated set of physical resources to each set of tasks, a pooled set of virtual resources can be quickly allocated as needed across all workloads. Reliance on the pool of virtual resources allows companies to improve latency. This increase in service delivery speed and efficiency is a function of the distributed nature of virtualized environments and helps to improve overall time-to-realize value. Using a distributed set of physical resources, such as servers, in a more flexible and efficient way delivers significant benefits in terms of cost savings and improvements in productivity. First, virtualization of physical resources (such as servers, storage, and networks) enables substantial improvement in the utilization of these resources. Second, virtualization enables improved control over the usage and performance of the IT resources. Third, virtualization provides a level of automation and standardization to optimize your computing environment. Fourth, consequently, virtualization provides a foundation for cloud computing. Virtualization increases the efficiency of the cloud that makes many complex systems easier to optimize. As a result, organizations have been able to achieve the performance and optimization to be able to access data that were previously either unavailable or very hard to collect. Big data platforms are increasingly used as sources of enormous amounts of data about customer preferences, sentiment, and behaviors (see Chapter 21, Section 21.1.1 "What Is Big Data?"). Companies can integrate this information with internal sales and product data to gain insight into customer preferences to make more targeted and personalized offers.

15.1.1 Characteristics of Virtualized Environment

In a virtualized environment, there are three major components: guest, host, and virtualization layer. The guest represents the system component that interacts with the virtualization layer rather than with the host, as would normally happen. The host represents the original environment where the guest is supposed to be managed. The virtualization layer is responsible for recreating the same or a different environment where the guest will operate.

Virtualization has three characteristics that support the scalability and operating efficiency required for big data environments:

1. *Partitioning*: In virtualization, many applications and operating systems are supported in a single physical system by partitioning (separating) the available resources.

2. *Isolation*: Each virtual machine is isolated from its host physical system and other virtualized machines. Because of this isolation, if one virtual instance crashes, the other virtual machines and the host system are not affected. In addition, data are not shared between one virtual instance and another.

3. *Encapsulation*: A virtual machines can be represented (and even stored) as a single file, so you can identify it easily based on the services it provides. For example, the file containing the encapsulated process could be a complete business service. This encapsulated virtual machine could be presented to an application as a complete entity. Thus, encapsulation could protect each application so that it does not interfere with another application.

Virtualization abstracts the underlying resources and simplifies their use, isolates users from one another, and supports replication, which, in turn, increases the elasticity of the system. Virtualization is a critical aspect of cloud computing, equally important to the providers and consumers of cloud services, and plays an important role in

- System security because it allows isolation of services running on the same hardware
- Portable performance and reliability because it allows applications to migrate from one platform to another
- Development and management of services offered by a provider
- Performance isolation

Virtualization—the process of using computer resources to imitate other resources—is valued for its capability to increase IT resource utilization, efficiency, and scalability. One obvious application of virtualization is server virtualization, which helps organizations to increase the utilization of physical servers and potentially save on infrastructure costs; companies are increasingly finding that virtualization is not limited only to servers but is valid and applicable across the entire IT infrastructure, including networks, storage, and software. For instance, one of the most important requirements for success with big data is having the right level of performance to support the analysis of large volumes and varied types of data. If a company only virtualizes the servers, they may experience bottlenecks from other infrastructure elements such as storage and networks; furthermore, they are less likely to achieve the latency and efficiency that they need and more likely to expose the company to higher costs and increased security risks. As a result, a company's entire IT environment needs to be optimized at every layer from the network to the databases, storage, and servers—virtualization adds efficiency at every layer of the IT infrastructure.

For a provider of IT services, the use of virtualization techniques has a number of advantages:

1. *Resource usage*: Physical servers rarely work to capacity because their operators usually allow for sufficient computing resources to cover peak usage. If virtual machines are used, any load requirement can be satisfied from the resource pool. In case the demand increases, it is possible to delay or even avoid the purchase of new capacities.
2. *Management*: It is possible to automate resource pool management. Virtual machines can be created and configured automatically as required.
3. *Consolidation*: Different application classes can be consolidated to run on a smaller number of physical components. Besides server or storage consolidation, it is also possible to include entire system landscapes, data and databases, networks, and desktops. Consolidation leads to increased efficiency and thus to cost reduction.
4. *Energy consumption*: Supplying large data centers with electric power has become increasingly difficult, and seen over its lifetime, the cost of energy required to operate a server is higher than its purchase price. Consolidation reduces the number of physical components. This, in turn, reduces the expenses for energy supply.

5. *Less space required*: Each and every square yard of data center space is scarce and expensive. With consolidation, the same performance can be obtained on a smaller footprint and the costly expansion of an existing data center might possibly be avoided.

6. *Emergency planning*: It is possible to move virtual machines from one resource pool to another. This ensures better availability of the services and makes it easier to comply with service-level agreements. Hardware maintenance windows are inherently no longer required.

Since the providers of cloud services tend to build very large resource centers, virtualization leads not only to a size advantage but also to a more favorable cost situation. This results in the following benefits for the customer:

1. *Dynamic behavior*: Any request can be satisfied just in time and without any delays. In case of bottlenecks, a virtual machine can draw on additional resources (such as storage space and I/O capabilities).

2. *Availability*: Services are highly available and can be used day and night without stop. In the event of technology upgrades, it is possible to hot-migrate applications because virtual machines can easily be moved to an up-to-date system.

3. *Access*: The virtualization layer isolates each virtual machine from the others and from the physical infrastructure. This way, virtual systems feature multitenant capabilities and, using a roles concept, it is possible to safely delegate management functionality to the customer. Customers can purchase IT capabilities from a self-service portal (customer emancipation).

The most direct benefit from virtualization is to ensure that MapReduce engines work better. Virtualization will result in better scale and performance for MapReduce. Each one of the map and reduce tasks needs to be executed independently. If the MapReduce engine is parallelized and configured to run in a virtual environment, you can reduce management overhead and allow for expansions and contractions in the task workloads. MapReduce itself is inherently parallel and distributed. By encapsulating the MapReduce engine in a virtual container, you can run what you need whenever you need it. With virtualization, you can increase your utilization of the assets you have already paid for by turning them into generic pools of resources (see Chapter 17, Section 17.2 "Google MapReduce").

There are side effects of virtualization, notably the performance penalty and the hardware costs. All privileged operations of a VM must be trapped and validated by the VMM, which ultimately controls system behavior; the increased overhead has a negative impact on performance. The cost of the hardware for a VM is higher than the cost for a system running a traditional operating system because the physical hardware is shared among a set of guest operating systems and it is typically configured with faster and/or multicore processors, more memory, larger disks, and additional network interfaces compared with a system running a traditional operating system.

A drawback of virtualization is the fact that the operation of the abstraction layer itself requires resources. Modern virtualization techniques, however, are so sophisticated that this overhead is not too significant: due to the particularly effective interaction of current multicore systems with virtualization technology, this performance loss plays only a minor role in today's systems. In view of possible savings and the quality benefits perceived by the customers, the use of virtualization pays off in nearly all cases.

15.1.2 Layering and Virtualization

A common approach to managing system complexity is to identify a set of layers with well-defined interfaces among them. The interfaces separate different levels of abstraction. Layering minimizes the interactions among the subsystems and simplifies the description of the subsystems. Each subsystem is abstracted through its interfaces with the other subsystems. Thus, we are able to design, implement, and modify the individual subsystems independently. The instruction set architecture (ISA) defines a processor's set of instructions. For example, the Intel architecture is represented by the ×86–32 and ×86–64 instruction sets for systems supporting 32-bit addressing and 64-bit addressing, respectively. The hardware supports two execution modes, a privileged, or kernel, mode and a user mode. The instruction set consists of two sets of instructions, privileged instructions that can only be executed in kernel mode and nonprivileged instructions that can be executed in user mode. There are also sensitive instructions that can be executed in kernel and in user mode but that behave differently.

Modern computing systems can be expressed in terms of the reference model described in Figure 15.1. The highest level of abstraction is represented by the application programming interface (API), which interfaces applications to libraries and/or the underlying operating system. The application binary interface (ABI) separates the operating system layer from the applications and libraries, which are managed by the OS. ABI covers details such as low-level data types, alignment, and call conventions and defines a format for executable programs. System calls are defined at this level. This interface allows portability of applications and libraries across operating systems that implement the same ABI. At the bottom layer, the model

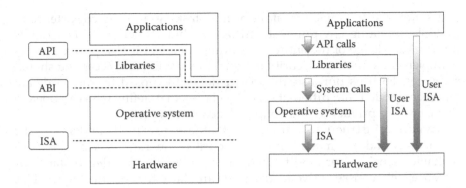

FIGURE 15.1
Layering and interfaces between layers of a computer system.

for the hardware is expressed in terms of the Instruction Set Architecture (ISA), which defines the instruction set for the processor, registers, memory, and interrupts management. ISA is the interface between hardware and software, and it is important to the operating system (OS) developer (system ISA) and developers of applications that directly manage the underlying hardware (user ISA).

The API defines the set of instructions the hardware was designed to execute and gives the application access to the ISA. It includes high-level languages (HLL) library calls, which often invoke system calls. A *process* is the abstraction for the code of an application at execution time; a *thread* is a lightweight process. The API is the projection of the system from the perspective of the HLL program and the ABI is the projection of the computer system seen by the process. Consequently, the binaries created by a compiler for a specific ISA and a specific operating system are not portable. Such code cannot run on a computer with a different ISA or on computers with the same ISA but different operating systems. However, it is possible to compile an HLL program for a VM environment, where portable code is produced and distributed and then converted dynamically by binary translators to the ISA of the host system. A dynamic binary translation converts blocks of guest instructions from the portable code to the host instruction and leads to a significant performance improvement as such blocks are cached and reused.

For any operation to be performed in the application level API, ABI and ISA are responsible for making it happen. The high-level abstraction is converted into machine-level instructions to perform the actual operations supported by the processor. The machine-level resources, such as processor registers and main memory capacities, are used to perform the operation at the hardware level of the central processing unit (CPU). This layered approach simplifies the development and implementation of computing systems and simplifies the implementation of multitasking and the coexistence of multiple executing environments. In fact, such a model not only requires limited knowledge of the entire computing stack, but it also provides ways to implement a minimal security model for managing and accessing shared resources. For this purpose, the instruction set exposed by the hardware has been divided into different security classes that define who can operate them, namely, privileged and nonprivileged instructions.

Privileged instructions are those that are executed under specific restrictions and are mostly used for sensitive operations, which expose (behavior-sensitive) or modify (control-sensitive) the privileged state. For instance, behavior-sensitive instructions are those that operate on the I/O, whereas control-sensitive instructions alter the state of the CPU registers. Nonprivileged instructions are those instructions that can be used without interfering with other tasks because they do not access shared resources. For instance, this category contains all the floating, fixed-point, and arithmetic instructions.

All the current systems support at least two different execution modes: supervisor mode and user mode. The first mode denotes an execution mode in which all the instructions (privileged and nonprivileged) can be executed without any restriction. This mode, also called master mode or kernel mode, is generally used by the operating system (or the hypervisor) to perform sensitive operations on hardware-level resources. In user mode, there are restrictions to control the machine-level resources. If code running in user mode invokes the privileged instructions, hardware interrupts occur and trap the potentially harmful execution of the instruction.

15.1.3 Virtual Machines

A virtual machine (VM) is an isolated environment that appears to be a whole computer but actually only has access to a portion of the computer resources. Each VM appears to be running on the bare hardware, giving the appearance of multiple instances of the same computer, though all are supported by a single physical system. Virtual machines have been around since the early 1970s, when IBM released its VM/370 operating system. There are two types of VM: process and system VMs. A process VM is a virtual platform created for an individual process and destroyed once the process terminates. Virtually, all operating systems provide a process VM for each one of the applications running, but the more interesting process VMs are those that support binaries compiled on a different instruction set. A system VM supports an operating system together with many user processes. When the VM runs under the control of a normal OS and provides a platform-independent host for a single application, we have an application virtual machine (e.g., Java Virtual Machine [JVM]).

A system virtual machine provides a complete system; each VM can run its own OS, which in turn can run multiple applications. Systems such as

Linux-VServer, OpenVZ (Open VirtualiZation), FreeBSD Jails, and Solaris Zones, based on Linux, FreeBSD, and Solaris, respectively, implement operating system-level virtualization technologies. Operating system-level virtualization allows a physical server to run multiple isolated operating system instances, subject to several constraints; the instances are known as containers, virtual private servers (VPSs), or virtual environments (VEs). For instance, OpenVZ requires both the host and the guest OS to be Linux distributions. These systems claim performance advantages over the systems based on a VMM such as Xen or VMware (there is only a 1%–3% performance penalty for OpenVZ compared to a stand-alone Linux server).

15.1.3.1 Virtual Machine Monitor (VMM)

A virtual machine monitor (VMM), also called a hypervisor, is the software that securely partitions the resources of a computer system into one or more virtual machines. A guest operating system is an operating system that runs under the control of a VMM rather than directly on the hardware: the VMM runs in kernel mode, whereas a guest OS runs in user mode. VMMs allow several operating systems to run concurrently on a single hardware platform; at the same time, VMMs enforce isolation among these systems, thus enhancing security. A VMM controls how the guest operating system uses the hardware resources. The events occurring in one VM do not affect any other VM running under the same VMM.

Thus, the VMM enables

- Multiple services to share the same platform
- The movement of a server from one platform to another, the so-called live migration
- System modification while maintaining backward compatibility with the original system

When a guest OS attempts to execute a privileged instruction, the VMM traps the operation and enforces the correctness and safety of the operation. The VMM guarantees the isolation of the individual VMs and thus ensures security and encapsulation, a major concern in cloud computing. At the same time, the VMM monitors system performance and takes corrective action to avoid performance degradation; for instance, the VMM may swap out a VM (copies all pages of that VM from real memory to disk and makes the real memory frames available for paging by other VMs) to avoid thrashing.

A VMM virtualizes the CPU and memory. For instance, the VMM traps interrupts and dispatches them to the individual guest operating systems. If a guest OS disables interrupts, the VMM buffers such interrupts until the guest OS enables them. The VMM maintains a shadow page table for each guest OS and replicates any modification made by the guest OS in its own shadow page table. This shadow page table points to the actual page frame and is used by the hardware component called the memory management unit (MMU) for dynamic address translation. Memory virtualization has important implications on performance. VMMs use a range of optimization techniques; for example, VMware systems avoid page duplication among different virtual machines; they maintain only one copy of a shared page and use copy-on-write policies, whereas Xen imposes total isolation of the VM and does not allow page sharing. VMMs control the virtual memory management and decide what pages to swap out; for example, when the ESX VMware server wants to swap out pages, it uses a balloon process inside a guest OS and requests it to allocate more pages to itself, thus swapping out pages of some of the processes running under that VM. Then it forces the balloon process to relinquish control of the free page frames.

There are two major types of hypervisors:

- Type I hypervisors run directly on top of the hardware. Therefore, they take the role of an operating system and interact directly with the ISA interface exposed by the underlying hardware, and they emulate this interface in order to allow the management of guest operating systems. These types of hypervisors are also called native virtual machines since they run natively on hardware.
- Type II hypervisors require the support of an operating system to provide virtualization services. This means that they are programs managed by the operating system, which interact with it through the ABI and emulate the ISA of virtual hardware for guest operating systems. These types of hypervisors are also called hosted virtual machines since they are hosted within an operating system.

15.1.3.2 VMM Solutions

A number of VMM solutions exist that are the basis of many utility or cloud computing environments.

a. VMWare ESXi: ESXi is a VMM from VMWare. VMware is a pioneer in the virtualization market. Its ecosystem of tools ranges from server and desktop virtualization to high-level management tools. It is a bare-metal hypervisor, meaning that it installs directly on the physical server, whereas others may require a host operating system. It provides advanced virtualization techniques of processor, memory, and I/O. Especially, through memory ballooning and page sharing, it can overcommit memory, thus increasing the density of VMs inside a single physical server.

b. Xen: Xen hypervisor started as an open-source project and has served as a base to other virtualization products, both commercial and open source. It has pioneered the paravirtualization concept, on which the guest operating system, by means of a specialized kernel, can interact with the hypervisor, thus significantly improving performance. In addition to an open-source distribution, Xen currently forms the base of commercial hypervisors of a number of vendors including Citrix XenServer and Oracle VM.

c. KVM: The kernel-based virtual machine (KVM) is a Linux virtualization subsystem. It has been part of the mainline Linux kernel since version 2.6.20, thus being natively supported by several distributions. In addition, activities such as memory management and scheduling are carried out by existing kernel features, thus making KVM simpler and smaller than hypervisors, which take control of the entire machine. KVM leverages hardware-assisted virtualization, which improves performance and allows it to support unmodified guest operating systems; currently, it supports several versions of Windows, Linux, and UNIX.

15.2 Types of Virtualization

15.2.1 Operating System Virtualization

The use of operating system virtualization or partitioning (such as IBM LPARs) in cloud environments may help to solve security and confidentiality problems, which would otherwise impair the acceptance of the cloud approach. For this type of virtualization, which is also called *container* or *jails*, the host operating system plays a major role. This is a concept where multiple identical system environments or runtime environments, which are completely isolated from each other, run under one operating system kernel. Seen from the outside, virtual environments appear as autonomous systems. All running applications use the same kernel, but they can only see the processes belonging to the same virtual environment.

Mainly, Internet service providers (ISPs), who offer (virtual) root servers, prefer this kind of virtualization because it is associated with a minor performance loss and a high degree of security. The drawback of operating system virtualization is its reduced flexibility: while multiple independent instances of the same operating system can be used simultaneously, it is not possible to run different operating systems at the same time. Popular examples of operating system virtualization are the container technology from Sun Solaris, OpenVZ for Linux, Linux-VServer, FreeBSD Jails, and Virtuozzo.

15.2.2 Platform Virtualization

Platform virtualization allows to run any desired operating systems and applications in virtual environments. There are two different models: full virtualization and paravirtualization. Both solutions are implemented on the basis of a virtual machine monitor or hypervisor. The hypervisor is a minimalistic meta-operating system used for distributing the hardware resources among the guest systems and for access coordination. A type-1 hypervisor is built directly on top of the hardware; a type-2 hypervisor runs under a traditional basic operating system.

Full virtualization is based on the simulation of an entire virtual computer with virtual resources, such as CPU, RAM, drives, and network adapters, including its own BIOS. Since the access to the most important resources, such as the processor and the RAM, is passed through, the processing speed of the guest operating systems nearly equals the speed to be expected if there was no virtualization. Other components, for example, drives or network adapters, are emulated. While this decreases the performance, it allows to run unmodified guest operating systems. Paravirtualization does not provide an emulated hardware layer to the guest operating systems, but only an application interface. For this purpose, the guest operating systems need to be modified because any direct access to hardware must be replaced by the corresponding hypervisor interface call. This is also referred to as hyper calls (just like system calls), which are used by the applications to call functions in the operating system kernel. Since this approach allows the guest system to participate actively
* in the virtualization (at least to some extent), a higher throughput than with full virtualization can be obtained, especially for I/O-intensive applications. Examples of full virtualization are the VMware products or, specifically for Linux, the Kernel-based Virtual Machine (KVM). Under Linux, mostly Xen-based solutions are used for paravirtualization. They play an important role, particularly in the realization of the Amazon Web Services.

15.2.3 Storage Virtualization

Cloud systems should also offer dynamically scalable storage space as a service. In this context, storage virtualization boasts a number of advantages. The fundamental idea of storage virtualization is to separate the data store from the classical file servers and to pool the physical storage systems. Applications use these pools to dynamically meet their storage requirements. For the data transfers, a special storage area network (SAN) or a local company network (LAN) is used. Data for cloud offerings are mostly available in the form of Web objects that can be retrieved or manipulated over the Internet. An additional abstract administration layer is interposed between the clients and the storage landscape so that the representation of a datum is decoupled from its physical storage. This has a variety of advantages with respect to data management and access scalability.

Central management also allows to operate the distributed storage systems at a lower cost. Moreover, different categories of data storage can be organized in storage hierarchies (tier concept). This makes it possible to implement an automated lifecycle management for data sets, from tier 0 with the most stringent availability and bandwidth requirements to lower and cheaper tier levels with a correspondingly lower quality of service. The data can be migrated between these levels without affecting the service. By using snapshots, even large data quantities can be backed up without a special backup window. A further advantage of storage virtualization is that distributed mirrors may be created and managed in order to avoid service disruptions in case of malfunctions. Amazon, for instance, creates up to three copies in different data centers when storing data.

15.2.4 Network Virtualization

Techniques such as load balancing are essential in cloud environments because it must be possible to dynamically scale the services offered. The resources are usually implemented as Web objects. For this reason, it is recommended to apply the procedures commonly used for Web servers: services can be accessed via virtual IP addresses. Through cluster technology, they realize load balancing as well as automatic failover in case of a failure. By forwarding DNS requests, it is also possible to integrate cloud resources into the customer's Internet namespace.

Network virtualization is also used for virtual local area networks (VLANs) and virtual switches. In this case, cloud resources appear directly in the customer's network. Internal resources can thus be replaced transparently by external resources. VLAN technology has the following advantages:

- *Transparency*: Distributed devices can be pooled together in a single logical network. VLANs are very helpful when designing the IT infrastructure for geographically disparate locations.

- *Security*: Certain systems that require particular protection can be hidden in a separate virtual network.

On the other hand, VLANs involve more overhead for network administration and for programming active network components (switches, etc.).

15.3 Service-Oriented Architecture (SOA)

SOA introduces a flexible architectural style that provides an integration framework through which software architects can build applications using a collection of reusable functional units (services) with well-defined interfaces, which it combines in a logical flow. Applications are integrated at the interface (contract) and not at the implementation level. This allows greater flexibility since applications are built to work with any implementation of a contract, rather than take advantage of a feature or idiosyncrasy of a particular system or implementation. For example, different service providers (of the same interface) can be dynamically chosen based on policies, such as price, performance, or other QoS guarantees, current transaction volume, and so on.

Another important characteristic of an SOA is that it allows many-to-many integration; that is, a variety of consumers across an enterprise can use and reuse applications in a variety of ways. This ability can dramatically reduce the cost/complexity of integrating incompatible applications and increase the ability of developers to quickly create, reconfigure, and repurpose applications as business needs arise. Benefits include reduced IT administration costs, ease of business process integration across organizational departments and with trading partners, and increased business adaptability.

SOA is a logical way of designing a software system to provide services to either end-user applications or to other services distributed in a network, via published and discoverable interfaces. To achieve this, SOA reorganizes a portfolio of previously siloed software applications and support infrastructure in an organization into an interconnected collection of services, each of which is discoverable and accessible through standard interfaces and messaging protocols. Once all the elements of an SOA are in place, existing and future applications can access the SOA-based services as necessary. This architectural approach is particularly applicable when multiple applications running on varied technologies and platforms need to communicate with each other.

The essential goal of an SOA is to enable general-purpose interoperability among existing technologies and extensibility to future purposes and architectures. SOA lowers interoperability hurdles by converting monolithic and static systems into modular and flexible components, which it represents as services that can be requested through an industry standard protocol. Much of SOA's power and flexibility derives from its ability to leverage standards-based functional services, calling them when needed on an individual basis or aggregating them to create composite applications or multistage business processes. The building-block services might employ preexisting components that are reused and can also be updated or replaced without affecting the functionality or integrity of other independent services. In this latter regard, the services model offers numerous advantages over large monolithic applications, in which modifications to some portions of the code can have unintended and unpredictable effects on the rest of the code to which it is tightly bundled. Simply put, an SOA is an architectural style, inspired by the service-oriented approach to computing, for enabling extensible interoperability.

SOA as a design philosophy is independent of any specific technology, for example, Web Services or J2EE. Although the concept of SOA is often

discussed in conjunction with Web Services, these two are not synonymous. In fact SOA can be implemented without the use of Web Services, for
example, using Java, C#, or J2EE. However, Web Services should be seen as a
primary example of a message delivery model that makes it much easier to
deploy an SOA. Web Services standards are key to enabling interoperability
as well as key issues including quality of system (QoS), system semantics,
security, management, and reliable messaging.

15.3.1 Operations in the SOA

SOA enables three primary operations; these are publication of the service
descriptions, finding the service descriptions, and binding or invocation of
services based on their service description. These three basic operations can
occur singly or iteratively. A logical view of the SOA is given in Figure 15.2.
This figure illustrates the relationship between the SOA operations and
roles. First, the Web Services provider publishes its Web Service(s) with the
discovery agency. Next, the Web Services client searches for desired Web
Services using the registry of the discovery agency. Finally, the Web Services
client, using the information obtained from the discovery agency, invokes
(binds to) the Web Services provided by the Web Services provider.

15.3.1.1 Publish Operation

Publishing a Web Service so that other users or applications can find it actually consists of two equally important operations. The first operation is
describing the Web Service itself; the other is the actual registration of the
Web Service.

The first requirement for publishing Web Services with the service registry is for a service provider to properly describe them in WSDL. For proper

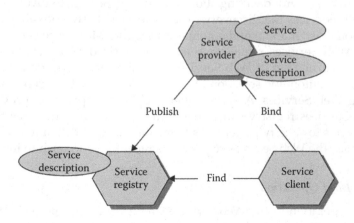

FIGURE 15.2
Web Services Operations.

description of a Web Service, three basic categories of information are necessary:

1. *Business information*: information regarding the Web Service provider or the implementer of the service
2. *Service information*: information about the nature of the Web Service
3. *Technical information*: information about implementation details and the invocation methods for the Web Service

Registration deals with storing the three basic categories of descriptive information about a service in the Web Services registry. For Web Services requestors to be able to find a Web Service, this service description information needs to be published with at least one discovery agency.

15.3.1.2 Find Operation

Finding the desired Web Services consists of first discovering the services in the registry of the discovery agency and then selecting the desired Web Service(s) from the search results.

Discovering Web Services involves querying the registry of the discovery agency for Web Services matching the needs of a Web Services requestor; a query is executed against the Web Service information in the registry entered by the Web Services provider. A query consists of search criteria such as type of service, preferred price range, what products are associated with this service, with which categories in company and product taxonomies this Web Service is associated, as well as other technical service characteristics. The find operation can be specified statically at design time to retrieve a service's interface description for program development or dynamically (at runtime) to retrieve a service's binding and location description for invocation.

Selection deals with deciding about which Web Service to invoke from the set of Web Services the discovery process returned. Two possible methods of selection exist: manual and automatic selection. Manual selection implies that the Web Services requestor selects the desired Web Service directly from the returned set of Web Services after manual inspection. The other possibility is automatic selection of the best candidate between potentially matching Web Services. A special client application program provided by the Web Services registry can achieve this. In this case, the Web Services requestor has to specify preferences to enable the application to infer which Web Service the Web Services requestor is most likely to wish to invoke.

15.3.1.3 Bind Operation

The final operation in the Web Services architecture and perhaps the most important one is the actual invocation of the Web Services. During the binding operation, the service requestor invokes the Web Service at runtime using the binding details in the service description to locate and contract to the service in either of these two ways:

1. Direct invocation of the Web Service by the Web Services requestor using the technical information included in the description of the service.
2. Mediation by the discovery agency for invoking the Web Service. In this case, all communication between the Web Services requestor and the Web Services provider goes through the Web Services registry of the discovery agency.

15.3.2 Roles in SOA

Corresponding to the three operations in the SOA, there are three primary roles, namely, service provider, the service registry, and the service requestor.

15.3.2.1 Web Services Provider

The Web Services provider is responsible for publishing the Web Services it provides in a service registry hosted by a service discovery agency. This involves describing the business, service, and technical information of the Web Service and registering that information with the Web Services registry in the format prescribed by the discovery agency.

From a business perspective, the Web Services provider is the organization that owns the Web Service and implements the business logic that underlies the service. From an architectural perspective, this is the platform that hosts and controls access to the service.

15.3.2.2 Web Services Registry

Web Services registry is a searchable directory where service descriptions can be published and searched. Service requestors find service descriptions in the registry and obtain binding information for services. This information is sufficient for the service requestor to contact, or bind to, the service provider and thus make use of the services it provides.

The Web Services discovery agency is responsible for providing the infrastructure required to enable the three operations in the Web Services architecture as described in the previous section: publishing the Web Services by the Web Services provider, searching for Web Services by the Web Services requestors, and invoking the Web Services.

15.3.2.3 Web Services Requestor

The next major role in the Web Services architecture is that of the Web Services requestor (or client). From a business perspective, this is the enterprise that requires certain functions to be satisfied. From an architectural perspective, this is the application that is looking for, and subsequently invoking, the service.

The Web Services requestor searches the service registry for the desired Web Services and using the information in the description to bind to the service. Two different kinds of Web Services requestors exist. The requestor role can be played either by another Web Service as part of an application (i.e., without a user interface) or a browser driven by an end user.

15.3.3 Layers in an SOA

On the basis of their requirements, enterprises may use SOA for the following:

- *Implementing end-to-end collaborative business processes*: The term *end-to-end business process* signifies that a succession of automated business processes and information systems in different enterprises (which are typically involved in intercompany business transactions) are successfully integrated. The aim is to provide seamless interoperation and interactive links between all the relevant members in an extended enterprise—ranging from product designers, suppliers, trading partners, and logistics providers to end customers. At this stage, an organization moves into the highest strategic level of SOA implementation. Deployment of services becomes ubiquitous, and federated services collaborate across enterprise boundaries to create complex products and services. Individual services in this extended enterprise may originate from many providers, irrespective of company-specific systems or applications.

- *Implementing enterprise service orchestrations*: This basic SOA entry point focuses on a typical implementation within a department or between a small number of departments and enterprise assets and comprises two steps: The first step is transforming enterprise assets and applications into an SOA implementation. This can start by service enabling existing individual applications or creating new applications using Web Services technology. This can begin by specifying a Web Service interface into an individual application or application element (including legacy systems). The next step after this basic Web Service implementation is implementing service orchestrations out of the service-enabled assets or newly created service applications.

- *Service enabling the entire enterprise*: The next stage in the SOA entry point hierarchy is when an enterprise seeks to provide a set of common services based on SOA components that can be used across the entire organization. Enterprise-wide service integration is achieved on the basis of commonly accepted standards. This results in achieving service consistency across departmental boundaries and is a precursor to integrating an organization with its partners and suppliers. Consistency is an important factor for this configuration as it provides both a uniform view to the enterprise and its customers as well as ensuring compliance with statutory or business policy requirements.

One problem when implementing an SOA at the enterprise level or implementing a cross-enterprise collaborative SOA is how to manage the SOA model, how to categorize the elements in this model, and how to organize them in such a way that the different stakeholders reviewing the model can understand it. Toward this end, it is often convenient to think of the SOA as comprising a number of distinct layers of abstraction that emphasize service interfaces, service realizations, and compositions of services into higher-level business processes. Each of these describes a logical separation of concerns by defining a set of common enterprise elements; each layer uses the functionality of the layer below it, adding new functionality, to accomplish its objective. The logical flow employed in the layered SOA development model may focus on a top-down development approach, which emphasizes how business processes are decomposed into a collection of business services and how these services are implemented in terms of preexisting enterprise assets.

SOA can considered to be comprised of the following six distinct layers:

1. *Domains*: A *business domain* is a functional domain comprising a set of current and future business processes that share common capabilities and functionality and can collaborate with each other to accomplish a higher-level business objective, such as loans, insurance, banking, finance, manufacturing, marketing, and human resources.

2. *Business processes*: This layer is formed by subdividing a business domain, such as distribution, into a small number of core business processes, such as purchasing, order management, and inventory, which are made entirely standard for use throughout the enterprise; having a large number of fine-grained processes leads to tremendous overhead and inefficiency, and hence, having a small collection of coarser-grained processes that are usable in multiple scenarios is a better option.

3. *Business services*: For any process, the right business services is to subdivide it into increasingly smaller subprocesses until the process cannot be divided any further. The resulting subprocesses then become candidate indivisible (singular) business services for implementation. *Business services* automate generic business tasks that provide value to an enterprise and are part of standard business

process. The more processes that an enterprise decomposes in this way, the more commonality across these subprocesses can be achieved. In this way, an enterprise has the chance of building an appropriate set of reusable business services.

This layer relies on the orchestration interface of a collection of business-aligned services to realize reconfigurable end-to-end business processes. Individual services or collections of services that exhibit various levels of granularity are combined and orchestrated to produce *new* composite services that not only introduce new levels of reuse but also allow the reconfiguration of business processes.

The interfaces get exported as service descriptions in this layer using a service description language, such as WSDL. The service description can be implemented by a number of service providers, each offering various choices of qualities of service based on technical requirements in the areas of availability, performance, scalability, and security.

> During the exercise of defining business services, it is also important to take existing utility logic, ingrained in code, and expose it as services, which themselves become candidate services that specify not the overall business process but rather the mechanism for implementing the process. This exercise should thus yield two categories of services: business functionality services that are reusable across multiple processes and a collection of fine-grained *utility* (or *commodity*) *services*, which provide value to and are shared by business services across the organization. Examples of utility services include services implementing calculations, algorithms, and directory management services.

4. *Infrastructure services*: Infrastructure services are subdivided into technical utility services, access services, management and monitoring services, and interaction services; these are not specific to a single line of business but are reusable across multiple lines of business. They also include mechanisms that seamlessly interlink services that span enterprises. This can, for example, include the policies, constraints, and specific industry messages and interchange standards (such as the need to conform to specific industry message and interchange standards like EDIFACT, SWIFT, xCBL, ebXML BPSS, or RosettaNet) that an enterprise, say within a particular vertical marketplace, must conform to in order to work with other similar processes. *Access services* are dedicated to transforming

data and integrating legacy applications and functions into the SOA environment. This includes the wrapping and service enablement of legacy functions.

5. *Service realizations*: This layer is the component realization layer that uses components for implementing services out of preexisting applications and systems found in the operational systems layer. Components comprise autonomous units of software that may provide a useful service or a set of functionality to a client (business service) and have meaning in isolation from other components with which they interoperate.

6. *Operational systems*: This layer is used by components to implement business services and processes. Layer 6 is shown to contain existing enterprise systems or applications, including customer relationship management (CRM) and ERP systems and applications, legacy applications, database systems and applications, and other packaged applications. These systems are usually known as enterprise information systems.

15.4 Web Services

Web Services are not implemented in a monolithic manner but rather represent a collection of several related technologies. The minimum infrastructure required by the Web Services paradigm is purposefully low to help ensure that Web Services can be implemented on and accessed from any platform using any technology and programming language. The development of open and accepted standards is a key strength of the coalitions that have been developing the Web Services infrastructure:

1. *Enabling technology standards.* Although not specifically tied to any specific transport protocol, Web Services build on ubiquitous Internet connectivity and infrastructure to ensure nearly universal reach and support. For instance, at the transport level, Web Services take advantage of HTTP, the same connection protocol used by Web servers and browsers. Web Services use Extensible Markup Language (XML) as the fundamental building block for nearly every other layer in the Web Services stack.

2. *Core services standards.* The core Web Services standards comprise the baseline standards SOAP, WSDL, and UDDI:

   a. *Simple Object Access Protocol*: SOAP is a simple XML-based messaging protocol on which Web Services rely to exchange information among themselves. It is based on XML and uses common Internet transport protocols like HTTP to carry its data. SOAP implements a request/response model for communication between interacting Web Services and uses HTTP to penetrate firewalls, which are usually configured to accept HTTP and FTP service requests.

b. *Service description*: Web Services can be used effectively when a Web Service and its client rely on standard ways to specify data and operations, to represent Web Service contracts, and to understand the capabilities that a Web Service provides. To achieve this, the functional characteristics of a Web Service are first described by means of a Web Services Description Language. WSDL defines the XML grammar for describing services as collections of communicating endpoints capable of exchanging messages.

c. *Service publication*: Web Service publication is achieved by UDDI, which is a public directory that provides publication of online services and facilitates eventual discovery of Web Services. Companies can publish WSDL specifications for services they provide and other enterprises can access those services using the description in WSDL. In this way, independent applications can advertise the presence of business processes or tasks that can be utilized by other remote applications and systems. Links to WSDL specifications are usually offered in an enterprise's profile in the UDDI registry.

3. *Service composition and collaboration standards.* These include the following standards:

a. *Service composition*: This describes the execution logic of Web Services-based applications by defining their control flows (such as conditional, sequential, parallel, and exceptional execution) and prescribing the rules for consistently managing their unobservable business data. In this way, enterprises can describe complex processes that span multiple organizations—such as order processing, lead management, and claims handling—and execute the same business processes in systems from other vendors. The Business Process Execution Language (BPEL) can achieve service composition for Web Services.

b. *Service collaboration*: This describes cross-enterprise collaborations of Web Service participants by defining their common observable behavior, where synchronized information exchanges occur through their shared contact points (when commonly defined ordering rules are satisfied). Service collaboration is materialized by the Web Services Choreography Description

Language (WS-CDL), which specifies the common observable behavior of all participants engaged in business collaboration. Each participant could be implemented not only by BPEL but also by other executable business process languages.

c. Coordination/transaction standards: Solving the problems associated with service discovery and service description retrieval is the key to success of Web Services. Currently, there are attempts underway toward defining transactional interaction among Web Services. The WS-Coordination and WS-Transaction initiatives complement BPEL to provide mechanisms for defining specific standard protocols for use by transaction processing systems, workflow systems, or other applications that wish to coordinate multiple Web Services. These three specifications work in tandem to address the business workflow issues implicated in connecting and executing a number of Web Services that may run on disparate platforms across organizations involved in e-business scenarios.

There are several vendors including companies such as IBM, Microsoft, BEA, and Sun Microsystems that supply products and services across the realm of Web Services functionality and implement Web Services technology stack. These vendors are considered as platform providers and provide both infrastructure, for example, WebSphere, .NET framework, and WebLogic, for building and deploying Web Services in the form of application servers and tools for orchestration and/or composite application development for utilizing Web Services within business operations.

15.5 Quality of Service (QoS)

The QoS offered by a Web Service is becoming the highest priority for service providers and their customers. QoS refers to the ability of the Web Service to respond to expected invocations and to perform them at the level commensurate with the mutual expectations of both its provider and its customers. Several quality factors that reflect customer expectations, such as constant service availability, connectivity, and high responsiveness, become key to keeping a business competitive and viable as they can have a serious impact upon service provision. QoS thus becomes an important criterion that determines the service usability and utility, both of which influence the popularity of a particular Web Service, and an important selling and differentiating point between Web Services providers. This requires not only focusing on the functional properties of services but also concentrating on describing the environment hosting the Web Service, that is, describing the nonfunctional capabilities of services. Each service hosting environment may offer various choices of QoS based on technical requirements regarding demands for around-the-clock levels of service availability, performance and scalability, security and privacy policies, and so on, all of which must be described.

In the Web Services' context, QoS can be viewed as providing assurance on a set of quantitative characteristics. These can be defined on the basis of important functional and nonfunctional service quality properties that include implementation and deployment issues as well as other important service characteristics such as service metering and cost, performance metrics (e.g., response time), security requirements, (transactional) integrity, reliability, scalability, and availability. These characteristics are necessary requirements to understand the overall behavior of a service so that other applications and services can bind to it and execute it as part of a business process.

Web Services QoS elements can be grouped under the following three broad categories:

1. *Performance and capacity*: This category considers such issues as transaction volumes, throughput rates, system sizing, utilization levels, whether underlying systems have been designed and tested to meet these peak load requirements, and, finally, how important are request/response times.

2. *Availability*: This category considers such issues as mean time between failure for all or parts of the system, disaster recovery mechanisms, mean time to recovery, whether the business can tolerate Web Services downtime and how much, and whether there is adequate redundancy built in so that services can be offered in the event of a system or network failure.

3. *Security/privacy*: This category considers such issues as response to systematic attempts to break into a system, privacy concerns, and authentication/authorization mechanisms provided.

15.6 Summary

This chapter presented an overview of technologies on which cloud computing depends: virtualization, service-oriented architectures (SOA), and Web Services. Virtualization technology is one of the fundamental components of cloud computing, Virtualization allows the creation of a secure, customizable, and isolated execution environment for running applications without affecting other users' applications. One of the primary reasons companies implement virtualization is to improve the performance and efficiency of processing of a diverse mix of workloads. Rather than assigning a dedicated set of physical resources to each set of tasks, a pooled set of virtual resources can be quickly allocated as needed across all workloads. Virtualization provides a great opportunity to build elastically scalable systems that can provision additional capability with minimum costs. SOA is an architectural style, inspired by the service-oriented approach to computing, for enabling extensible interoperability. Much of SOA's power and flexibility derives from its ability to leverage standards-based functional services, calling them when needed on an individual basis, or aggregating them to create composite applications or multistage business processes. Although the concept of SOA is often discussed in conjunction with Web Services, these two are not synonymous. Web Services standards are key to enabling interoperability as well as key issues including quality of service (QoS), system semantics, security, management, and reliable messaging.

16

Cloudware Vendor Solutions

Cloud computing is on-demand access to a shared pool of computing resources. It helps consumers to reduce costs, reduce management responsibilities, and increase business agility. For this reason, it is becoming a popular paradigm, and increasingly more companies are shifting toward IT cloud computing solutions. Advantages are many, but being a new paradigm, there are also challenges and inherent issues. These relate to data governance, service management, process monitoring, infrastructure reliability, information security, data integrity, and business continuity.

16.1 Infrastructure as a Service (IaaS) Solutions

IaaS provides developers with on-demand infrastructure resources such as compute, storage, and communication as virtualized services in the cloud. The provider actually manages the entire infrastructure and operates data centers large enough to provide seemingly unlimited resources. The client is responsible for all other aspects of deployment, which can include the operating system itself, together with programming languages, Web servers, and applications. IaaS normally employs a pay-as-you-go model with vendors typically charging by the hour. Once connected, the developers work with the resources as if they owned them. IaaS has been largely facilitated by the advances in operating system virtualization, which enables a level of indirection or abstraction with regard to direct hardware usage. The virtual machine (VM) is the most common form for providing computational resources, and users normally get superuser access to their virtual machines. Virtualized forms of fundamental resources such as computing power, storage, or network bandwidth are provided and can be composed in order to construct new cloud software environments or applications. Virtualization enables the IaaS provider to control and manage the efficient utilization of the physical resources and allows users unprecedented flexibility in configuration while protecting the physical infrastructure of the data center. The IaaS model allows for existing applications to be directly migrated from an organization's servers to the cloud supplier's hardware, potentially with minimal or no changes to the software.

16.1.1 Amazon

In the mid-2000, Amazon introduced Amazon Web Services (AWS), based on the IaaS delivery model. In this model, the cloud service provider offers an infrastructure consisting of compute and storage servers interconnected by high-speed networks that support a set of services to access these resources. An application developer is responsible for installing applications on a platform of his or her choice and managing the resources provided by Amazon. Launched in July 2002, AWS provides online services for websites or client-side applications. Amazon S3 was launched in March 2006, and Amazon EC2 was built in August 2006 with the Amazon infrastructure and developers base available worldwide. Since then, AWS became the market leader in cloud computing, by virtue of its early entry, rapid innovation, and flexible cloud services. In June 2007, Amazon claimed that more than 330,000 developers had signed up to use AWS. As a core part of AWS, EC2 provides the computing facility for organizations and is capable of supporting a variety of applications. In November 2010, Amazon made the switch of its flagship retail website itself to EC2 and AWS.

16.1.1.1 Elastic Compute Cloud (EC2)

EC2 is a Web Service with a simple interface for launching instances of an application under several operating systems, such as several Linux distributions, Microsoft Windows Server 2003 and 2008, OpenSolaris, FreeBSD, and NetBSD.

An instance is created either from a predefined Amazon Machine Image (AMI) digitally signed and stored in S3 or from a user-defined image. The image includes the operating system, the runtime environment, the libraries, and the application desired by the user. AMIs create an exact copy of the original image but without configuration-dependent information such as the hostname or the MAC address.

A user can

1. Launch an instance from an existing AMI and terminate an instance
2. Start and stop an instance
3. Create a new image
4. Add tags to identify an image
5. Reboot an instance

EC2 is based on the Xen virtualization strategy discussed in detail in Section 15.1.3.2. In EC2, each virtual machine or instance functions as a virtual private server. An instance specifies the maximum amount of resources available to an application, the interface for that instance, and the cost per hour.

A user can interact with EC2 using a set of SOAP messages and can list available AMIs, boot an instance from an image, terminate an image, display the running instances of a user, display console output, and so on. The user has root access to each instance in the elastic and secure computing environment of EC2. The instances can be placed in multiple locations in different regions and availability zones.

EC2 allows the import of virtual machine images from the user environment to an instance through a facility called VM import. It also automatically distributes the incoming application traffic among multiple instances using the elastic load-balancing facility. EC2 associates an elastic IP address with an account; this mechanism allows a user to mask the failure of an instance and remap a public IP address to any instance of the account without the need to interact with the software support team.

16.1.1.2 Simple Storage Service (S3)

Simple storage service (S3) is a storage service designed to store large objects. It supports a minimal set of functions: write, read, and delete. S3 allows an application to handle an unlimited number of objects ranging in size from 1 byte to 5 TB. An object is stored in a bucket and retrieved via a unique developer-assigned key. A bucket can be stored in a region selected by the user. S3 maintains the name, modification time, an access control list, and up to 4 kB of user-defined metadata for each object. The object names are global. Authentication mechanisms ensure that data are kept secure; objects can be made public, and rights can be granted to other users.

S3 supports PUT, GET, and DELETE primitives to manipulate objects but does not support primitives to copy, rename, or move an object from one bucket to another. Appending to an object requires a read followed by a write of the entire object.

16.1.1.3 Elastic Block Store (EBS)

EBS provides persistent block-level storage volumes for use with Amazon EC2 instances. A volume appears to an application as a raw, unformatted, and reliable physical disk; the size of the storage volumes ranges from 1 GB to 1 TB. The volumes are grouped together in availability zones and are automatically replicated in each zone. An EC2 instance may mount multiple volumes, but a volume cannot be shared among multiple instances. The EBS supports the creation of snapshots of the volumes attached to an instance and then uses them to restart an instance. The storage strategy provided by EBS is suitable for database applications, file systems, and applications using raw data devices.

It is important to mention that an EBS volume can only be mounted to one single instance, which, in turn, must be located in the same availability zone. An EBS volume implements persistent storage, that is, it preserves the data after termination of the instance.

16.1.1.4 SimpleDB

SimpleDB is a nonrelational data store that allows developers to store and query data items via Web Service requests. It supports store-and-query functions traditionally provided only by relational databases. SimpleDB creates multiple geographically distributed copies of each data item and supports high-performance Web applications; at the same time, it automatically manages infrastructure provisioning, hardware and software maintenance, replication and indexing of data items, and performance tuning.

Amazon SimpleDB is not designed for complex database schemes or transactional properties but intended to provide simply structured, yet highly reliable data storage, which is considered to be sufficient for a wide range of applications. The database administration and optimization tasks are thus reduced to a minimum. For applications that depend on the performance and the comprehensive functionality of today's commercial relational database systems (RDBMS), Amazon RDS is the better choice (see Section 16.2.1 below).

In line with the limited functionality of SimpleDB, its interface is also restricted to a few simple Web Service calls. This should ensure both ease of learning and a user-friendly behavior:

- *CreateDomain, ListDomains, DeleteDomain*: Create, list, or delete domains. Domains correspond to the tables existing in relational databases. Each command can only address one single domain at a time.
- *DomainMetadata*: Reads metadata of a domain, such as the current storage space requirements.
- *PutAttributes*: Adds or updates a record based on a record identifier and attribute/value pairs.
- *BatchPutAttributes*: Simultaneously triggers multiple insert operations to increase the performance.
- *DeleteAttributes*: Deletes records, attributes, or values.
- *GetAttributes*: Reads an identified (partial) record.
- *Select*: Queries the database using an SQL-like syntax, but without being applied to multiple domains (as with Join).

16.1.1.5 Simple Queue Service (SQS)

SQS is a hosted message queue. SQS is a system for supporting automated workflows; it allows multiple Amazon EC2 instances to coordinate their activities by sending and receiving SQS messages. Any computer connected to the Internet can add or read messages without any installed software or special firewall configurations.

Applications using SQS can run independently and asynchronously and do not need to be developed with the same technologies. A received message is *locked* during processing; if processing fails, the lock expires and the message is available again. The time-out for locking can be changed dynamically via the ChangeMessageVisibility operation. Developers can access SQS through standards-based SOAP and query interfaces. Queues can be shared with other AWS accounts and anonymously; queue sharing can also be restricted by IP address and time of day.

The SQS interface provides the following services that are not usually started by a user by entering a command but by corresponding Web Service calls issued by the associated components:

- *CreateQueue* creates a queue in the AWS user context.
- *ListQueues* lists the existing queues.
- *DeleteQueue* deletes a queue.
- *SendMessage* places a message in a queue.
- ReceiveMessage reads one (or more) message(s) from a queue.
- *ChangeMessageVisibility* explicitly sets the visibility of a read message for other potential readers.
- *DeleteMessage* deletes a read message.
- *SetQueueAttributes* sets queue attributes, for example, the interval between two read operations of the same message.
- *GetQueueAttributes* reads queue attributes, for example, the number of messages currently in the queue.
- *AddPermission* enables shared access to a queue from multiple user contexts.
- *RemovePermission* disables the shared access by other user contexts.

The SQS message queue has special importance within the range of AWS cloud offerings because it can be used effectively to scale applications. A sender (publisher) can place messages in a queue, which can then be read out and processed by a registered recipient. To dissociate different components in an application, a service-consuming component can place its jobs as requests in the queue from where they are fetched by service-providing components. Skillful programming allows to operate critical components simultaneously on multiple EC2 instances using the path thus defined. This way, bottlenecks existing with certain components can be eliminated flexibly at runtime, and the system's overall performance is no longer limited by the bottleneck. The example in Section 4.1.7 illustrates the resulting system architecture.

16.1.1.6 CloudWatch

CloudWatch is a monitoring infrastructure used by application developers, users, and system administrators to collect and track metrics important for optimizing the performance of applications and for increasing the efficiency of resource utilization. Without installing any software, a user can monitor approximately a dozen preselected metrics and then view graphs and statistics for these metrics.

When launching an Amazon Machine Image (AMI), a user can start the CloudWatch and specify the type of monitoring. Basic Monitoring is free of charge and collects data at 5 min intervals for up to 10 metrics; Detailed Monitoring is subject to a charge and collects data at 1 min intervals. This service can also be used to monitor the latency of access to EBS volumes, the available storage space for RDS DB instances, the number of messages in SQS, and other parameters of interest for applications.

16.1.1.7 Auto Scaling

Auto Scaling exploits cloud elasticity and provides automatic scaling of EC2 instances. The service supports grouping of instances, monitoring of the instances in a group, and defining triggers and pairs of CloudWatch alarms and policies, which allow the size of the group to be scaled up or down. Typically, a maximum, a minimum, and a regular size for the group are specified.

An Auto Scaling group consists of a set of instances described in a static fashion by launch configurations. When the group scales up, new instances are started using the parameters for the RunInstances EC2 call provided by the launch configuration. When the group scales down, the instances with older launch configurations are terminated first. The monitoring function of the Auto Scaling service carries out health checks to enforce the specified policies; for example, a user may specify a health check for Elastic Load Balancing, and then Auto Scaling will terminate an instance exhibiting a low performance and start a new one. Triggers use CloudWatch alarms to detect events and then initiate specific actions; for example, a trigger could detect when the CPU utilization of the instances in the group goes above 90% and then scale up the group by starting new instances. Typically, triggers to scale up and down are specified for a group.

16.1.1.8 Elastic Beanstalk

Elastic Beanstalk, a service that interacts with other AWS services, including EC2, S3, SNS, Elastic Load Balancing, and Auto Scaling, automatically handles the deployment, capacity provisioning, load balancing, Auto Scaling, and application monitoring functions [356]. The service automatically scales the resources as required by the application, either up or down, based on the default Auto Scaling settings.

Some of the management functions provided by the service are

1. Deployment of a new application version (or rollback to a previous version)
2. Access to the results reported by CloudWatch monitoring service
3. E-mail notifications when application status changes or application servers are added or removed
4. Access to server login files without needing to log in to the application servers

The Elastic Beanstalk service is available to developers using a Java platform, the PHP server-side description language, or .NET Framework. For example, a Java developer can create the application using any Integrated Development Environment (IDE) such as Eclipse and package the code into a Java Web application archive (a file of type .war) file. The .war file should then be uploaded to the Elastic Beanstalk using the management console and then deployed, and in a short time, the application will be accessible via a URL.

16.1.1.9 *Regions and Availability Zones*

Today, Amazon offers cloud services through a network of data centers on several continents. In each region, there are several availability zones interconnected by high-speed networks; regions communicate through the Internet and do not share resources. An availability zone is a data center consisting of a large number of servers. A server may run multiple virtual machines or instances, started by one or more users; an instance may use storage services, S3, EBS, and SimpleDB, as well as other services provided by AWS. A cloud interconnect allows all systems in an availability zone to communicate with one another and with systems in other availability zones of the same region.

Storage is automatically replicated within a region; S3 buckets are replicated within an availability zone and between the availability zones of a region, whereas EBS volumes are replicated only within the same availability zone. Critical applications are advised to replicate important information in multiple regions to be able to function when the servers in one region are unavailable due to catastrophic events. A user can request virtual servers and storage located in one of the regions. The user can also request virtual servers in one of the availability zones of that region. The Elastic Compute Cloud (EC2) service allows a user to interact and to manage the virtual servers.

The billing rates in each region are determined by the components of the operating costs, including energy, communication, and maintenance costs. Thus, the choice of the region is motivated by the desire to minimize costs, reduce communication latency, and increase reliability and security.

An instance is a virtual server. The user chooses the region and the availability zone where this virtual server should be placed and selects from a limited menu of instance types: the one that provides the resources, CPU cycles, main memory, secondary storage, communication, and I/O bandwidth needed by the application. When launched, an instance is provided with a DNS. This name maps to a private IP address for internal communication within the internal EC2 communication network and a public IP address for communication outside the internal Amazon network (e.g., for communication with the user that launched the instance). Network Address Translation (NAT) maps external IP addresses to internal ones. The public IP address is assigned for the lifetime of an instance, and it is returned to the pool of available public IP addresses when the instance is either stopped or terminated. An instance can request an elastic IP address rather than a public IP address. The elastic IP address is a static public IP address allocated to an instance from the available pool of the availability zone. An elastic IP address is not released when the instance is stopped or terminated and must be released when no longer needed.

16.1.1.10 Charges for Amazon Web Services

Amazon charges a fee for EC2 instances, EBS storage, data transfer, and several other services. The charges differ from one region to another and depend on the pricing model (see http://aws.amazon.com/ec2/pricing for the current pricing structure). EC2 has flexible and multiple price models, which allow cloud consumers to reduce costs based upon workloads. The costs are calculated based on factors such as the tenant model, regions, and computing usage, instance type and operating system of the instances. The tenant model includes On-Demand, Reserved (Light, Medium, and Heavy for 1-Year, 3-Year), and Spot.

There are three pricing models for EC2 instances:

1. The On-Demand Instances model allows consumers to pay for computing capacity by the hour without long-term commitments. On-demand instances use a flat hourly rate, and the user is charged for the time an instance is running; no reservation is required for this most popular model.

2. The Reserved Instances model gives consumers the option to make a low, one-time payment for each instance they reserve and in turn receive a significant discount on the hourly charge for that instance. For reserved instances a user pays a one-time fee to lock in a typically lower hourly rate. This model is advantageous when a user anticipates that the application will require a substantial number of CPU cycles and this amount is known in advance. Additional capacity is available at a larger standard rate.

3. The Spot Instances model enables consumers to bid for Amazon EC2 computing capacity. Spot price, which varies in real time based on supply and demand. In case of spot instances, users bid on unused capacity and their instances are launched when the market price reaches the threshold specified by the user.

There are three pricing models for EC2 instances: on-demand, reserved, and spot. On-demand instances use a flat hourly rate, and the user is charged for the time an instance is running; no reservation is required for this most popular model. For reserved instances, a user pays a one-time fee to lock in a typically lower hourly rate. This model is advantageous when a user anticipates that the application will require a substantial number of CPU cycles, and this amount is known in advance. Additional capacity is available at the larger standard rate. In case of spot instances, users bid on unused capacity, and their instances are launched when the market price reaches a threshold specified by the user.

16.2 Platform as a Service (PaaS) Solutions

For many businesses, specific rather than general requirements may mean that a generic SaaS application will not suffice. PaaS creates a managed environment in the cloud where complex, tailor-made applications can be constructed, tested, and deployed. The provider supplies a hardware platform together with software environments specifically designed to support cloud application development. The target users are developers who build, test, deploy, and tune applications on the cloud platform (for the end user, the result is still a browser-based application). Automatic scalability, monitoring, and load balancing are provided so that an increase in demand for a resource such as a Web application will not result in degradation in performance (of course, this may mean an increased charge is incurred). PaaS vendors will also ensure the availability of applications, so, for example, should there be any problems on the underlying hardware, the application will be automatically redeployed to a working environment, without a detrimental effect on the end user's experience. PaaS should accelerate development and deployment and result in a shorter time to market when compared with traditional software development using an organization's own data center.

The developer does not control and has no responsibility for the underlying cloud infrastructure but does have control over the deployed applications. The vendor is responsible for all operational and maintenance aspects of the service and will also provide billing and metering information. The vendor commonly offers a range of tools and utilities to support the design, integration, testing, monitoring, and deployment of the application.

Typically, the tool set will include services to support

- Collaboration and team working
- Data management—sometimes referred to as data as a service (DaaS)
- Authentication or identification—sometimes referred to as authentication as a service (AaaS)
- Performance monitoring and management
- Testing
- Queue services
- E-mail and messaging
- User interface components

Often the supporting tools are configured as RESTful services, which can readily be composed when building applications. It is important that developers are able to write applications without needing to know the details of the underlying technology of the cloud to which they will be deployed. Standard programming languages are often available, although normally with some restrictions based on security and scalability concerns. Additional support for programming and configuration tasks is often available from a developer community specific to the PaaS offering.

16.2.1 Amazon Relational Database Service

Amazon Relational Database Service (Amazon RDS) is a PaaS that makes it easy to set up, operate, and scale a relational database in the cloud. Similar to the EC2 operating model, this service provides elastic capacities and at the same time handles time-consuming database administration tasks. Thanks to automated database backups and snapshots, the Amazon RDS is highly reliable: a database instance can be recovered for any point in time or recovery point that lies within the agreed retention period.

Amazon CloudWatch allows users to monitor the utilization of the computing and storage capacities of their database instances and to scale the available resources vertically using a simple API call as needed. In connection with highly demanding applications involving many read operations, it is possible to scale out by launching the so-called Read Replica instances. A corresponding high-availability offering allows the provisioning of synchronously replicated database instances without additional costs in multiple availability regions as a safeguard against failures at a single location. This way, it is possible to mask maintenance windows as RDS switches the database services transparently between the locations.

Amazon RDS enables access to all MySQL database functions so that it is a nobrainer to migrate existing applications while maintaining the preferred database tools and programming languages. If an existing application already uses a MySQL database, the data can be exported with mysqldump and then be piped directly into Amazon RDS. For larger databases of 1 GB or more, we recommend to create a database schema in RDS first, then convert the data into a flat file, and finally import it into the RDS instance using the mysqlimport utility. The same method can be used when exporting data from the database services.

Amazon RDS selects the optimum configuration parameters for database instances, taking the relevant computing resources and storage capacity requirements into account. However, it is also possible to change the default setting through configuration management APIs. Since RDS is implemented as a PaaS offering, it is not possible to set the database parameters by directly accessing the servers through the SSH.

For the management of its database services, Amazon not only offers command line tools and libraries for various programming languages but also a convenient web-based management console.

16.2.2 Google App Engine (GAE)

Google is also a leader in the Platform-as-a-Service (PaaS) space. App Engine is a developer platform hosted on the cloud. Initially it supported only Python, but support for Java was added later, and detailed documentation for Java is now available. The database for code development can be accessed with Google Query Language (GQL) with an SQL-like syntax.

Google App Engine is a PaaS that includes a programming environment, tool support, and an execution environment. This *instrumentation* can be used to develop Web applications for the scalable Google infrastructure. Google App Engine virtually frees Web application designers from any tasks involving server administration so that they can focus on developing the required application functionality.

GAE applications are built using three different programming languages:

1. The Python programming language was the first to appear when GAE was initially made available to developers. Python System Development Kit (SDK) is still available and widely used.

2. Java has been added together with integration with Eclipse. Using the Eclipse plug-in, developers can run and test their Java applications locally and deploy to the cloud with a single click. Google uses the Java virtual machine with the Jetty servlet engine and a standard WAR file structure. Any programming language that can run on a JVM-based interpreter, such as JRuby, Groovy, JavaScript (Rhino) and Scala, can also be run on GAE, although there may be a little more work in terms of initial configuration. GAE also supports many of the Java standards and frameworks such as servlets, the Spring Framework, and Apache Struts.

3. Go concurrent programming language developed by Google.

Java, Python, and Go runtime environments are provided with APIs (application program interface) to interact with Google's runtime environment. Google App Engine provides developers with a simulated environment to build and test applications locally with any operating system that supports a suitable version of the Python, Java, or Go language environments but with a number of important restrictions. Recently, a number of projects such as AppScale have been able to run Python, Java, and Go GAE applications on EC2 and other cloud vendors.

The GAE platform also allows developers to write code and integrate custom-designed applications with other Google services. GAE also has a number of composable supporting services including

- Integrated Web Services
- Authentication using Google Accounts
- Scalable nonrelational, schema-less storage using standard Java persistence
- models
- Fast in-memory storage using a key-value cache (memcache)
- Task queues and scheduling

Google App Engine not only supports Web application developers by providing the local runtime environment mentioned earlier and the associated automation of transfer and deployment on the Google servers. What is more, many other tools, for example, Google Plugin for Eclipse, are available for the development of Google Web Toolkit (GWT) applications. Even when it comes to the App Engine pricing system, Google has the needs of potential developers in mind: each developer benefits from free quotas for CPU load, storage use, data transfer, etc., which can be used up on a daily basis and which are usually sufficient to run developer systems and basic Web applications. If it is predictable that more resources will be needed, they can be purchased as an option. Prices for commercial use are comparable to those for the Amazon Web Services.

16.2.3 Google Cloud Print

Google Cloud Print is another interesting cloud computing concept. This service allows any application to print to any output device on the Internet. Especially users of mobile end devices might greatly benefit from this option. While modern Internet-enabled devices, such as notebooks, touchpads, and mobile phones, are becoming more and more widespread, it is often difficult or even impossible to set up a local printer that can be used by these devices. The lack of suitable printer drivers, the partly insufficient device resources, and the variety of operating systems in use add to this issue.

With Google Cloud Print, print jobs are sent to a service that directly forwards them to a Cloud Print–compatible network printer; for this purpose, a special authorization or accounting procedure is used. As long as a printer is connected to the Internet, the service can be set up for worldwide use. If the printer is incompatible with the Cloud Print technology, it must be connected to a server where a suitable proxy is installed. In this case, the print job data are first converted to a compatible printer language. For this purpose, the required set of printer drivers is made available by the Cloud Print provider. Then the print job is transmitted to the server that finally forwards the preprocessed print jobs to the selected printer. Google Cloud Print represents an unconstrained standard that can be implemented freely by the industry. Many manufacturers, such as HP, meanwhile offer a number of low-priced, compatible printers.

This development has the potential to allow a broad range of service offerings to spread around this technology, including services that go far beyond the simple replacement of centralized print job management in companies and universities. In the future, university students will have the possibility, for example, to directly upload their lecture notes from their touchpad to a print provider's site to have them printed and bound. Due to economies of scale, the provider is able to offer this service at a very good price, including free delivery. The standardized interface provides the ability to greatly automate and streamline certain tasks, particularly the cross-company processing of bulk mail, quality prints, etc. What is more, a new distribution model for publishing houses is beginning to materialize: The installation of machines in the public on which—controlled by a mobile device—electronic magazines, newspapers, or books can be printed on the spot.

16.2.4 Windows Azure

Windows Azure is Microsoft's cloud computing platform for the execution of software in Microsoft's data centers. Windows Azure is an operating system, SQL Azure is a cloud-based version of the SQL Server, and Azure AppFabric (formerly .NET Services) is a collection of services for cloud applications.

Windows Azure has three core components:

1. Compute, which provides a computation environment.
2. Storage for scalable storage.
3. Fabric controller, which deploys, manages, and monitors applications; it interconnects nodes consisting of servers, high-speed connections, and switches.

The Windows Azure platform comprises a compute service for running applications, a storage service for storing data, and an SQL service for providing highly available relational databases in the cloud. The storage service can be used to store large objects containing text or binary data. Based on this service, Windows Azure Drive allows to format a binary data object to be used as an NTFS volume. A queue service ensures a reliable data exchange between the components. Finally, the virtual network service constitutes the basis for a transparent communication between local and remote resources. It can be used to integrate Windows Azure services into a local Active Directory. In addition, the platform includes the Windows Azure AppFabric that uses secure connectivity to bridge traditional IT systems to cloud applications. With the Azure service platform, software products can be installed as cloud services on the Internet or alternatively as applications in the in-house data center. The two methods can also be combined to implement a flexibly scalable hybrid cloud. For this purpose, Web and business developers have the choice of a variety of established tools, such as Microsoft.NET, Visual Studio, or many other products that are available as commercial or open-source software. Applications can be developed and tested locally before they are finally uploaded and published to the Azure cloud.

Scaling, load balancing, memory management, and reliability are ensured by a fabric controller, a distributed application replicated across a group of machines that owns all of the resources in its environment—computers, switches, load balancers—and it is aware of every Windows Azure application.

The fabric controller decides where new applications should run; it chooses the physical servers to optimize utilization using configuration information uploaded with each Windows Azure application. The configuration information is an XML-based description of how many Web role instances, how many worker role instances, and what other resources the application needs. The fabric controller uses this configuration file to determine how many VMs to create.

Blobs, tables, queues, and drives are used as scalable storage. A blob contains binary data; a container consists of one or more blobs. Blobs can be up to a terabyte, and they may have associated metadata (e.g., the information about where a JPEG photograph was taken). Blobs allow a Windows Azure role instance to interact with persistent storage as though it were a local NTFS6 file system. Queues enable Web role instances to communicate asynchronously with worker role instances.

The Content Delivery Network (CDN) maintains cache copies of data to speed up computations. The connect subsystem supports IP connections between the users and their applications running on Windows Azure. The API to Windows Azure is built on REST, HTTP, and XML. The platform includes five services: Live Services, SQL Azure, AppFabric, SharePoint, and Dynamics CRM. A client library and tools are also provided for developing cloud applications in Visual Studio.

The computations carried out by an application are implemented as one or more roles; an application typically runs multiple instances of a role. For the use of Windows Azure services, three different roles have been defined:

1. The Web role supports the development and execution of Web applications with Internet Information Server 7.
2. The worker role provides supporting services to the Web role, for example, in multitier applications.
3. The VM role allows the execution of a virtual machine on Windows Server 2008 R2. The image is stored on a virtual hard disk in the Azure storage service. This role further provides functionality for granting privileges and using remote desktop connections.

The usage-oriented Windows Azure pricing model is based on what is actually used so that IT solutions can be deployed without up-front investment in hardware and software and more compute power can be added later as required. Pricing is in line with Amazon's or Google's model (Table 16.1).

The Microsoft Azure platform currently does not provide or support any distributed parallel computing frameworks, such as MapReduce, Dryad, or MPI, other than the support for implementing basic queue-based job scheduling.

Windows Azure was announced in October 2008 followed by SQLAzure Relational Data's announcement in March 2009. In October 2010, full IIS support was added to Windows Azure.

The general characteristics of Windows Azure are as follows:

- *Model*: Windows Azure provides both PaaS and IaaS types of cloud services.

- *Compatibility*: Several Microsoft products have been integrated with Windows Azure to help cloud consumers better use and manage cloud services. For example, consumers can use the Microsoft SQL Server2 to access and operate their SQL Azure database.

- *Deployment and interface*: The Windows Azure interface includes a management portal and powerful command-line tools. The management portal and command-line tools are easy to use, and help cloud consumers and end users manage applications and cloud services. In addition, several Microsoft products (e.g., Microsoft Visual

TABLE 16.1

Fees for Windows Azure Compute Resources (USD per h)

Instance Type	CPU (GHz)	Memory	Disk (GB)	I/O	Costs
Extra small	1.0	768 MB	20	Low	0.05
Small	1.6	1.75 GB	225	Moderate	0.12
Medium	2 × 1.6	3.5 GB	490	High	0.24
Large	4 × 1.6	7 GB	1000	High	0.48
Extra large	8 × 1.6	14 GB	2040	High	0.96

Studio and Microsoft WebMatrix) can help deploy and manage user applications and cloud resources. Windows Azure also has a REST-based service API for managements and deployments.

- *Hypervisor*: Microsoft does not provide Windows Azure Hypervisor as a single product for cloud consumers or end users.
- *Reliability*: Windows Azure guarantees a 99.95% computing reliability and a 99.9% role instance and storage reliability.
- *OS Supports*: Windows Azure currently supports both Windows and Linux Operation Systems for Virtual Machines (VMs).
- *Scalability*: In Windows Azure, computing resources can be separated from storage to achieve independent scalability of computing and storage. This mechanism also provides isolation and multitenancy. In 2012, the Windows Azure team built a new flat network system that could create Flat Network Storage (FNS) crossing all Windows Azure data centers. The FNS system resulted in bandwidth improvements of network connectivity to support Windows Azure VMs. The new networking design better supports high-performance computing (HPC) applications that require massive communications and significant bandwidth between computing nodes.
- *Cost*: Windows Azure does not have upfront costs. The cost for each component is based on the cloud configuration. Windows Azure payment plans include pay-as-you-go, 6-month plans, or 12-month plans. Microsoft provides an official cost calculator2 for Windows Azure so that users can easily estimate the cost.

Cost is calculated based on computing usage. The cost includes

- Website cost
- Virtual machines cost
- Cloud service cost
- Mobile service cost
- Data management cost

16.3 Software as a Service (SaaS) Solutions

Software as a Service (SaaS) is a hosted application that is available over the Internet via a Web browser. SaaS, sometimes referred to as *on-demand software*, is the most complete of the cloud services. Computing hardware, software, and the solution are offered by a vendor. The cloud application layer is the only layer visible to end users who are the target users for this layer.

The end user does not manage or control the underlying infrastructure. Their only responsibility is for entering and managing their data based on interactions with the software. The user interacts directly with the hosted software via the browser. We should note that SaaS existed well before the concept of cloud computing emerged; nevertheless, it is now an integral part of the cloud model.

Creating and delivering software via the SaaS layer is an attractive alternative to the more traditional desktop applications, which must be installed on the user's machine. With SaaS, the application is deployed in the cloud, so the work of testing, maintaining, and upgrading software is greatly simplified since it can all occur in one place rather than being rolled out to the desktops of potentially thousands of users. Configuration testing is reduced in complexity due to centralization and the preset restrictions in the deployment environment. Developers can also use a simplified strategy when applying upgrades and fixes. Furthermore, composition, as discussed previously, becomes a straightforward option as soon as the cloud services are developed. Last but not least, the providers also benefit from greater protection to their intellectual property as the application is not deployed locally and pirated versions of the software will be much harder to obtain and distribute.

A number of typical characteristics of SaaS are listed as follows:

- Software is available globally over the Internet either free or paid for by subscription based on customer usage.
- Collaborative working is easily provided and generally encouraged.
- Automatic upgrades are handled by the vendor with no required customer input.
- All users have the same version of the software.
- The software will automatically scale on demand.
- Distribution and maintenance costs are significantly reduced.

There are a huge variety of SaaS applications already available, and their number appears to be growing at an exponential rate. A small selection of prominent examples of SaaS is discussed later, which hopefully gives a good illustration of the SaaS approach. We recommend that you investigate for yourself by viewing the sites mentioned, even if only briefly.

16.3.1 Google

Services such as Gmail, Google Drive, Google Calendar, Picasa, and Google Groups are free of charge for individual users and available for a fee for organizations. These services are running on a cloud and can be invoked from a broad spectrum of devices, including mobile ones, such as iPhones, iPads, and Blackberrys; laptops; and tablets. The data for these services are stored in data centers on the cloud.

The Gmail service hosts e-mails on Google servers and provides a Web interface to access them and tools for migrating from Lotus Notes and Microsoft Exchange. Google Docs is a web-based software for building text documents, spreadsheets, and presentations. It supports features such as tables, bullet points, basic fonts, and text size; it allows multiple users to edit and update the same document and view the history of document changes; and it provides a spell checker. The service allows users to import and export files in several formats, including Microsoft Office, PDF, text, and OpenOffice extensions.

Google Calendar is a browser-based scheduler; it supports multiple calendars for a user, the ability to share a calendar with other users, the display of daily/weekly/monthly views, and the ability to search events and synchronize with the Outlook Calendar. Google Calendar is accessible from mobile devices. Event reminders can be received via SMS, desktop pop-ups, or e-mails. It is also possible to share your calendar with other Google Calendar users. Picasa is a tool to upload, share, and edit images; it provides 1 GB of disk space per user free of charge. Users can add tags to images and attach locations to photos using Google Maps. Google Groups allows users to host discussion forums to create messages online or via e-mail.

16.3.2 Salesforce.com

Salesforce.com is a leading cloud provider for customer relationship management (CRM) software. In the sense of Chapter 3, Salesforce.com is an SaaS offering that has been complemented by a PaaS offering where independent third parties can develop and offer add-on software. The Salesforce.com portfolio consists of four major parts:

1. The central part is a CRM SaaS offering called Salesforce. It provides a web-based solution for sales, marketing, customer service, partner management, and others. The central component is available in several bundles that reflect different capabilities and numbers of users; it is usually paid for on a monthly or yearly basis. As a classic SaaS offering, this multitenant application runs on the Salesforce.com servers and does not have to be installed locally.

2. Force.com is the name of a PaaS offering that allows customers or independent software vendors (ISVs) to develop their own web-based business applications and run them on the salesforce.com infrastructure. For more technical details on Force.com, see the following discussion.

3. The business applications developed on Force.com can be obtained on the AppExchange marketplace: the choice includes free and paid apps. Via Force.com, the applications are preintegrated with the Salesforce.com CRM, and their functionality often complements the latter or is fine-tuned for particular industries.

4. Both Force.com and Salesforce.com are accompanied by organized user communities named Developer Force and Salesforce.com Community, which provide both user networking and professional consulting services offered by Salesforce.com and their partners.

The Force.com platform includes development tools and a programming environment suitable, for example, to develop the logic of an application in Apex, a programming language with a Java-like syntax. Other features are the support of user interface development with Visualforce, the integration of software testing procedures, or the connection of external Web Services through a dedicated API.

For the development of custom Web applications, a programming model with a highly data-centric approach is used: The development of an application usually starts with the creation of an object model that will later hold the application data. For the data elements, constraints can be defined that improve the data quality. Just like workflows and acceptance processes, these two initial steps can be defined directly from within the development environment, using the available data. Use of the Apex programming language is only required if a more complex application logic must be implemented. The developer tools are not only helpful in the essential steps mentioned earlier required to meet functional and nonfunctional requirements but also in the context of the organizational workflows that are necessary to pack the developed application and deploy it as an offering to potential customers. Developers who wish to learn how to write programs on Force.com can obtain comprehensive material in the form of tutorials, manuals, reference documents, and code samples on the web-based community portals.

16.4 Open Source Cloud Solutions

In addition to commercial cloud services, many open-source cloud computing solutions can be flexibly tailored to build private cloud services for the specific demands of a user. This chapter introduces four major open-source cloud computing solutions: Nimbus, OpenNebula, Eucalyptus, and CloudStack.

16.4.1 Nimbus

Started in 2003 and developed by the Argonne National Laboratory at the University of Chicago, Nimbus focuses on the computing demands of scientific users. Nimbus has two variations: Nimbus Infrastructure and Nimbus Platform. Nimbus Infrastructure provides solutions for IaaS and has been designed to support the needs of data-intensive scientific research projects.

Nimbus Platform is a set of tools that assists consumers to leverage IaaS cloud computing. The toolset comprises functions for application installation, configuration, monitoring, and repair. Nimbus supports Xen and KVM hypervisors (see Section 15.1.3 "Virtual Machines"). Nimbus Platform enables consumers to create hybrid clouds of Nimbus Infrastructure, Amazon AWS, and other clouds.

When a consumer is subscribed to Nimbus service, a virtual workspace is created. The workspace comprises the front end, the workspace service, the back end, and the VM workspace. The Virtual Machine (VM) workspace is deployed onto the Virtual Machine Monitor (VMM) node, which is a physical node. Once the deployment has been done, consumers can access the cloud service node via the HTTP interface. Cumulus is a crucial component of Nimbus, serving as the front end to the Nimbus VM image repository. Any VM image must be loaded into the Cumulus repository before booting.

The general characteristics of Nimbus are as follows:

- *Cloud model*: Nimbus is a solution for IaaS.
- *Compatibility*: Cumulus storage extends the Amazon S3 REST API, and is S3 compatible.
- *Deployment and interface*: Users directly interact with VMs in the node pool in almost the same way as interacting with a physical machine. Nimbus publishes information about the VM such as the IP address of each VM so that users can know information about each VM. Users deploy applications to Nimbus clouds by using a cloudkit configuration that includes a manager service hosting and an image repository.
- *Hypervisors*: Nimbus supports KVM and Xen.
- *Reliability*: To achieve the same level of reliability as S3, the hardware configuration of Cumulus needs to be at the same level as S3, that is, the reliability of Nimbus partially depends on the hardware infrastructure the Cumulus builds on.
- *OS support*: Nimbus supports various Linux distributions.
- *Scalability*: The Cumulus Redirection module of Nimbus manages scalability by keeping track of the workload of the service. As Cumulus is compatible with the Amazon S3 REST API, it can be configured to run as a set of replicated hosts to support horizontal scalability.
- *Cost*: Nimbus is a free, open-source solution.

16.4.2 OpenNebula

OpenNebula was started as a research project in 2005 and the first version was released in 2008. The latest version of OpenNebula, released in 2013, is 3.8.3. OpenNebula can use multiple hypervisors such as VMware ESXi, Xen, and KVM. It is capable of managing private, public, or hybrid IaaS clouds. It can flexibly adapt cloud interfaces from EC2 Query, OGF's Open Cloud Computing Interface (OCCI), and vCloud.

OpenNebula is designed to provide a solution for building enterprise-level data centers and IaaS clouds. Its modular-based architecture allows cloud builders to configure and implement a diverse range of cloud services while maintaining a high level of stability and quality. OpenNebula includes a core module, a set of plug-in drivers, and multiple tools. The core module manages and monitors virtual resources such as VMs, virtual networks, virtual storage, and images. It also handles client requests and invokes corresponding drivers to perform operations on resources. The plug-in drivers serve as adapters to interact with middleware. Core functions are exposed to end users through a set of tools and APIs including REST API (e.g., EC2-Query API), the OpenNebula Cloud API (OCA), and APIs for native drivers, for example, for connecting to AWS.

The general characteristics of OpenNebula are as follows:

- *Cloud model*: Designed to support IaaS clouds by leveraging existing infrastructure; OpenNebula usually does not have specific requirements on the infrastructure.
- *Compatibility*: OpenNebula can be deployed to existing infrastructure and integrated with various cloud services.
- *Deployment and interfaces*: Interfaces are available for cloud consumers and providers. Cloud providers can also develop customized tools with cloud interfaces. Consumers can use either the Command Line Interface (CLI) or the SunStone Web Portal to perform most operations, especially the management of resources. In addition, the latest release provides interfaces to cloud providers such as Amazon.
- *Hypervisors*: OpenNebula supports three major hypervisors: KVM, Xen, and VMware. Because the hypervisor driver can be shifted between different hypervisors, it provides a solution for a multihypervisor platform.
- *Reliability*: The OpenNebula system has designed a specialized quality check module, OpenNebula QA, to ensure the quality of every release.
- *OS support*: All major Linux and Windows versions are supported.
- *Scalability*: OpenNebula has been employed in building large-scale infrastructure as well as highly scalable databases; virtualization drivers can be adjusted to achieve maximum scalability.
- *Cost*: While OpenNebula is a completely free solution, its enterprise version, OpenNebulaPro, is distributed on an annual subscription basis.

OpenNebula is one of the most popular open-source cloud computing solutions because of its focus on openness, excellence, cooperation, and innovation; cloud providers and consumers especially value the openness that allows an appreciable degree of customization to applications. OpenNebula has been widely deployed for enterprise-level private clouds and data centres including IBM Global Business Services, High Performance Computing (HPC), science communities such as NASA, and also cloud integrators and product developers like KPMG.

16.4.3 Eucalyptus

Eucalyptus (Elastic Utility Computing Architecture Linking Your Programs to Useful Systems) was developed by researchers at the University of California, Santa Barbara. The first version of the software was released in 2008. The latest version is Eucalyptus 3.2.11 released in February 2013. Eucalyptus supports enterprise private and hybrid cloud computing. It also supports hybrid cloud services by integrating Amazon Web Services (AWS) API.

Eucalyptus provides an IaaS solution to build private or hybrid clouds. By virtualization of physical machines in the data center, cloud providers can provide virtualized computer hardware resources, including computing, network, and storage to cloud consumers. Consumers can access the cloud through command-line tools (like euca2ools) or through a web-based dashboard. Euca2ools enable interaction with Web services compatible with Eucalyptus and Amazon cloud services. Eucalyptus also supports AWS-compatible APIs on top of Eucalyptus for consumers to communicate with AWS.

The general characteristics of Eucalyptus are as follows:

- *Cloud model*: Eucalyptus provides access to collections of virtualized computer hardware resources for computing, network, and storage to provide IaaS cloud services. Users can assemble their own virtual cluster where they can install, maintain, and execute their own application stack using Eucalyptus.
- *Compatibility*: Eucalyptus is compatible with Amazon AWS cloud services; it provides compatibility with a range of AWS features including Amazon EC2, AMI, Amazon S3, Amazon IAM, and Amazon EBS.
- *Deployment and interface*: Eucalyptus supports the Amazon AWS APIs for EC2 and S3.
- *Hypervisors*: Eucalyptus is compatible with Xen, KVM, and VMware Hypervisors
- *Reliability*: Eucalyptus 3 can be deployed with high availability as it has improved the reliability of the IaaS cloud using automatic failover and failback mechanisms

- *OS support*: Eucalyptus 3.2 supports Windows Server 2003 and 2008, Windows 7, and all modern Linux distributions such as RedHat.
- *Scalability*: Eucalyptus supports scalability starting with Eucalyptus 2.0 at two levels: front-end transactional scalability and back-end resource scalability
- *Cost*: Users can choose between the open-source free Eucalyptus Cloud and the priced Eucalyptus Enterprise Cloud.

IT-savvy organizations across the globe run Eucalyptus clouds for their agility, elasticity, and scalability required by highly demanding applications; these include NASA, Cornell University, George Mason University, and so on.

16.4.4 CloudStack

First released in 2010, CloudStack was initially developed by cloud.com, a start-up supported by venture capital. Citrix acquired cloud.com in late 2011, and donated CloudStack to the Apache Software Foundation in 2012. Becoming a subproject of Apache Software Foundation dramatically increased the development speed of CloudStack and solidified its leadership in open-source cloud solutions. Currently, CloudStack is licensed under Apache License, Version 2. CloudStack can also use multiple hypervisors such as VMware ESXi, Xen, and KVM. In addition to its own API, it also implements the Amazon EC2 and S3 APIs to support interoperability with Amazon cloud services. CloudStack is designed to manage large networks of Virtual Machines (VMs) for enabling a highly available and scalable IaaS platform. It includes the entire "stack" of features, which most organizations expect within an IaaS cloud.

The CloudStack architecture envisions an IaaS platform with computing, network, and storage resources managed through a shared architecture comprising of at least one hypervisor solution. This model provides a multitenant mode for supporting tenancy abstraction ranging from departmental to public cloud reseller modes; it encompasses core functions such as the user interface and image management, and allows cloud providers to provide advanced services such as high availability and load balancing. All services are tied together through a series of Web service APIs, which enable CloudStack to support the unique needs of consumers.

The general characteristics of CloudStack are as follows:

- *Cloud model*: CloudStack is a solution for IaaS-level cloud platforms. It pools, manages, and configures network, storage, and computing nodes to support public, private, and hybrid IaaS clouds.
- *Compatibility*: In addition to its own APIs, CloudStack is compatible with Amazon EC2 and S3 APIs as well as the vCloud APIs for hybrid cloud applications.

- *Deployment and interface*: Users can manage their clouds with a user-friendly Web-based interface, command-line tools, or through a RESTful API. CloudStack provides a feature-rich out-of-the-box user interface that implements the CloudStack API to manage the infrastructure; it is an AJAX-based solution compatible with most Web browsers. A snapshot view of the aggregated storage, IP pools, CPU, memory, and other resources in use gives users comprehensive status information about the cloud.

- *Hypervisors*: CloudStack is compatible with a variety of hypervisors including VMware vSphere, KVM, Citrix XenServer, and Xen Cloud Platform (XCP).

- *Reliability*: CloudStack is a highly available and scalable IaaS solution, in that no single component failure can cause cluster or cloud-wide outage. It enables downtime-free management for server maintenance and reduces the workload of managing a large-scale cloud deployment.

- *OS support*: CloudStack supports Linux for the management server and for underlying computing nodes. Depending on the employed hypervisor, CloudStack supports a wide range of guest operating systems including Windows, Linux, and various versions of Berkeley Software Distributions (BSDs).

- *Scalability*: CloudStack is capable of managing large numbers of servers across geographically distributed data centres through a linearly scalable and centralized management server. This capability eliminates the need for intermediate cluster-level management servers. It supports integration with both software and hardware firewalls and load balancers to provide additional security and scalability to a user's cloud environment (such as F5 load balancer and Netscaler).

- *Cost*: CloudStack itself is a free software licensed under the Apache License; however, using a commercial hypervisor will incur corresponding costs.

CloudStack is one of the leading open-source cloud solutions with its rich functionality and user-friendly Web-based graphical user interface. Many leading telecommunication companies are using Citrix Cloud Platform powered by Apache CloudStack, including British Telecom (BT), Nippon Telegraph and Telephone (NTT), Tata Communications, and Korea Telecom.

16.4.5 Apache Hadoop

Hadoop is an open-source software platform that allows to easily process and analyze very large data sets in a computer cluster. Hadoop can, for example, be used for Web indexing, data mining, log file analyses, machine learning, finance analyses, scientific simulations, or research in the bioin-formatics field.

The general characteristics of the Hadoop system are as follows:

- *Scalability*: It is possible to process data sets with a volume of several petabytes (PB) by distributing them to several thousand nodes of a computer cluster.
- *Efficiency*: Parallel data processing and a distributed file system allow to manipulate the data quickly.
- *Reliability*: Multiple copies of the data can be created and managed. In case a cluster node fails, the workflow reorganizes itself without user intervention. Hence, automatic error-correction is possible.

Hadoop has been designed with scalability in mind so that cluster sizes of up to 10,000 nodes can be realized. The largest Hadoop cluster at Yahoo! currently comprises 32,000 cores in 4,000 nodes, where 16 PB of data are stored and processed. It takes about 16 h to analyze and sort a 1 PB data set on this cluster.

16.4.5.1 MapReduce

Hadoop implements the MapReduce programming model, which is also of great importance in the Google search engine and applications (see Section 17.3 Hadoop). Even though the model relies on massive parallel processing of data, it has a functional approach.

In principle, it has two functions:

1. *Map function*: Reads key/value pairs to generate intermediate results, which are then output in the form of new key/value pairs.
2. *Reduce function*: Reads all intermediate results, groups them by keys, and generates aggregated output for each key.

Usually, the procedures generate lists or queues to store the results of the individual steps. As an example, let is look at how the vocabulary in a text collection can be acquired: The Map function extracts the individual words from the text, the Reduce function reads them, counts the number of occurrences, and stores the result in a list. In parallel processing, Hadoop distributes the texts or text fragments to the available nodes of a computer cluster. The Map nodes process the fragments assigned to them and output the individual words. These outputs are available to all nodes via a distributed file system. The Reduce nodes then read the word lists and count the number of words. Since counting can only start after all words have been processed by the Map function, a bottleneck might arise here.

16.5 Summary

This chapter details some of the prominent cloud vendor services. For the IaaS delivery model, we describe the various cloud services provided by Amazon, including Elastic Compute Cloud (EC2), Simple Storage System (S3), Elastic Block Store (EBS), Simple DB, Simple Queue Service (SQS), CloudWatch, Auto Scaling, and Elastic Beanstalk. For the PaaS delivery model, we discuss the examples of Amazon Relational Database Service, Google Apps Engine (GAE), Google Cloud Print, and Windows Azure. For the SaaS delivery model, we describe cloud services from Google and Salesforce.com. Finally, we describethe open source cloud solutions, namely, Nimbus, OpenNebula, Eucalyptus, OpenStack, and, Apache Hadoop.

17

Cloudware Application Development

17.1 Reliability Conundrum

The cloud, with its tendency to use commodity hardware and virtualization and with the potential for enormous scale, presents many additional challenges to designing reliable applications. In all engineering disciplines, reliability is the ability of a system to perform its required functions under stated conditions for a specified period of time. In software, for application reliability, this becomes the ability of a software application and all the components it depends on (operating system, hypervisor, servers, disks, network connections, power supplies, etc.) to execute without faults or halts all the way to completion. But completion is defined by the application designer. Even with perfectly written software and no detected bugs in all underlying software systems, applications that begin to use thousands of servers will run into the mean time to failure in some piece of hardware, and some number of those instances will fail. Therefore, the application depending on those instances will also fail.

Many design techniques for achieving high reliability depend upon redundant software, hardware, and data. For redundant software components, this may consist of double- or triple-redundant software components (portions of your application) running in parallel with common validation checks. One idea is to have the components developed by different teams based on the same specifications. This approach costs more, but extreme reliability may require it. Because each component is designed to perform the same function, the failures of concurrent identical components are easily discovered and corrected during quality-assurance testing.

Although redundant software components provide the quality-assurance process with a clever way to validate service accuracy, certain applications may want to deploy component redundancy into the production environment. In such conditions, multiple parallel application processes can provide validity checks on each other and let the majority rule. Although the redundant software components consume extra resource consumption, the trade-off between reliability and the cost of extra hardware may be worth it.

Another redundancy-based design technique is the use of services such as clustering (linking many computers together to act as a single faster computer), load balancing (workloads kept balanced between multiple computers), data replication (making multiple identical copies of data to be processed independently and in parallel), and protecting complex operations with transactions to ensure process integrity. Naturally, when one is using cloud provider services, many of these services are inbuilt in the base infrastructure and services.

Redundant hardware is one of the most popular strategies for providing reliable systems. This includes redundant arrays of independent disks (RAID) for data storage, redundant network interfaces, and redundant power supplies. With this kind of hardware infrastructure, individual component failures can occur without affecting the overall reliability of the application. It is important to use standardized commodity hardware to allow easy installation and replacement.

In 2008, Google had to solve the massive search problem across all content on the Web, which was bordering on one trillion unique URLs. They ended up employing loosely coupled distributed computing on a massive scale: clusters of commodity (cheap) computers working in parallel on large data sets. Even with individual server with excellent reliability statistics, with hundreds of thousands of servers, there were still multiple failures per day as one machine or another reached its mean time to failure. Google had no choice but to give up on reliability of the hardware and switch things over to achieve the same with the reliability of the software. The only way to build a reliable system across a group of large number of unreliable computers is to employ suitable software to address those failures. MapReduce was the software framework invented by Google to address this issue; the name MapReduce was inspired by the *map* and *reduce* functions of the functional programming language Lisp.

Parallel programming on a massive scale has the potential to not only address the issue of reliability but also deliver a huge boost in performance. This is opportune because, given the problems with large data sets of the Web, without massive parallelism, leave aside reliability, the processing itself may not be achievable.

17.1.1 Functional Programming Paradigm

The functional programming paradigm treats computation as the evaluation of mathematical functions with zero (or minimal) maintenance of states or data updates. As opposed to procedural programming in languages such as C or Java, it emphasizes that the application be written completely as functions that do not save any state. Such functions are called pure functions. This is the first similarity with MapReduce abstraction. All input and output values are passed as parameters, and the map and reduce functions are not expected to save state. However, the values can be input and output from a file system or a database to ensure persistence of the computed data. Programs written using pure functions eliminate side effects. So the output of a pure function depends solely on the inputs provided to it. Calling a pure function twice with the same value for an argument will give the same result both times. Lisp is one such popular functional programming language where two powerful recursion schemes—called map and reduce—enable powerful decomposition and reuse of code.

17.1.1.1 Parallel Architectures and Computing Models

MapReduce provides a parallel execution platform for data parallel applications. This and the next section describe core concepts involved in understanding such systems.

17.1.1.1.1 Flynn's Classification

Michael J. Flynn in 1966 created a taxonomy of computer architectures that support parallelism based on the number of concurrent control and data streams the architecture can handle. This classification is used extensively to characterize parallel architectures. They are briefly described here:

1. *Single instruction, single data (SISD) stream*: This is a sequential computer that exploits no parallelism, like a PC (single core).
2. *Single instruction, multiple data (SIMD) stream*: This architecture supports multiple data streams to be processed simultaneously by replicating the computing hardware. Single instruction means that all the data streams are processed using the same compute logic. Examples of parallel architectures that support this model are array processors or Graphics Processing Unit (GPU).
3. *Multiple instruction, single data (MISD) stream*: This architecture operates on a single data stream but has multiple computing engines using the same data stream. This is not a very common architecture and is sometimes used to provide fault tolerance with heterogeneous systems operating on the same data to provide independent results that are compared with each other.
4. *Multiple instruction, multiple data (MIMD) stream*: This is the most generic parallel processing architecture where any type of distributed application can be programmed. Multiple autonomous processors executing in parallel work on independent streams of data. The application logic running on these processors can also be very different. All distributed systems are recognized to be MIMD architectures.

A variant of SIMD is called single program, multiple data (SPMD) model, where the same program executes on multiple compute processes.

While SIMD can achieve the same result as SPMD, SIMD systems typically execute in lock step with a central controlling authority for program execution. As can be seen, when multiple instances of the map function are executed in parallel, they work on different data streams using the same map function. In essence, though the underlying hardware can be a MIMD machine (a compute cluster), the MapReduce platform follows a SPMD model to reduce programming effort. Of course, while this holds for simple use cases, a complex application may involve multiple phases, each of which is solved with MapReduce—in which case the platform will be a combination of SPMD and MIMD.

17.1.1.2 Data Parallelism versus Task Parallelism

Data parallelism is a way of performing parallel execution of an application on multiple processors. It focuses on distributing data across different nodes in the parallel execution environment and enabling simultaneous subcomputations on these distributed data across the different compute nodes. This is typically achieved in SIMD mode (Single Instruction Multiple Data mode) and can either have a single controller controlling the parallel data operations or multiple threads working in the same way on the individual compute nodes (SPMD). In contrast, task parallelism focuses on distributing parallel execution threads across parallel computing nodes. These threads may execute the same or different threads. These threads exchange messages either through shared memory or explicit communication messages, as per the parallel algorithm. In the most general case, each of the threads of a Task Parallel system can be doing completely different tasks but coordinating to solve a specific problem. In the most simplistic case, all threads can be executing the same program and differentiating based on their node IDs to perform any variation in task responsibility. Most common Task Parallel algorithms follow the master–worker model, where there is a single master and multiple workers. The master distributes the computation to different workers based on scheduling rules and other task-allocation strategies. MapReduce falls under the category of data-parallel SPMD architectures.

Due to the functional programming paradigm used, the individual mapper processes processing the split data are not aware (or dependent) upon the results of the other mapper processes. Also, since the order of execution of the mapper function does not matter, one can reorder or parallelize the execution. Thus, this inherent parallelism enables the mapper function to scale and execute on multiple nodes in parallel. Along the same lines, the reduce functions also run in parallel; each instance works on a different output key. All the values are processed independently, again facilitating implicit data parallelism. The extent of parallel execution is determined by the number of map and reduce tasks that are configured at the time of job submission.

17.2 Google MapReduce

The MapReduce architecture and programming model pioneered by Google is an example of a modern system architecture designed for processing and analyzing large data sets and is being used successfully by Google in many applications to process massive amounts of raw Web data. The MapReduce system runs on top of the Google File System, within which data are loaded and partitioned into chunks and each chunk is replicated. Data processing is colocated with data storage: when a file needs to be processed, the job scheduler consults a storage metadata service to get the host node for each chunk and then schedules a *map* process on that node, so that data locality is exploited efficiently.

> Google engineers designed MapReduce to solve a specific practical problem. Therefore, it was designed as a programming model combined with the implementation of that model—in essence, a reference implementation. The reference implementation was used to demonstrate the practicality and effectiveness of the concept and to help ensure that this model would be widely adopted by the computer industry. Over the years, other implementations of MapReduce have been created and are available as both open source and commercial products.

The MapReduce architecture allows programmers to use a functional programming style to create a map function that processes a key–value pair associated with the input data to generate a set of intermediate key–value pairs and a reduce function that merges all intermediate values associated with the same intermediate key.

Users define a map and a reduce function:

1. The map function processes a (key, value) pair and returns a list of intermediate (key, value) pairs:

   ```
   map (in _ key,in _ value)- > list(out _ key,
   intermediate _ value).
   ```

2. The reduce function merges all intermediate values having the same intermediate key:

   ```
   reduce (out _ key, list(intermediate _ value)) → list
   (out _ value).
   ```

The former processes an input key–value pair, producing a set of intermediate pairs. The latter is in charge of combining all of the intermediate values related to a particular key, outputting a set of merged output values (usually just one). MapReduce is often explained illustrating a possible solution to the problem of counting the number of occurrences of each word in a large collection of documents. The following pseudocode refers to the functions that need to be implemented:

```
map(String input_key, String input_value):
//input_key: document name
//input_value: document contents
for each word w in input_value:
EmitIntermediate(w, "1");
reduce(String output_key,
Iterator intermediate_values):
//output_key: a word
//output_values: a list of counts
int result = 0;
for each v in intermediate_values:
result + = ParseInt(v);
Emit(AsString(result));
```

The map function emits in output each word together with an associated count of occurrences (in this simple example just one). The reduce function provides the required result by summing all of the counts emitted for a specific word. MapReduce implementations (e.g., Google App Engine and Hadoop) then automatically parallelize and execute the program on a large cluster of commodity machines. The runtime system takes care of the details of partitioning the input data, scheduling the program's execution across a set of machines, handling machine failures, and managing required intermachine communication.

The programming model for MapReduce architecture is a simple abstraction where the computation takes a set of input key–value pairs associated with the input data and produces a set of output key–value pairs. The overall model for this process is shown in Figure 17.1. In the map phase, the input data are partitioned into input splits and assigned to map tasks associated with processing nodes in the cluster. The map task typically executes on the same node containing its assigned partition of data in the cluster. These map tasks perform user-specified computations on each input key–value pair from the partition of input data assigned to the task and generate a set of intermediate results for each key. The shuffle and sort phase then takes the intermediate data generated by each map task, sorts these data with intermediate data from other nodes, divides these data into regions to be processed by the reduce tasks, and distributes these data as needed to nodes where the reduce tasks will execute. All map tasks must complete prior to the shuffle and sort and reduce phases. The number of reduce tasks does not need to be the same as the number of map tasks. The reduce tasks perform additional user-specified operations on the intermediate data possibly merging values associated with a key to a smaller set of values to produce the output data.

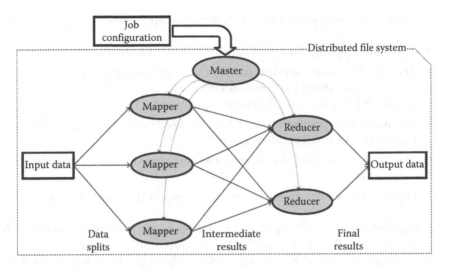

FIGURE 17.1
Execution phases in a generic MapReduce application.

For more complex data processing procedures, multiple MapReduce calls may be linked together in sequence.

The MapReduce programs can be used to compute derived data from documents such as inverted indexes, and the processing is automatically parallelized by the system that executes on large clusters of commodity-type machines, highly scalable to thousands of machines. Since the system automatically takes care of details like partitioning the input data, scheduling and executing tasks across a processing cluster, and managing the communications between nodes, programmers with no experience in parallel programming can easily use a large distributed processing environment.

17.2.1 Google File System (GFS)

Google File System (GFS) is the storage infrastructure that supports the execution of distributed applications in Google's computing cloud. GFS was designed to be a high-performance, scalable distributed file system for very large data files and data-intensive applications providing fault tolerance and running on clusters of commodity hardware. GFS is oriented to very large files dividing and storing them in fixed-size chunks of 64 Mb by default, which are managed by nodes in the cluster called chunkservers. Each GFS consists of a single master node acting as a nameserver and multiple nodes in the cluster acting as chunkservers using a commodity Linux-based machine (node in a cluster) running a user-level server process. Chunks are stored in plain Linux files, which are extended only as needed and replicated on multiple nodes to provide high availability and improve performance.

GFS has been designed with the following assumptions:

- The system is built on top of commodity hardware that often fails.
- The system stores a modest number of large files; multi-GB files are common and should be treated efficiently, and small files must be supported, but there is no need to optimize for that.
- The workloads primarily consist of two kinds of reads: large streaming reads and small random reads.
- The workloads also have many large sequential writes that append data to files.
- High sustained bandwidth is more important than low latency.

The architecture of the file system is organized into a single master, which contains the metadata of the entire file system, and a collection of chunk servers, which provide storage space. From a logical point of view, the system is composed of a collection of software daemons, which implement either the master server or the chunk server. A file is a collection of chunks for which the size can be configured at file system level. Chunks are replicated on multiple nodes in order to tolerate failures. Clients look up the master server and identify the specific chunk of a file they want to access. Once the chunk is identified, the interaction happens between the client and the chunk server. Applications interact through the file system with a specific interface supporting the usual operations for file creation, deletion, read, and write. The interface also supports snapshots and records append operations that are frequently performed by applications. GFS has been conceived by considering that failures in a large distributed infrastructure are common rather than a rarity; therefore, specific attention has been given to implementing a highly available, lightweight, and fault-tolerant infrastructure. The potential single point of failure of the single-master architecture has been addressed by giving the possibility of replicating the master node on any other node belonging to the infrastructure.

17.2.2 Google's BigTable

BigTable is the distributed storage system designed to scale up to petabytes of data across thousands of servers. BigTable provides storage support for several Google applications that expose different types of workload: from throughput-oriented batch-processing jobs to latency-sensitive serving of data to end users. BigTable's key design goals are wide applicability, scalability, high performance, and high availability. To achieve these goals, BigTable organizes the data storage in tables of which the rows are distributed over the distributed file system supporting the middleware, which is the Google File System. From a logical point of view, a table is a multidimensional sorted map indexed by a key that is represented by a string of arbitrary length.

A table is organized into rows and columns; columns can be grouped in column family, which allow for specific optimization for better access control, the storage, and the indexing of data. A simple data access model constitutes the interface for client applications that can address data at the granularity level of the single column of a row. Moreover, each column value is stored in multiple versions that can be automatically time stamped by BigTable or by the client applications.

Google's BigTable solution's objective was to develop a relatively simple storage management system that could provide fast access to petabytes of data, potentially redundantly distributed across thousands of machines. Physically, BigTable resembles a B-tree index-organized table in which branch and leaf nodes are distributed across multiple machines. Like a B-tree, nodes split as they grow, and—because nodes are distributed—this allows for high scalability across large numbers of machines. Data elements in BigTable are identified by a primary key, column name, and, optionally, a time stamp. Lookups via primary key are predictable and relatively fast. BigTable provides the data storage mechanism for Google App Engine.

Data are stored in BigTable as a sparse, distributed, persistent multidimensional sorted map structure, which is indexed by a row key, column key, and a time stamp. Rows in a BigTable are maintained in order by row key, and row ranges become the unit of distribution and load balancing called a tablet. Each cell of data in a BigTable can contain multiple instances indexed by the time stamp. BigTable uses GFS to store both data and log files. The API for BigTable is flexible, providing data management functions like creating and deleting tables and data manipulation functions by row key including operations to read, write, and modify data. Index information for BigTables utilizes tablet information stored in structures similar to a B-tree. MapReduce applications can be used with BigTable to process and transform data, and Google has implemented many large-scale applications that utilize BigTable for storage including Google Earth (Tables 17.1 and 17.2).

17.3 Hadoop

Hadoop is an open-source software project sponsored by the Apache Software Foundation. Following the publication in 2004 of the research paper describing Google MapReduce, an effort was begun in conjunction

TABLE 17.1

MapReduce Cloud Implementations

Owner	Imp Name and Website	Start Time	Last Release	Distribution Model
Google	Google MapReduce, http://labs.google.com/ papers/inapreduce.html	2004	—	Internal use by Google
Apache	Hadoop, http://hadoop. apache.org/	2004	Hadoop0.20.0, April 22, 2009	Open source
GridGain	GridGain http://www. gridgain.com.	2005	GridGain 2.1.1, February 26, 2009	Open source
Nokia	Disco ht, tp:// discoproject.org/	2008	Disco 0.2.3, September 9, 2009	Open source
Geni.com	SkyNet, http://skynet. rubyforge.org	2007	SkynetO.9.3, May 31, 2008	Open source
Manjrasoft	MapReduce.net (optional service of Aneka), http://www.manjrasoft. com products.html	2008	Aneka 1.0, March 27, 2009	Commercial

TABLE 17.2

Comparison of MapReduce Implementations

	Google MapReduce	Hadoop	Disco
Focus	Data intensive	Data intensive	Data intensive
Architecture	Master–slave	Master–slave	Master–slave
Platform	Linux	Cross-platform	Linux, Mac OS X
Storage system	GFS	HDFS, CloudStore, S3	GlusterFS
Implementation technology	C++	JAVA	Erlang
Programming environment	JAVA and Python	JAVA, shell utilities using Hadoop streaming, C++ using Hadoop pipes	Python
Deployment	Deployed on Google clusters	Private and public cloud (EC2)	Private and public cloud (EC2)
Some users and applications	Google	Baidu, NetSeer, A9.com, Facebook	Nokia Research Center

with the existing Nutch project to create an open-source implementation of the MapReduce architecture. It later became an independent subproject of Lucene, was embraced by Yahoo! after the lead developer for Hadoop became an employee, and became an official Apache top-level project in February of 2006. Hadoop now encompasses multiple subprojects in addition to the base core, MapReduce, and Hadoop distributed file system (HDFS). These additional subprojects provide enhanced application processing capabilities to the base Hadoop implementation and currently include Avro, Chukwa, HBase, Hive, Pig, and ZooKeeper.

The Apache Hadoop project develops open-source software for reliable, scalable, distributed computing. Hadoop includes these subprojects:

- *Hadoop Common*: The common utilities that support the other Hadoop subprojects
- *Avro*: A data serialization system that provides dynamic integration with scripting languages
- *Cassandra*: A scalable multimaster database with no single point of failure
- *Chukwa*: A data collection system for managing large distributed systems
- *HBase*: A scalable, distributed database that supports structured data storage for large tables
- *HDFS*: A distributed file system that provides high-throughput access to application data
- *Hive*: A data warehouse infrastructure that provides data summarization and adhoc querying
- *MapReduce*: A software framework for distributed processing of large data sets on compute clusters
- *Mahout*: A scalable machine learning and data mining library
- *Pig*: A high-level dataflow language and execution framework for parallel computation
- *ZooKeeper*: A high-performance coordination service for distributed applications

The Hadoop MapReduce architecture is functionally similar to the Google implementation except that the base programming language for Hadoop is Java instead of C++. The implementation is intended to execute on clusters of commodity processors utilizing Linux as the operating system environment but can also be run on a single system as a learning environment. Hadoop clusters also utilize the *shared nothing* distributed processing paradigm linking individual systems with local processor, memory, and disk resources using high-speed communication switching capabilities typically in rack-mounted configurations. The flexibility of Hadoop configurations allows small clusters to be created for testing and development using desktop systems or any system running Unix/Linux providing a JVM environment; however, production clusters typically use homogeneous rack-mounted processors in a data center environment.

The Hadoop MapReduce architecture is similar to the Google implementation creating fixed-size input splits from the input data and assigning the splits to map tasks. The local output from the map tasks is copied to reduce nodes where it is sorted and merged for processing by reduce tasks that produce the final output.

The Hadoop execution environment supports additional distributed data processing capabilities that are designed to run using the Hadoop MapReduce architecture. Several of these have become official Hadoop subprojects within the Apache Software Foundation. These include a distributed file system called HDFS, which is analogous to GFS in the Google MapReduce implementation. HBase is a distributed column-oriented database that provides similar random access read/write capabilities and is modeled after BigTable was implemented by Google. HBase is not relational, and does not support SQL, but provides a Java API and a command-line shell for table management. Hive is a data warehouse system built on top of Hadoop that provides SQL-like query capabilities for data summarization, ad hoc queries, and analysis of large data sets. Other Apache-sanctioned projects for Hadoop include Avro, a data serialization system that provides dynamic integration with scripting languages; Chukwa, a data collection system for managing large distributed systems; Pig, a high-level dataflow language and execution framework for parallel computation; and ZooKeeper, a high-performance coordination service for distributed applications.

17.3.1 Hadoop Distributed File System (HDFS)

HDFS is a highly fault-tolerant, scalable, and distributed file system architected to run on commodity hardware. The HDFS architecture was designed to solve two known problems experienced by the early developers of large-scale data processing. The first problem was the ability to break down the files across multiple systems and process each piece of the file independent of the other pieces and finally consolidate all the outputs in a single result set. The second problem was the fault tolerance both at the file processing level and the overall system level in the distributed data processing systems.

The three principle goals of HDFS architecture are

1. Process extremely large files ranging from multiple gigabytes to petabytes
2. Streaming data processing to read data at high-throughput rates and process data on read
3. Capability to execute on commodity hardware with no special hardware requirements

The HDFS design is based on the following assumptions:

- *Redundancy*—Hardware will be prone to failure, and processes can run out of infrastructure resources, but redundancy built into the design can handle these situations.

- *Scalability*—Linear scalability at a storage layer is needed to utilize parallel processing at its optimum level. Designing for 100% linear scalability.

- *Fault tolerance*—The automatic ability to recover from failure and complete the processing of data.

- *Cross-platform compatibility*—The ability to integrate across multiple architecture platforms.

- *Compute and storage in one environment*—Data and computation colocated in the same architecture removing redundant I/O and excessive disk access.

HDFS is analogous to GFS in the Google MapReduce implementation. A block in HDFS is equivalent to a chunk in GFS and is also very large, 64 Mb by default, but 128 Mb is used in some installations. The large block size is intended to reduce the number of seeks and improve data transfer times. Each block is an independent unit stored as a dynamically allocated file in the Linux local file system in a DataNode directory. If the node has multiple disk drives, multiple DataNode directories can be specified. An additional local file per block stores metadata for the block. HDFS also follows a master–slave architecture, which consists of a single master server that manages the distributed file system namespace and regulates access to files by clients called the NameNode. In addition, there are multiple DataNodes, one per node in the cluster, which manage the disk storage attached to the nodes and assigned to Hadoop. The NameNode determines the mapping of blocks to DataNodes. The DataNodes are responsible for serving read and write requests from file system clients such as MapReduce tasks, and they also perform block creation, deletion, and replication based on commands from the NameNode.

HDFS is a file system, and like any other file system architecture, it needs to manage consistency, recoverability, and concurrency for reliable operations. These requirements have been addressed in the architecture by creating image, journal, and checkpoint files.

17.3.1.1 HDFS Architecture

1. *NameNode (master node)*: The NameNode is a single master server that manages the file system namespace and regulates access to files by clients. Additionally, the NameNode manages all the operations like opening, closing, moving, naming, and renaming of files and directories. It also manages the mapping of blocks to DataNodes.

2. *DataNodes (slave nodes)*: DataNodes represent the slaves in the architecture that manage data and the storage attached to the data. A typical HDFS cluster can have thousands of DataNodes and tens

of thousands of HDFS clients per cluster since each DataNode may execute multiple application tasks simultaneously. The DataNodes are responsible for managing read and write requests from the file system's clients, block maintenance, and perform replication as directed by the NameNode. The block management in HDFS is different from a normal file system. The size of the data file equals the actual length of the block. This means if a block is half full, it needs only half of the space of the full block on the local drive, thereby optimizing storage space for compactness, and there is no extra space consumed on the block unlike a regular file system.

3. *Image*: An image represents the metadata of the namespace (inodes and lists of blocks). On startup, the NameNode pins the entire namespace image in memory. The in-memory persistence enables the NameNode to service multiple client requests concurrently.

4. *Journal*: The journal represents the modification log of the image in the local host's native file system. During normal operations, each client transaction is recorded in the journal, and the journal file is flushed and synced before the acknowledgment is sent to the client. The NameNode upon startup or from a recovery can replay this journal.

5. *Checkpoint*: To enable recovery, the persistent record of the image is also stored in the local host's native files system and is called a checkpoint. Once the system starts up, the NameNode never modifies or updates the checkpoint file. A new checkpoint file can be created during the next startup, on a restart, or on demand when requested by the administrator or by the CheckpointNode.

17.3.2 HBase

HBase is an open-source, nonrelational, column-oriented, multidimensional, distributed database developed on Google's BigTable architecture. It is designed with high availability and high performance as drivers to support storage and processing of large data sets on the Hadoop framework. HBase is not a database in the purist definition of a database. It provides unlimited scalability and performance and supports certain features of an ACID-compliant database. HBase is classified as a NoSQL database due to its architecture and design being closely aligned to Base (Being Available and Same Everywhere). Why do we need HBase when the data are stored in the HDFS file system, which is the core data storage layer within Hadoop? HBase is very useful for operations other than MapReduce execution and operations that are not easy to work with in HDFS and when you need random access to data. First, it provides a database-style interface to Hadoop, which enables developers to deploy programs that can quickly read or write to specific subsets of data in an extremely voluminous data set, without having to search and process through the entire data set.

Second, it provides a transactional platform for running high-scale, real-time applications as an ACID-compliant database (meeting standards for atomicity, consistency, isolation, and durability) while handling the incredible volume, variety, and complexity of data encountered on the Hadoop platform. HBase supports the following properties of ACID compliance:

1. *Atomicity*: All mutations are atomic within a row. For example, a read or write operation will either succeed or fail.

2. *Consistency*: All rows returned for any execution will consist of a complete row that existed or exists in the table.

3. *Isolation*: The isolation level is called *read committed* in the traditional DBMS.

4. *Durability*: All visible data in the system are durable data. For example, to phrase durability, a read will never return data that have not been made durable on disk.

17.3.2.1 HBase Architecture

Data are organized in HBase as rows and columns and tables, very similar to a database; however, here is where the similarity ends.

HBase architecture is described as follows:

1. Tables

    a. Tables are made of rows and columns.

    b. Table cells are the intersection of row and column coordinates. Each cell is versioned by default with a time stamp. The contents of a cell are treated as an uninterpreted array of bytes.

    c. A table row has a sortable row key and an arbitrary number of columns.

2. Rows

    a. Table row keys are also byte arrays. In this configuration, anything can serve as the row key as opposed to strongly typed data types in the traditional database.

    b. Table rows are sorted, byte-ordered, by row key, the table's primary key, and all table accesses are via the table's primary key.

    c. Columns are grouped as families, and a row can have as many columns as loaded.

3. Columns and column groups (families)

   a. In HBase, row columns are grouped into column families.

   b. All column family members will mandatorily have a common prefix, for example, the columns person:name and person:comments are both members of the person column family, where as e-mail:identifier belongs to the e-mail family.

   c. A table's column families must be specified upfront as part of the table schema definition.

   d. New column family members can be added on demand.

17.3.3 Hive

Apache Hive is a data warehouse infrastructure built on top of Hadoop provided by Facebook. Similar to Pig, Hive was initially designed as an in-house solution for large-scale data analysis. As the company expanded, the parallel RDBMS infrastructure originally deployed at Facebook began to choke at the amount of data that had to be processed on a daily basis. Following the decision to switch to Hadoop to overcome these scalability problems in 2008, the Hive project was developed internally to provide the high-level interface required for a quick adoption of the new warehouse infrastructure inside the company. Since 2009, Hive is also available for the general public as an open-source project under the Apache umbrella. Inside Facebook, Hive runs thousands of jobs per day on different Hadoop clusters ranging from 300 to 1200 nodes to perform a wide range of tasks including periodical reporting of click counts, ad hoc analysis, and training machine learning models for ad optimization. Other companies working with data in the petabyte magnitude like Netflix are reportedly using Hive for the analysis of website streaming logs and catalog metadata information.

The fundamental goals of designing Hive were the following:

- Build a system for managing and querying data using structured techniques on Hadoop
- Use native MapReduce for execution at HDFS and Hadoop layers
- Use HDFS for storage of Hive data
- Store key metadata in an RDBMS
- Extend SQL interfaces, a familiar data warehousing tool in use at enterprises
- *High extensibility*: User-defined types, user-defined functions, formats, and scripts
- Leverage extreme scalability and performance of Hadoop
- Interoperability with other platforms

The main difference between Hive and the other languages previously discussed comes from the fact that Hive's design is more influenced by classic relational warehousing systems, which is evident both at the data model and at the query language level. Hive thinks of its data in relational terms—data sources are stored in tables, consisting of a fixed number of rows with predefined data types. Similar to Pig and Jaql, Hive's data model provides support for semistructured and nested data in the form of complex data types like associative arrays (maps), lists, and structs, which facilitates the use of denormalized inputs. On the other hand, Hive differs from the other higher-level languages for Hadoop in the fact that it uses a catalog to hold metadata about its input sources. This means that the table schema must be declared and the data loaded before any queries involving the table are submitted to the system (which mirrors the standard RDBMS process). The schema definition language extends the classic DDL CREATE TABLE syntax. Currently, Hive does not provide support for updates, which means that any data load statement will enforce the removal of any old data in the specified target table or partition. The standard way to append data to an existing table in Hive is to create a new partition for each append set. Since appends in an OLAP environment are typically performed periodically in a batch manner, this strategy is a good fit for most real-world scenarios.

The Hive Query Language (HiveQL) is an SQL dialect with various syntax extensions. HiveQL supports many traditional SQL features like from clause subqueries, various join types, group bys and aggregations, as well as many useful built-in data processing functions that provide an intuitive syntax for writing Hive queries to all users familiar with the SQL basics. In addition, HiveQL provides native support for in-line MapReduce job specification. The semantics of the mapper and the reducer are specified in external scripts, which communicate with the parent Hadoop task through the standard input and output streams (similar to the streaming API for user-defined functions (UDFs) in Pig).

17.3.4 Pig

As data volumes and processing complexities increase, analyzing large data sets introduces dataflow complexities that become harder to implement in a MapReduce program. There was a need for an abstraction layer over MapReduce: a high-level language that is more user friendly, is SQL-like in terms of expressing dataflows, has the flexibility to manage multistep data transformations, and handles joins with simplicity and easy program flow. Apache Pig was the first system to provide a higher-level language on top of Hadoop. Pig started as an internal research project at Yahoo (one of the early adopters of Hadoop) but due to its popularity subsequently was promoted to a production-level system and adopted as an open-source project by the

Apache Software Foundation. Pig is widely used both inside and outside Yahoo for a wide range of tasks including ad hoc data analytics, ETL tasks, log processing, and training collaborative filtering models for recommendation systems.

The fundamental goals of designing Pig were as follows:

- *Programming flexibility*: The ability to break down complex tasks comprised of multiple steps and interprocess-related data transformations should be encoded as dataflow sequences that are easy to design, develop, and maintain.
- *Automatic optimization*: Tasks are encoded to let the system optimize their execution automatically. This allows the user to have greater focus on program development, allowing the user to focus on semantics rather than efficiency.
- *Extensibility*: Users can develop user-defined functions (UDFs) for more complex processing requirements.

Pig queries are expressed in a declarative scripting language called Pig Latin, which provides SQL-like functionality tailored toward big data's specific needs. Most notably from the syntax point of view, Pig Latin enforces implicit specification of the dataflow as a sequence of expressions chained together through the use of variables. This style of programming is different from SQL, where the order of computation is not reflected at the language level, and is better suited to the ad hoc nature of Pig as it makes query development and maintenance easier due to the increased readability of the code.

Unlike traditional SQL systems, the data do not have to be stored in a system-specific format before it can be used by a query. Instead, the input and output formats are specified through storage functions inside the load and store expressions. In addition to ASCII and binary storage, users can implement their own storage functions to add support for other custom formats. Pig uses a dynamic type system to provide native support for nonnormalized data models. In addition to the simple data types used by relational databases, Pig defines three complex types—tuple, bag, and map—which can be nested arbitrary to reflect the semistructured nature of the processed data. For better support of ad hoc queries, Pig does not maintain a catalog with schema information about the source data. Instead, input schema is defined at the query level either explicitly by the user or implicitly through type inference. At the top level, all input sources are treated as bags of tuples; the tuple schema can be optionally supplied as part of the load expression.

17.4 Summary

The chapter discusses the functional programming paradigm followed by the Google MapReduce algorithm and its reference implementation. We discuss the Hadoop ecosystem including Hadoop Distributed File System (HDFS), HBase NoSQL database, Hive data warehouse solution, and the Pig query language for ad hoc analytical requirements. In the end, we discuss CADM for the development of cloudware applications.

18

Cloudware Operations and Management

Cloud computing is on-demand access to a shared pool of computing resources. It helps users to reduce costs, reduce management responsibilities, and increase business agility. For this reason, it is becoming a popular paradigm, and increasingly more companies are shifting toward IT cloud computing solutions. Advantages are many but, being a new paradigm, there are also challenges and inherent issues. These relate to data governance, service management, process monitoring, infrastructure reliability, information security, data integrity, and business continuity.

18.1 Characteristics of Cloud Operations

Cloud computational resources can be scaled up and down on demand and paid for on a metered usage basis. The usage characteristics have multiple options based on a host of parameters. These options have to be evaluated on a case-by-case basis.

This ability to configure the various parameters per requirements (or changing requirements) provides tremendous advantages for clients in that they do not have to maintain internal computing systems designed for peak loads that may occur only a small percentage of the time. Though cloud offers the ability to provision massive amounts of computing power and storage, these quantities are not unlimited; like any other physical system, cloud computation must operate within physical limits imposed by practical boundary conditions. Cloud users might have to fit their applications into one set of resource usage categories defined by the cloud provider. The cloud paradigm also supports innovation in that a variety of new, advanced applications can be used in an affordable manner while reducing the total cost of ownership. Some applications that are of long duration and have stable computational requirements might be better served by in-house or leased computers and storage than by paying cloud fees over a long period of time.

As introduced in the earlier chapters, the major benefits of cloud computing can be summarized as follows:

- Means to move from operating in a capital expenditure environment to an operational expenditure environment.
- Ability to rapidly deploy innovative business and research applications in a cost-effective manner.
- Use of virtualization to detach business services from the underlying execution infrastructure.
- Disaster recovery and business continuity capabilities are intrinsic in the cloud paradigm.
- Ability of the cloud provider to apply security safeguards more effectively and efficiently in a centralized environment.
- Ability to select among a variety of cloud vendors that provide reliable scalable services, metered billing, and advanced development resources.
- Scalable infrastructure that can rapidly provision and de-allocate substantial resources on an as-needed basis.

18.2 Core Services

18.2.1 Discovery and Replication

Service discovery promotes reusability by allowing service users to find the existing services. RESTful services support discovery and reuse at design time. Replication can be used to create and maintain copies of an enterprise's data at these sites. When events affecting an enterprise's primary location occur, key application services can effectively be restarted and run at the remote location incurring no capital expenditure, only operational expenditure, until such time as the primary site is brought back online. Replication keeps all replicas as a part of one atomic transaction. Replication technology is available in storage arrays and network-based appliances and through host-based software.

18.2.2 Load Balancing

Load balancing prevents system bottlenecks due to unbalanced loads. It also considers implementing failover for the continuation of a service after failure of one or more of its components. This means that a load balancer provides a mechanism by which instances of applications can be provisioned and deprovisioned automatically without changing network configuration.

This is an inherited feature from grid-based computing for cloud-based platforms. Energy conservation and resource consumption are not always a focal point when discussing cloud computing; however, with proper load balancing in place, resource consumption can be kept to a minimum. This not only serves to keep costs low and enterprises *greener*; it also puts less stress on the hardware infrastructure of each individual component, making them potentially last longer. Load balancing also enables other important features such as scalability.

18.2.3 Resource Management

Cloud computing provides a way of deploying and accessing massively scalable shared resources on demand, in real time, and at affordable cost. Cloud resource management protocols deal with all kinds of homogeneous and heterogeneous resource environment. Management of virtualized resources, workload and resource scheduling, cloud resource provisioning with QoS, and scalable resource management solutions are the concerning points. Dynamic resource scheduling across a virtualized infrastructure for those environments is another issue for cloud.

18.2.4 Data Governance

When data begin to move out of organizations, they are vulnerable to disclosure or loss. The act of moving sensitive data outside the organizational boundary may also violate national regulations for privacy. In Germany, passing data across national boundaries can be a federal offence. Governance in the cloud *who and how* is the big challenge for enterprise clouds. Governance places a layer of processes and technology around services (location of services, service dependencies, service monitoring, service security, and so on) so that anything occurring will be quickly known. There are some questions that need to be solved before mission-critical data and functionality can be moved outside a controllable environment.

18.2.4.1 Interoperability

Interoperability means easy migration and integration of applications and data between different vendors' clouds. Owing to different hypervisors (KVM, Hyper-V, ESX, ESXi), VM technologies, storage, configuring operating systems, various security standards, and management interfaces, many cloud systems are not interoperable. However, many enterprises want interoperability between their in-house infrastructure and the cloud. The issue of interoperability needs to be addressed to allow applications to be ported between clouds or to use multiple cloud infrastructures before critical business applications are delivered from the cloud. Most clouds are completely opaque to their users. Most of the time, users are fine with this until there is an access issue. In such situations, frustration increases exponentially with time, partly because of the opacity. Is a mechanism like a network weather map, in other words, some form of monitoring solution like autonomous agents, required?

18.2.4.2 Data Migration

Data migration between data centers or cloud systems are important concerns of taxonomy. While migrating data, some considerations should be taken into account like no data loss, availability, scalability, cost efficiency, and load balancing. Users should be able to move their data and applications any time from one to another seamlessly, without any one vendor controlling it. Seamless transfer, as in mobile communication, is required for cloud computing to work. Many enterprises do not move their mission-critical data and applications to the cloud because of vendor lock-in, security, governance, and many more complications.

18.2.5 Management Services

The management services contain deployment, monitoring, reporting, service-level agreement, and metering billing. We discuss these in detail.

18.2.5.1 Deployment and Configuration

To reduce the complexity and administrative burden across the cloud deployment, we need the automation process life cycle. RightScale Cloud Management Platform addresses three stages of the cloud application deployment lifecycle, namely, design, manage, and deploy. Automated configuration and maintenance of individual or networked computers, from the policy specification, is very important in the computing arena; it improves robustness and functionality without sacrificing the basic freedoms and self-repairing concepts. That is why, to handle complex systems like cloud environment and data center, we need such configuration management. Configuration management framework tools help software developers and engineers to manage server and application configuration by writing code, rather than running commands by hand.

18.2.5.2 Monitoring and Reporting

Developing, testing, debugging, and studying the performance of cloud systems are quite complex. Management cost increases significantly as the number of sites increases. To address such problems, we need monitoring and reporting mechanisms. Monitoring basically monitors the SLA lifecycle. It also determines when an SLA completes and reports to the billing services. There are some services that monitor the cloud systems and produce health reports such as Hyperic HQ, which monitors SimpleDB, Simple Queue Service, and Flexible Payment Service, all offered by Amazon. It collects the matrix and provides a rich analysis and reporting.

18.2.5.3 Service-Level Agreements (SLAs) Management

Users always want stable and reliable system service. Cloud architecture is considered to be highly available, up and running 24 h × 7 days. Many cloud service providers have made huge investments to make their system reliable. However, most cloud vendors today do not provide high availability assurances. If a service goes down, for whatever reason, what can a user do? How can users access their documents stored in the cloud? In such a case, the provider should pay a fine to the user as compensation to meet SLAs. An SLA specifies the measurement, evaluation, and reporting of the agreed service-level standards such as the following:

1. How raw quality measures will be used to evaluate agreed service component
2. How the raw quality measures will be qualified as a service quality measure
3. How the qualified quality measures will be used to estimate the service quality levels
4. How the results of service evaluation will be reported
5. How disputes on service-level evaluation will be resolved

Currently, Amazon offers a *99.9% monthly uptime percentage* SLA for Simple Storage Service (Amazon S3) and credit is limited to 10%. Amazon credits 25% of charges if the availability drops below 99.0%, whereas 3Tera Virtual Private Data Center (VPDC) service will include a 99.999% availability SLA that is supposed to help assure customers about putting mission-critical apps and services in the cloud.

18.2.5.4 Metering and Billing

Transparent metering and billing will increase the trust level of users toward cloud services. Pay-as-you-go subscription or pay-as-you-consume model of billing and metering are popular for cloud. This service gets the status of the SLA and invokes the credit service, which debits the user credit card or account and informs the user. There are many pricing strategies such as RAM hours, CPU capacity, bandwidth (inbound/outbound data transfer), storage space (gigabytes of data), software license fee, and subscription-based pricing. There are some interesting new billing models such as GoGrid prepaid cloud hosting plan and IDC cloud billing research, which are great examples of moving cloud pricing models toward telecom models.

18.2.5.5 Authorization and Authentication

In public clouds, safeguards must be placed on machines to ensure proper authentication and authorization. Within the private cloud environment, one can track, pinpoint, control, and manage users who try to access machines with improper credentials. Single sign-on is the basic requirement for a customer who accesses multiple cloud services.

18.2.6 Fault Tolerance

In case of failure, there will be a hot backup instance of the application, which is ready to take over without disruption. Cloud computing outages extend into the more refined version of cloud service platforms. Some outages have been quite lengthy. For example, Microsoft Azure had an outage that lasted 22 h on March 13–14, 2008. Cloud reliance can cause significant problems if downtime and outages are removed from your control. Table 18.1 shows failover records from some cloud service provider systems. These are significant downtime incidents. Reliance on the cloud can cause real problems when time is money.

Google has also had numerous difficulties with its Gmail and application services. These difficulties have generated significant interest in both traditional media and the blogosphere owing to deep-seated concerns regarding service reliability. The incidents mentioned here are just the tip of the iceberg. Every year, thousands of websites struggle with unexpected downtime and hundreds of networks break or have other issues. So, the major problem for cloud computing is how to minimize outage/failover to provide reliable services. It is important to adopt the well-known Recovery-Oriented Computing (ROC) paradigm in large data centers. Google uses Google File System (GFS) or distributed disk storage; every piece of data is replicated three times. If one machine dies, a master redistributes the data to a new server.

18.3 Core Portfolio of Functionality

This section discusses the functional components that have to be provided by the service provider. All the cloud service providers may not provide all of these components; on the contrary, additional components may be added if the need arises. The delivery of these functional components may be based on different optimization criteria like cost optimization or performance optimization and fulfill different user constrains like budget, performance, instance types, load balancing, instance prices, and service workload.

TABLE 18.1

Outages in Different Cloud Services

Cloud Service and Outage	Duration	Date	Implications
Google GMAil	30 h	October 16, 2008	Users could not access their emails
Google Gmail and Google Apps	24 h	August 15, 2008	Those affected by the outage received a 502 server error when trying to log in to Gmail and Google Apps
FlexiScale: core network failure	18 h	October 31, 2008	All services were unavailable to customers
Amazon S3	6–8 h	July 20, 2008	Users could not access the storage due to single bit error leading to gossip protocol blowup
Google Network	3 h	May 14, 2009	The vast majority of Google services became unavailable, including Gmail, YouTube, Google News, and even the google.com home page. The outage affected about 14% of Google users worldwide
Google News	1.5 h	May 18, 2009	Users saw a 503 server error, along with a message to try their requests again later
Google News	2 h	September 22, 2009	Many users experienced difficulties accessing Google News
Google Gmail	2 h	September 1, 2009	Users could not access their emails
Amazon EC2	8 h	December 10, 2009	Customers experienced a loss of connectivity to their service instances
Microsoft Sidekick	6 days	March 13, 2009	The massive outage left Sidekick customers without access to their calendar, address book, and other key aspects of their service
Microsoft Azure	22 h	March 13, 2009	The outage occurred before the service came out of beta. The outage left people without access to their applications
Netsuite	30 min	April 27, 2010	The company's cloud applications were inaccessible to customers worldwide

The following components are required:

1. *Service management*: This is the most critical component of the cloud service provider and performs the tasks related to service discovery, service selection (in conjunction with other functions like semantic engine and SLA and QoS management), service provisioning, and service deprovisioning. The allocation of services is done in a manner to preserve on-demand and elastic nature of cloud services provisioning.

2. *Metering/billing*: This functionality keeps track of the services consumed along with any aggregations and discounts and other pricing-related information (in conjunction with rules management and monitoring engine). This module may have integration with external third-party payment gateways and will meter data from various resources and aggregate them so that they can be rated and billed.

3. *Data management*: Also, an important functionality dealing with data and their storage and security (in conjunction with integration, transformation, security management, and SLA and QoS management modules).

4. *Monitoring*: Monitors the business activities, SLAs, holistic service status, outstanding alerts, and policy violations. This module interfaces with most other modules and provides them with relevant information.

5. *Security management*: May include identity and access management functionality for handling user roles and access issues and security services like authentication, authorization, auditing, encryption, wire-level security, and other conventional security measures required in a distributed environment. Privacy-related issues are also handled by this module.

6. *SLA and QoS management*: Makes use of metrics in the relevant areas, some of them being legal metrics, SLA, and QoS requirements pertaining to regulatory, privacy, data security, and penalties management; interfaces with the security management, policy management, rules management, logging/audit trails, monitoring, and performance management modules; defines metrics for usage and assessment of charge-backs, promotions, and discount-related information management.

7. *Performance management*: Handles the performance-related aspects of business processes and services and also of the underlying resources. It interfaces with certain other modules like monitoring and support and incident management.

8. *Policy management*: Policy handling including policy creation and assessment, mapping, attachment, and deployment is performed as

part of this module. Policy enforcement and escalations must also be applied as appropriate as part of the functionality of this module.

9. *Self-service*: Provides the customer with the ability to self-register and perform self-service functions including provisioning and management tasks and also administration. This module ties in with the security management module and may tie in with other modules like rules management, SLA and QoS management, and policy management.

10. *Support and incident management*: Performance and utilization of diagnostic information at multiple levels to troubleshoot and resolve issues. Cloud management infrastructure must provide diagnostics capabilities for the full stack. Incident management aims to restore normal cloud operation as quickly as possible and minimize the adverse effect on business operations. This may include resolution of the root causes of incidents and thus minimizes the adverse impact of incidents and problems on business and prevents recurrence of incidents related to these errors. If resolution is not possible, then alternative service deployment may be needed in conjunction with service management module.

11. *Analytics*: This module collects and makes use of historical data and provides analytical information for both internal usages of the cloud service provider but also for the cloud users. It provides insight into business trends and user preferences, transaction visibility, business key performance indicator (KPI) monitoring, reporting functionality, and dashboards.

12. *Orchestration*: Business transactions are often executed by coordinating or arranging existing applications and infrastructure in the cloud to implement business processes. This involves usage of technologies and tools like ESBs, process engines, middleware, legacy, and packaged applications. Event processing may be a useful functionality to have as it provides asynchronous resource coordination and automated policy-based orchestration.

13. *Transformation*: Involves changing an entity in one form to another at runtime, for example, transformation of an entity from one data model to another or transformation of message from one protocol to another.

14. *Logging/audit trails*: Performs creation of logs and audit trails. The module is essential for fulfilling regulatory compliance and also for interfacing to incident management and essential for security management, SLA and QoS management, and support and incident management modules.

15. *Mediation*: This module helps to resolve the differences between two or more systems in order to integrate them seamlessly. Mediation

can happen at different levels such as security, transport, message, API, and protocol.

16. *Integration*: Is used to facilitate the combination or aggregation of separately produced components or services to perform a larger task, ensuring that the problems in their interactions are addressed by using some intermediary tool, say, mediation. This module is necessary for interfacing to the multiple service providers who do not follow the same standards in terms of protocols, technology, APIs, etc.

17. *Semantic engine*: A specialized entity (could be optional) that will support the creation of a common understanding of and relationships among entities in a domain by means of creation and usage of ontologies. This module thus helps in easier mapping and understanding of services provided by different cloud vendors and helps in creating ease of interoperability and common understanding among cloud services provided by different vendors.

18. *Rules management*: Is a support functionality that is utilized by many other modules to perform complex decision making and evaluation functionality and also to map the business requirements in a declarative, easy-to-use manner that allows easy update and changes.

18.4 Metrics for Interfacing to Cloud Service Providers

The cloud service providers provide mediation, optimization, and value-addition functionality based on either business or technical requirements of the corresponding cloud users. It is based on these business requirements, specified by means of defined metrics, that the cloud service providers provide the best-fitting service to the users and/or perform necessary selection and filtering of the cloud service providers.

The metrics relevant for interfacing with the cloud service providers are as follows:

1. Business metrics
   a. *Cost of service*: Is one of the most important criteria and may be quite difficult to specify and calculate. Cost of service may be divided into one-time and ongoing fixed costs. The cost may be based on per hour/per transaction/volume of transfer, quota, time of request, availability of resources, and so on. Other criteria may be commercial versus noncommercial provider usage,

time, and/or cost optimization scheduling. It is one of the most important but also the most difficult parameter to monitor and regulate as it involves comparison of costing data for existing services to newly available services as well. Costing data are complicated and may involve multiple subparameters like fixed one-time costs and variable costs depending on time and demand.

b. *Regulatory criteria-driven requirements*: Regulatory requirements that need to be fulfilled by the cloud user are further taken over to the cloud service provider; examples of such data are geographical constraints like location of data, logging, audit trails, and privacy.

c. *Maximum tolerable latency*: This is typically a technical metrics, but in case a business process has some specific response time-related constraint, then it forms part of the business parameter.

d. Business SLA requirements, including availability guarantees, availability penalties, and penalty exclusions.

e. Data-related requirements, for example, data security, data location, backup, and storage.

f. *Role-based access control*: These are business-related access control metrics pertaining to hierarchy of the user organization.

g. On-demand business process deployment and user or application constraint management like management of budget and deadline.

h. Governance criteria like quality of service, policy enforcement, and management.

i. Environmental constraints, for example, preference for green providers and incentives for green providers.

2. Technical metrics
   a. Virtual machine requirements, say, those related to central processing unit (CPU), memory, and operating system
   b. Security-related requirements like details related to encryption, digital key management, and identity and access management
   c. Technical SLA requirements, including alerts and notifications when SLA terms reach a critical point
   d. Maximum provisioning time
   e. Redundancy and disaster recovery metrics for critical applications or functionality
   f. Environment safety and protection-related metrics like carbon emission rate and CPU power efficiency

18.5 Selection Criteria for Service Provider(s)

Planning for the involvement of a cloud service provider as part of the cloud strategy is necessary and important for an enterprise when there is a need for using more than one cloud service providers.

There are a number of generic criteria for the selection of a cloud service provider. The criteria are given as follows:

- *Cost/price*: This is one of the main deciding factors for the users of the service. This should include all the components of the cost, including one-time and ongoing costs. The transparency of the costs is also important to ensure that there are no hidden costs and so is the review of the historical price trend to see how the cost of services has changed in the past.

- *Viability and reputation of the provider*: Since this is a nascent field and the users build their business processes based on the services provided by the cloud service provider, it is very important to look at the reputation of the cloud service provider and assess their references, the number of successful projects delivered, and the duration of time for which they have been in business. It is also important to assess the long-term stability of the provider's business.

- *Security and privacy*: Security provision as a value addition is one of the reasons for using the cloud service providers in the first place. It is important to ensure that the service providers provide an end-to-end security (including physical, application, and personnel related) for the services they provide. Equally important (also from the point of view of regulatory requirements fulfillment) is the review of how a service provider manages privacy identity and access management, and single sign-on feature provision is important. It is important that the service provider provides details of security certifications that their services may fulfill.

- *Regulatory requirements fulfillment*: Depending on the domain, the cloud users have to fulfill varied requirements (say, audit trail provision, logging, etc.). The cloud service provider on behalf of the user must ensure that the compliance requirements are met.

- *Transparency of operational information and data*: It is necessary to get the insight into the basic operational information and data related to one's operation from the point of view of the cloud service user, and the service provider that provides this information should be preferred.

- *Data management*: Data are an important asset of the enterprise. Functions related to data like data storage and backup, confidentiality and privacy of data, and geographical location of data are met. Data availability is important for the user's business, and, thus, it is important to review the contingency plan that the service provider has in the event of a data center/service failure (either own or of the service provider's).

- *SLA management*: The service provider's readiness in SLA to provide some transparency into the vendor's operations in the areas of audits and compliances and its flexibility to meet SLA requirements including penalties in case of noncompliance is a key deciding factor as it impacts the fulfillment of many other requirements. The flexibility of the SLA to reflect the service user's business requirement, including a match to inputted translated business metric of the cloud service user, must be reviewed.

- *Contingency and disaster recovery plans*: It is important to review the contingency and disaster recovery plans that a service provider provides.

- *Features of services provided*: Ultimately, the distinguishing factor among different service providers may be the features of the service they offer including the ability to find the best services for the user based on the user's business and its businesses location.

- *Service provider references*: A review of the references provided by the service provider is a necessary step. This should ideally include the possibility to communicate with actual users of the services of the service provider.

18.6 Service-Level Agreements (SLAs)

Most customers are willing to move their own premise setup to a hosted environment only if their data are kept securely and privately as well as nonfunctional properties such as availability or performance are guaranteed. Providing cloud services to customers requires not only managing the resources in a cost-efficient way but also running these services in certain quality satisfying the needs of the customers. The quality of delivered services is usually formally defined in an agreement between the provider and the customer, which is called service-level agreement (SLA). In the following sections, we introduce the basic ideas of such agreements from the technical perspective of cloud computing and give an overview on techniques to achieve the technical goals of service quality.

Service-level agreements define the common understanding about services, guarantees, and responsibilities. They consist of two parts:

1. *Technical part*: The technical part of an SLA (so-called service-level objectives) specifies measurable characteristics such as performance goals like response time, latency or availability, and the importance of the service.

2. *Legal part*: The legal part defines the legal responsibilities as well as fee/revenue for using the service (if the performance goals are met) and penalties (otherwise).

Supporting SLAs requires both the monitoring of resources and service providing as well as resource management in order to minimize the penalty cost while avoiding over-provisioning of resources.

There are two types of SLAs from the perspective of application hosting. These are described in detail here.

1. *Infrastructure SLA*: The infrastructure provider manages and offers guarantees on availability of the infrastructure, namely, server machine, power, and network connectivity. Enterprises manage themselves, their applications that are deployed on these server machines. The machines are leased to the users and are isolated from machines of other users. In such dedicated hosting environments, a practical example of service-level guarantees offered by infrastructure providers is shown in Table 18.2.

2. *Application SLA*: In the application colocation hosting model, the server capacity is available to the applications based solely on their resource demands. Hence, the service providers are flexible in allocating and de-allocating computing resources among the colocated applications. Therefore, the service providers are also responsible for ensuring to meet their user's application SLOs. For example, an enterprise can have the following application SLA with a service provider for one of its application, as shown in Table 18.3.

TABLE 18.2

Key Contractual Elements of an Infrastructural SLA

Hardware availability	99% uptime in a calendar month
Power availability	99.99% of the time in a calendar month
Data center network availability	99.99% of the time in a calendar month
Backbone network availability	99.999% of the time in a calendar month
Service credit for unavailability	Refund of service credit prorated on downtime period
Outage notification guarantee	Notification of customer within 1 h of complete downtime
Internet latency guarantee	When latency is measured at 5 min intervals to an upstream provider, the average doesn't exceed 60 ms
Packet loss guarantee	Shall not exceed 1% in a calendar month

TABLE 18.3

Key Contractual Elements of an Application SLA

Service-level parameter metric	• Website response time (e.g., max of 3.5 s per user request) • Latency of Web server (WS) (e.g., max of 0.2 s per request) • Latency of DB (e.g., max of 0.5 s per query)
Function	• Average latency of WS = (latency of Web server 1 + latency of Web server 2)/2 • Website response time = average latency of Web server + latency of database
Measurement directive	• DB latency available via http://mgmt server/em/latency • WS latency available via http://mgmtserver/ws/instanceno/latency
Service-level objective	• Service assurance • Website latency <1 s when concurrent connection <1000
Penalty	• 1000 USD for every minute while the SLO was breached

It is also possible for a customer and the service provider to mutually agree upon a set of SLAs with different performance and cost structure rather than a single SLA. The customer has the flexibility to choose any of the agreed SLAs from the available offerings. At runtime, the customer can switch between the different SLAs.

Table 18.1 describes the amount of acceptable downtime per year for the corresponding level of availability.

18.6.1 Quality of Service (QoS)

Quality of Service (QoS) is a well-known concept in other areas. For example, in networking, QoS is defined in terms of error rate, latency, or bandwidth and implemented using flow control, resource reservation, or prioritization.

In classic database system operation, QoS and SLAs are mostly limited to provide reliable and available data management. Query processing typically aims at executing each query as fast as possible, but not to guarantee given response times. However, for database services hosted on a cloud infrastructure and provided as multitenant service, more advanced QoS concepts are required. Important criteria or measures are the following:

1. *Availability*: The availability measure describes the ratio of the total time the service and the data are accessible during a given time interval and the length of this interval. For example, Amazon EC2 guarantees an availability of 99.95% for the service year per region, which means downtimes in a single region up to 4.5 h per year are acceptable. Availability can be achieved by introducing redundancies: data

are stored at multiple nodes and replication techniques (see Section 3.4) are used to keep multiple instances consistent. Then, only the number of replicas and their placement affect the degree of availability. For instance, Amazon recommends to deploy to EC2 instances in different availability zones to increase availability by having geographically distributed replicas, which is done for data in SimpleDB automatically.

2. *Consistency*: Consistency as service guarantee depends on the kind of service provision. In case of a hosted database like Amazon RDS or Microsoft Azure SQL, full ACID guarantees are given, whereas scalable distributed data stores such as Amazon SimpleDB guarantee only levels of eventual consistency. Techniques for ensuring different levels of consistency are standard database techniques that can be found in the concerned textbooks.

3. *(Query) response time*: The response time of a query can either be defined in the form of deadline constraints, for example, response time of a given query is ≤10 s, or not per query but as percentile constraints. The latter means, for example, that 90% of all requests need to be processed within 10 s; otherwise, the provider will pay a penalty charge. Though response time guarantees are not the domain of standard SQL database systems, there exist some techniques in real-time databases.

18.6.2 Pricing Models for Cloud Systems

One of the central ideas and key success factors of the cloud computing paradigm is the pay-per-use price model. Ideally, customers would pay only for the amount of the resources they have consumed. Looking at the current market, services differ widely in their price models and range from free or advertisement-based models over time or volume-based models to subscription models. Based on a discussion of the different cost types for cloud services, we present some fundamentals of pricing models in the following.

18.6.2.1 Cost Types

For determining a pricing model, all direct and indirect costs of a provided service have to be taken into account. The total costs typically comprise capital expenditures (CapEx) and operational expenditures (OpEx). Capital expenditures describes all costs for acquiring assets such as server and network hardware, software licenses, but also facilities, power, and cooling infrastructure. Operational expenditures (OpEx) includes all costs for running the service, for example, maintenance costs for servers, facilities, infrastructure, payroll, but also legal and insurance fees. One of the economical promises of cloud computing is to trade CapEx for OpEx by outsourcing IT hardware and services. Furthermore, for customers, there is no need for long-term commitment to resources. However, this typically comes with higher OpEx. A good analogy is the rental car business: relying on rental cars instead of maintaining a company-owned fleet of cars avoids the big investment for purchasing cars (CapEx) but requires to pay the rates for each day of use (OpEx). Cloud-based data management has some specific characteristics that have to be taken into account for pricing models, because data are not as elastic as pure computing jobs. The data sets have to be uploaded to the cloud and stored there for a longer time and usually require additional space for efficient access (indexing) and high availability (backups).

Thus, the following types of operational costs are included:

1. *Storage costs*: Database applications require to store data persistently on disk. In order to guarantee a reliable and high available service, backup and archiving have to be performed, which need additional storage space. In addition, for efficient access, index structures are required, which are also stored on disk. The disk space occupied by all these data (application data, backup, indexes) is considered part of the storage costs.

2. *Data transfer costs*: This type of costs covers the (initial and regularly) transfer of application data to the service provider over the network as well as the costs for delivering requests and results between clients and the database service.

3. *Computing costs*: This represents typically the main cost type for processing database services and includes processing time of running a data management system (computing time, license fee) or processing a certain number of requests (queries, transactions, read/write requests).

Further costs can be also billed for certain guarantees and service-level agreements, for example, availability, consistency level, or provided hardware (CPU type, disk type).

18.6.2.2 Subscription Types

Based on the individual cost types, different pricing models can be built. Currently available models are often inspired by other commercial services (such as mobile phone plans) and range from a pure pay-per-use approach where each type of cost is billed individually to flat-rate like subscription models for longer time periods. In principle, these models can be classified according to two dimensions: the unit for billing and the degree of flexibility.

The first dimension describes which unit is used for billing. Typical units are as follows:

1. *Time of usage*: Examples are Amazon's EC2 and RDS instances or Rackspace instances that are billed in $ per hour for specific hardware configurations. Further parameters such as system load, compute cycles, bandwidth usage, or number or requests are not taken into account. However, this means that it is up to the customer to choose an appropriate configuration needed to handle the envisaged load.

2. *Volume-based units*: Examples are GB per month for storage services or the number of requests per time unit. Volume-based pricing models simplify resource provisioning and SLAs for the service provider: based on the requested volume, the provider can estimate the required resources (disk space, computing resources to handle the requests).

The second dimension characterizes the flexibility of service usage:

- *On demand or pay per use*: This is the most flexible approach—a customer can use the service at any time as well as when he really wants to use it. Apart from the consumed units (time, volume), no additional costs are billed.

- *Auction based*: In this model, which is, for instance, offered by Amazon with their so-called *spot instances* in EC2, customers can bid for unused capacities. The prices for such machine instances vary over time depending on supply and demand. As long as the price for these instances is below the customer's bid (e.g., a maximum price), the customer can use such instances. Of course, this makes sense only for applications that are very flexible in their usage times. Furthermore, such applications have to deal with possible interruptions that occur when prices of instances exceed the bid.

- *Reservation based*: This means that capacity is reserved at the provider's side for which the customer has to pay a one-time or regular fee. In return, the customer receives a discount for the regular charges (time or volume).

Currently, cloud service providers use different combinations of these subscription and cost types. This allows a flexible choice for customers but makes it sometimes difficult to select the most appropriate offer. Some providers offer basic online pricing calculators for determining the price of given configurations, for example, Amazon3 and Microsoft.

Reason	Respondents Who Agree
Improved system reliability and availability	50%
Pay only for what you use	50%
Hardware savings	47%
Software license savings	46%
Lower labor costs	44%
Lower maintenance costs	42%
Reduced IT support needs	40%
Ability to take advantage of the latest funtinality	40%
Less pressure on internal resources	39%
Solve problems related to updating/upgrading	39%
Rapde deployment	39%
Ability to scale up resources to meet needs	39%
Ability to focus on core competencies	38%
Take advantage of the improved economies of scale	37%
Reduced infrastructure management needs	37%
Lower energy costs	29%
Reduced space requirements	26%
Create new revenue streams	23%

A broad set of concerns identified by the NIST working group on cloud security includes the following:

- Potential loss of control/ownership of data
- Data integration, privacy enforcement, data encryption
- Data remanence after deprovisioning
- Multitenant data isolation
- Data location requirements within national borders
- Hypervisor security
- Audit data integrity protection
- Verification of subscriber policies through provider controls
- Certification/accreditation requirements for a given cloud service

The top workloads mentioned by the users involved in this study are data mining and other analytics (83%), application streaming (83%), help desk services (80%), industry-specific applications (80%), and development environments (80%).

The study also identified workloads that are not good candidates for migration to a public cloud environment:

- Sensitive data such as employee and health-care records
- Multiple codependent services (e.g., online transaction processing)

- Third-party software without cloud licensing
- Workloads requiring auditability and accountability
- Workloads requiring customization

18.6.3 Software Licensing

Software licensing for cloud computing is an enduring problem without a universally accepted solution at this time. The license management technology is based on the old model of computing centers with licenses given on the basis of named users or as site licenses. This licensing technology, developed for a centrally managed environment, cannot accommodate the distributed service infrastructure of cloud computing or of grid computing.

Only very recently, IBM reached an agreement allowing some of its software products to be used on EC2. Furthermore, MathWorks developed a business model for the use of MATLAB in grid environments. The Software-as-a-Service (SaaS) deployment model is gaining acceptance because it allows users to pay only for the services they use.

There is significant pressure to change the traditional software licensing model and find nonhardware-based solutions for cloud computing. The increased negotiating power of users, coupled with the increase in software piracy, has renewed interest in alternative schemes such as those proposed by the SmartLM research project (www.smartlm.eu). SmartLM license management requires a complex software infrastructure involving SLA, negotiation protocols, authentication, and other management functions.

A commercial product based on the ideas developed by this research project is elasticLM, which provides license and billing for web-based services. The architecture of the elasticLM license service has several layers: co-allocation, authentication, administration, management, business, and persistency. The authentication layer authenticates communication between the license service and the billing service as well as the individual applications; the persistence layer stores the usage records. The main responsibility of the business layer is to provide the licensing service with the license prices, and the management coordinates various components of the automated billing service.

When a user requests a license from the license service, the terms of the license usage are negotiated and they are part of an SLA document. The negotiation is based on application-specific templates and the license cost becomes part of the SLA. The SLA describes all aspects of resource usage, including the ID of application, duration, number of processors, and guarantees, such as the maximum cost and deadlines. When multiple negotiation steps are necessary, the WS-Agreement negotiation protocol is used.

To understand the complexity of the issues related to software licensing, we point out some of the difficulties related to authorization. To verify the authorization to use a license, an application must have the certificate of an authority. This certificate must be available locally to the application because the application may be executed in an environment with restricted network access. This opens up the possibility for an administrator to hijack the license mechanism by exchanging the local certificate.

18.7 Summary

This chapter discussed issues related to cloudware operations and management. It described the portfolio of core services (including discovery and replication, load balancing, resource management, data governance, management services, and fault tolerance) and functionality (including Service management, Metering/billing, Data management, monitoring, security management, SLAs and QoS management, performance management, policy management, self-service, support and incident management, analytics, orchestration, transformation, logging/audit trails, mediation, integration, rules management, and semantic engine). It described metrics for interfacing with service providers and, hence, criteria for their selection. In the last part of the chapter, we discussed SLAs (Service Level Agreements) that assist in evaluating the normalcy of ongoing operations.

19

Cloudware Security

19.1 Governance

We have seen a number of factors that need to be considered when thinking about a move to a cloud infrastructure. The number of potential issues arising from neglecting any aspect of cloud security is large, and the implications can be catastrophic. While some of the items are mostly applicable during the preparation for a switchover to cloud, many of the pertinent activities remain afterward and require careful management. It is the overall management of the business assets that should provide the focus for how security (among other things) should be considered, and this management is generally referred to as governance.

Governance is oft-talked about in the context of corporate governance, rules, practices, customs, policies, and processes, which define and steer how an organization conducts its daily business. Typically, the act of governance results in two key functions:

1. Providing defined chains of authority, responsibility, and communication to empower the appropriate staff to take decisions
2. Defining mechanisms for policy and control, together with the associated measurements, to facilitate roles for staff to undertake

As you would expect, corporate governance is all encompassing when it comes to the operations, strategy, and future trajectory of an organization. It has tended to have been regarded as applicable to more traditional demarcations within business, such as the accounting, sales, human resource management, or production functions, to name a few. However, as we have seen in this book and in Chapter 9 specifically (Enterprise Cloud Computing), the opportunities created by consuming IT as a utility have the potential to completely transform how business is conducted.

As such, there is emerging recognition that the impact of IT on a business is crucial to its survival, and as more and more legislative and policy controls now assume the use of IT, the need to govern the IT function is now paramount. As we have found elsewhere in this book, cloud computing presents a convergence of the IT and business functions, and thus the governance of IT is a logical next step for organizations who choose to embrace the cloud.

19.1.1 IT Governance

The practice of IT governance needs to understand the different ways in which IT services are delivered and maintained in the presence of a cloud infrastructure. For instance, for each of the issues identified, the governance body needs to understand the implications of noncompliance with regulatory, ethical, business, and political constraints, as well as be suitably prepared for unforeseen circumstances in the future. The management of this activity requires delegation to a body that can oversee and marshal the holistic perspective while also having the authority to intervene when required. However, it is unlikely that the governing body can understand the totality of detail that increasingly complex technology involves, so consideration of the following items is needed:

1. While IT governance is concerned with the technical capabilities and requirements of technology, the adoption of cloud services, together with the associated delegation of authority (but not responsibility) to external partners, reinforces the need to understand the issues that affect the business. Governing IT that includes clouds means that an intimate understanding of the responsibility boundaries is as important as whether there is sufficient bandwidth at each point of access.

2. A governance perspective usefully tends to bring to the fore the potential myriad of services that may already exist. This list of services is likely to increase in the future, so it is important to empower the governing board to be able to rationalize and standardize where practicable. This in itself demands a blend of technical and business expertise that the governing board can call upon to inform its operations.

3. The pace of change is rapid, and cloud adoption will only accelerate this change. Meaningful data are required to inform the decisions taken, and therefore it is necessary to not only understand what metrics need to be monitored and reported, but the mechanisms that do this must be automated.

IT governance should be seen as an opportunity to manage the emerging organization that aspires to align its IT and business needs to generate value. It is likely that this will require a risk-based approach to management, which of course is hugely dependent on the quality of information used to assess and monitor risk. Standards such as COBIT assist the creation of benchmarks upon which monitoring can take place, but the resulting actions that are derived from this approach should not be underestimated. How many in-house applications are subject to security auditing, for instance? What are the implications of executing tested but not audited code on public clouds?

The ability to govern effectively means that an organization must ensure that it can attribute cause to effect, be flexible in its operations, and have the capability to accurately monitor its activities.

19.1.2 Security

The first release of the Cloud Security Alliance (CSA) report in 2010 identified seven top threats to cloud computing:

1. *Abuse of the cloud* refers to the ability to conduct nefarious activities from the cloud—for instance, using multiple *AWS* instances or applications supported by *IaaS* to launch Distributed Denial-of-Service (DDoS) attacks (which prevent legitimate users from assessing cloud services) or to distribute spam and malware.

2. *Shared technology* considers threats due to multitenant access supported by virtualization. VMMs can have flaws allowing a guest operating system to affect the security of the platform shared with other virtual machines.

3. *Insecure APIs* may not protect users during a range of activities, starting with authentication and access control to monitoring and application control during runtime mode.

4. *Malicious insiders* risk arises because the cloud service providers do not disclose their hiring standards and policies; potential harm due to this particular form of attack is quite substantial.

5. *Data loss* or *leakage* risks arise because proprietary or sensitive data maybe permanently lost when cloud data replication fails and is also followed by a storage media failure; similarly, inadvertent or unauthorised access to such information by third parties can have severe consequences. Since, maintaining copies of the data outside the cloud is often unfeasible due to the sheer volume of data, both of these risks can have devastating consequences for an individual or an organization using cloud services.

6. *Account* or service *hijacking refers to stealing of credentials* and is a significant threat.

7. *Unknown risk profile* refers to exposure to the ignorance or underestimation of the very risks of cloud computing.

According to this report, the IaaS delivery model can be affected by all threats. PaaS can be affected by all but the shared technology, whereas SaaS is affected by all but abuse and shared technology.

19.1.3 Privacy

The advent of the Internet has transformed the familiar issues of security and privacy beyond recognition. This is primarily because of the following:

- Every computer can be accessed and influenced by any other computer anywhere on the Internet; this effectively eliminates the concept of locality.
- Businesses that are supported on such an extended and open IT-substrate are vulnerable to scrutiny and monitoring by interested parties.
- Browsing and visitations of websites to gather information in turn also exposes the visitor's behavior itself to interpretation and analyses.

Consumer privacy issues and concerns have a drastic effect on the enterprise's ability to market to, connect to, and create an ongoing relationship with your customers. Consumers are gravely concerned regarding the abuse of their privacy, but gathering a certain amount of information is necessary for companies to personalize and to serve their customers better. Thus, there is a need for a balance between protecting a consumer's privacy and the need for enterprises to target and personalize their offerings to the customers. In 2000, US Federal Trade Commission recommended fair information practices of *notice, choice, access,* and *security.* Enterprises need to adhere to these practices when creating and implementing their privacy policies.

Consumer-oriented commercial websites that collect personal identifying information from or about consumers online would be required to comply with the four widely accepted fair information practices:

1. *Notice*: Websites would be required to provide consumers clear and conspicuous notice of their information practices, including what information they collect; how they collect it (e.g., directly or through nonobvious means such as cookies); how they use it; how they provide Choice, Access, and Security to consumers; whether they disclose the information collected to other entities; and whether other entities are collecting information through the site.
2. *Choice*: Websites would be required to offer consumers choices as to how their personal identifying information is used beyond the reason for which the information was provided (e.g., to consummate a transaction). Such choice would encompass both internal secondary uses (such as marketing back to consumers) and external secondary uses (such as disclosing data to other entities).
3. *Access*: Websites would be required to offer consumers reasonable access to the information a Website has collected about them,

including a reasonable opportunity to review information and to correct inaccuracies or delete information.

4. *Security*: Websites would be required to take reasonable steps to protect the security of the information they collect from consumers. The Commission recognizes that the implementation of these practices may vary with the nature of the information collected and the uses to which it is put, as well as with technological developments. For this reason, the Commission recommends that any legislation be phrased in general terms and be technologically neutral. Thus, the definitions of fair information practices set forth in the statute should be broad enough to provide flexibility to the implementing agency in promulgating its rules or regulations.

Third-party privacy seals are a good way to gain the trust of consumers because these third-party consumer privacy protection organizations certify the enterprise's privacy policy. There are two types of privacy seal programs. One has strict guidelines that prohibit sites from sharing consumer information they collect with other business partners or from using it for direct marketing programs. Secure Assure, which has 200 member companies, offers an audit program but requires members to adhere to a stringent privacy guideline stating that the e-business will never share a consumer's private information with a third party. Other privacy seal programs award stamps-of-approval to sites that simply stick to whatever privacy policy promises they have made. The oldest and most well known privacy seal program is Electronic Frontier Foundation's TRUSTe (www.truste.org), which was started in 1997 and has more than 1300 companies as members.

The threat to information privacy and security can never be eliminated, but controls and technologies can be applied to reduce the risks to acceptable levels. The challenges faced by the various enterprises in this regard are described below.

19.1.4 Trust

Trust in the context of cloud computing is intimately related to the general problem of trust in online Activities. The Internet offers individuals the ability to obscure or conceal their identities. The resulting anonymity reduces the cues normally used in judgments of trust. The identity is critical for developing trust relations; it allows us to base our trust on the past history of interactions with an entity. Anonymity causes mistrust because identity is associated with accountability, and, in the absence of identity, accountability cannot be enforced. The opacity extends immediately from identity to personal characteristics. It is impossible to infer whether the entity or individual we transact with is who it pretends to be, since the transactions occur between entities separated in time and distance. Finally, there are no guarantees that the entities we transact with fully understand the role they have assumed.

To compensate for the loss of clues, we need security mechanisms for access control, transparency of identity, and surveillance. The mechanisms for access control are designed to keep intruders and mischievous agents out. Identity transparency requires that the relationship between a virtual agent and a physical person be carefully checked through methods such as biometric identification. Digital signatures and digital certificates are used for identification. Credentials are used when an entity is not known. Credentials are issued by a trusted authority and describe the qualities of the entity using the credential. A Doctor of Dental Surgery diploma hanging on the wall of a dentist's office is a credential that the individual has been trained by an accredited university and hence is capable of performing a set of dental procedures; similarly, a digital signature is a credential used in many distributed applications. Surveillance could be based on intrusion detection or on logging and auditing. The first option is based on real-time monitoring, the second on off-line sifting through audit records.

There are primarily two ways of determining trust, namely, Policies and Reputation. Policies reveal the conditions to obtain trust and the actions to take when some of the conditions are met. Policies require the verification of credentials. Reputation is a quality attributed to an entity based on a relatively long history of interactions with or possibly observations of the entity. Recommendations are based on trust decisions made by others and filtered through the perspective of the entity assessing the trust.

19.2 Security Risks

It is important to understand the risks of inadequate security so that an enterprise can make an informed judgment about what, if any, information should be trusted to the cloud.

Since the actual risks to a system are varied, an enterprise typically takes a generalized approach to security and then manages exceptions separately, for instance, identity management; it is usual for an employee to require a user identity for access to a system when on the organization's premises. But this access has a different set of potential vulnerabilities if the employee is working at home or in the field. These specific situations might not apply to the cloud provider, who will by default create security strategies that are relevant for that type of business.

The individual security mechanisms that a number of applications use may not transfer easily to a cloud environment, and therefore, a detailed understanding of the approach taken toward security is required if further, unintended vulnerabilities are not to be introduced. This situation is not new; organizations have been outsourcing data storage and telephone call centers for some time now. What is different about cloud, however, is the depth of the infrastructure that is being entrusted into the cloud. Both data storage and customer care management are discrete, vertical functions that have been devolved to third parties. The devolvement of infrastructure/applications/services is a horizontal function that contains the heart of the organization's operations.

Such is the potential complexity of the situation that enterprises adopt a risk-based approach to security. This is where controls are prioritized toward areas where security risks are the most damaging. One part of a risk-based approach is to ensure that service-level agreements (SLAs) are in place. However, SLAs are often used to protect the supplier, not the customer, again underlining the importance of understanding the security detail so that the requirements are catered for properly. So, even though you may be impressed with the physical security during the sales tour of the cloud provider's premises, it is still your responsibility to ensure that all of the other aspects of security are assured as well.

Access control is an example of perimeter security. Without an account and a password, you cannot penetrate the perimeter of the network. This assumes though that those who have an account are honest and trustworthy. Unfortunately, most breaches in security are the result of employees who have legitimate access, and they are rarely detected. These internal threats don't go away if you move to the cloud, unless the cloud provider offers a more secure service that you can utilize, which adds protection over and above what you are currently using.

When dealing with security, it helps to be paranoid. Migrating systems to the cloud might increase the headcount of people who have access to your data, so your security strategy must have a provision to deal with threats from the inside. While virtualization is seen as an example of a specific technology that has enabled the cloud delivery model to become workable, it also complicates the demands placed upon a security strategy as servers, storage, and even networks are now executing in virtual environments. Rather than the traditional risk of an employee divulging a password or snooping around a system, an employee of a cloud provider might only have to provide access to the virtualization layer for havoc to be wreaked. In fact, if we consider the elasticity function of a cloud resource, any hacker would have plenty of compute resource to use for nefarious purposes if granted some access to a cloud.

The provision of IT security is a challenge, and the effort required to do it can make the migration to cloud appear an attractive one if it reduces the hassle. However, the enterprise is placing trust in its provider and needs to assure itself that the provider's capabilities are at least as good as the current architecture. Another factor is that today's IT systems are complex and becoming even more intricate and bespoke. The traditional model of security is to define a hard security perimeter (usually around the data center) and monitor all inbound and outbound traffic. The problem with multitenant clouds is that all sorts of traffic will be present at the access point, and in fact another enterprise's traffic might be deemed as hostile, even though the public cloud architecture is designed to have multiple sets of data coexisting in one virtual appliance. This arrangement means that it is meaningless to have systems in place that can both monitor and proactively protect against potential breaches. The system has to be secure at the point of access.

The reality is that breaches happen, and then enterprises need to quickly plug the hole, trying to understand what went wrong afterward. In the case of serious breaches, the default behavior might be to shut a system down completely, which is massively disruptive for a business. Finally, when an enterprise is choosing a potential cloud provider, it may find it difficult to assess a candidate supplier, since they don't release details of their internal services for the purposes of maintaining security.

19.3 Dimensions of Security

As described earlier, a move to a cloud provider means that an enterprise will have to establish a level of trust with the provider. The process of building trust involves open communication and sufficient understanding on the part of the enterprise, to ask the pertinent questions. We shall now consider six key functional areas of security, in order to derive a checklist of security fundamentals that must be present within a cloud system before migrating to a public cloud:

1. Identity management
2. Network security
3. Data security
4. Instance security
5. Application architecture
6. Patch management

19.3.1 Identity Management

The first area is that of identity management. The nature of cloud services means that identity management is paramount if end users can securely access the services that they need to do their jobs. Since we anticipate a user to be interacting with a business process that is composed of one or more cloud services (see), we wouldn't expect the user to have to manage separate access details for each separate service that was invoked. In fact, in a Web browser environment (the default interface for cloud services), this would be particularly dangerous as users would simply let their passwords be saved for convenience. So, along comes another person, who does not have access to the payroll reports, and they use a computer where the passwords have been saved in the Web browser. Chaos ensues! Single sign-on (SSO) is one example of a system where user identities are managed across a number of separate systems. The user signs in with their account, and they are automatically authorized to access business processes that they have been granted permission for. Behind the scenes, this needs a role-based permission system so that access rights can be quickly assembled, maintained, and revoked, for individuals and en masse.

An additional benefit is that all users of the services benefit from the simplification of SSO, which in an service-oriented environment (SOE) setting means the suppliers as well. The IT department also finds that SSO assists the management of user profiles in that, firstly, they won't have to manage individual accounts on a per-service basis (a lot of work) and, secondly, much of the account maintenance can be automated. For example, a default set of identities can be created for a number of job functions, which can be automatically provisioned for new users. These may either be suitable already or can be augmented with other capabilities quickly. This of course, reinforces the need to have a comprehensive understanding of a service-based architecture, in relation to the business that is being conducted. Once an identity management mechanism is in place, account migration to the cloud is simplified.

19.3.2 Network Security

Network security is of course an important concern for any corporate, distributed system. The one issue that a move to a cloud brings is the fact that an enterprise's application network traffic is transported along with every other application's network traffic. This means that packets that are exchanged between secure access points are mingling with packets that are exchanged between less secure applications. While cloud providers appear to segregate network traffic by utilizing virtual local area networks (VLAN), the separation is virtual (as the name implies), and therefore, at packet level, the traffic is still mixed and shares the same cable. So, the sensitive accounts data for payroll is present on the network, along with the (relatively) less sensitive sales figures for the last quarter.

These data can't be accessed without the correct permissions, but it does mean that an employee with network administration rights could, with a bit of work, have sight of the confidential information, even though they would not have any operational need to do so. The traditional model of security in this case has relied on trust, but in the context of a cloud environment (where the network administrator is not directly on your payroll), there is a need to actively control exactly who has access to what.

Since cloud adoption means that an enterprise has devolved all responsibility for the infrastructure, it is not possible to build a private, physical network, nor is it good practice to trust the honesty of a VLAN administrator. The solution in this case is to provide end-to-end encryption of data packets, between authorized applications. The fear of packets coexisting with other packets has much greater ramifications when the owner of the other packets might be a competitor. This is bound to occur at some point if access to the cloud services is made over publicly shared Internet connections (which they are). It is possible to acquire a private leased line to a cloud provider, but this creates an architecture that is more like a traditional data center.

At some point, an enterprise is going to need access to a service that is outside of this infrastructure. The more comprehensive alternative is to create a Virtual Private Network (VPN) connection to the cloud provider. This can be terminated upon entry to the cloud provider, or for maximum security, it can be terminated at the application. Of course, it is rare to get something for nothing. Data encryption is costly in terms of processing overhead and directly increases network latency, reducing throughput performance. It is therefore prudent to consider what data need encrypting, and what the risks would be if it ever got released publicly. You would expect an enterprise's salary information to be kept private, but a lot of the standard transactional reporting data might be safe enough in a VLAN scenario. It follows that a security risk assessment should be an integral part of the planning for a security strategy.

19.3.3 Data Security

There is no question that data are the key asset of an organization. Data security is therefore a fundamental component of the security strategy for a cloud migration. If data are lost, or inaccessible, then the effect can disarm an enterprise. There are many instances where an enterprise suffers a major infrastructure failure and the data are securely backed up somewhere, but the systems cannot be reconstructed quickly enough afterward, causing revenue losses. One of the apparent comforts of owning your hardware is that you feel in control when disaster strikes. If you delegate this control to a cloud provider, what measures should be taken?

In terms of responsibility, the cloud provider is wholly accountable for the provision and maintenance of hardware. Since a consumer of cloud services no longer has the responsibility for the infrastructure, they need to find a way of dealing with the overall responsibilities associated with managing data security, such as protecting against data theft or malicious change and compliance with legislative measures for the transport, storage, and expunging of data.

In terms of data transport, this can be dealt with by utilizing VPN mechanisms as described earlier. In terms of data storage, a move to the cloud means that by default external parties now have access to the infrastructure that the data reside on and, by implication, also have access to the data themselves. If an enterprise assumes a paranoid stance, then the only option is to encrypt the data and concentrate upon developing a secure mechanism for encryption key management, stored in a place where they cannot be accessed by external parties. The administrators of the cloud provider can of course still see data; it's just that they can't make sense of it.

While the discussion has separated network security from data security, the reality is that they should be considered together when planning the security strategy. A risk assessment will determine which parts of the system must be encrypted, and if transport encryption is required, then the data must be sensitive enough to protect, so storage encryption will be required also. The extent of this security will be determined by what the likely risk of data leakage will be, as a trade-off against reduced network performance.

Another major shift in security thinking for cloud deployments is created by the dynamic environment of virtualization. Traditional approaches to IT security assume a static infrastructure that expands in a planned, orderly way. If more storage is required, it is designed, incorporated into the overall security strategy, and then implemented. Security policies are amended if need be, and new procedures commissioned accordingly. Usually, the data store is deep within a hardened security perimeter. In terms of off-premise data centers, this is certainly the case.

19.3.4 Instance Security

However, the agile, collaborative environment of the cloud, which exposes internal services for external consumption, which dynamically provisions extra compute and storage resources on demand, is a more challenging beast to tame. An enterprise must now be more concerned with instance security. The secure data store is now a virtual entity, composed of a number of secure repositories that are associated with individual service instances. Whereas the scope of a traditional security model was that of the system to be secured, the scope is now limited to a particular instance but also is multiplied by the number of instances that are executing at any one time. It follows that centralized management of security is more complex, and more of the security controls require delegating to the individual services themselves. Instance security is provided in the following ways:

1. *Instance-level firewall*: Typically, the cloud provider will provide a firewall for each VLAN that is present. This firewall serves to virtually separate the traffic between user's respective VLANs. Bearing in mind the caveat mentioned earlier that the physical separation of traffic is not implemented with VLANs, it is necessary to ensure that each instance has a firewall to marshal only authorized traffic into the associated virtual machine. As this relates to application security, it is clearly not the responsibility of the cloud provider and therefore may require extra in-house expertise developing on the part of the consuming enterprise. Individual instance firewalls are controlled on a fine-grained basis, and it is likely that different instances will present different requirements. However, this is the maintenance cost of ensuring that only the appropriate data are passed onto each instance.

2. *Daemons/background services*: Anyone who has set up an externally exposed server and then *hardened* the build will be aware of background services that can be exploited by those with malicious intent. Each instance must be assessed to understand what operating system services are required to complete the job and ensure that nothing else is enabled.

3. *Penetration testing*: There are two parts to this activity. First, an audit mechanism should identify if the existing security measures have been properly implemented. The outcome of this might be a list of items that need attention to ensure that the overall security strategy is maintained. Secondly, a series of invasive tests simulate the effects of the instances in response to external attack. These tests place the system under load, which may expose hitherto undetected vulnerabilities. In practice, such auditing and testing is at the fringe of the relationship between an enterprise and a cloud provider, since the provision of shared services to a number of enterprises potentially puts all of them at risk if one particular enterprise starts doing penetration tests. Cloud providers have responded in two ways. The first is to have vulnerability testing as a cloud service (SaaS), which can perform some of the work required. The second approach is to develop in-house penetration capabilities at the cloud provider, who will conduct tests and present a list of recommendations for the enterprise to consider.

4. *Intrusion detection/prevention*: Intrusion Detection Systems (IDS) monitor network traffic and report anomalies in relation to predetermined security policies. The logs generated can then be used to identify areas that may need hardening, or they may be used as part of a forensic investigation after a breach has occurred. An Intrusion Prevention System (IPS) takes any anomalous behavior and proactively alters a firewall to prevent a recurrence of the behavior by stopping the traffic from accessing the instance. For cloud environments, host-based variants of IDS/IPS are required (HIDS/HIPS) for each application instance that consumes external traffic. Again, the extent to which this security measure is deployed will be dependent upon the risk profile produced by the initial assessment.

5. *Application auditing*: There are still cases where unwanted intruders can circumvent network security and gain access to applications. Automated application auditing monitors the applications that are installed and raises an alarm when files are changed. This is often referred to as file change monitoring and can be applied to any application or system files where changes would not normally be expected.

6. *Antivirus*: One approach to prevent viruses or malware being installed is to robustly prevent users from installing applications themselves. However, this does not stop e-mails being opened nor, in the case of

the cloud-based SOE, does it mean that the business partners of an enterprise enforce such policies. In the same way that other vulnerabilities at instance level need to be protected, the same applies to the prevention of malicious programs being allowed to penetrate and infect the service. Each instance should ideally be checked with the latest signature file every time that it is executed to maintain maximum protection. This does create a significant overhead for services that are in frequent use, and therefore a cloud-based antivirus/malware service may be more appropriate. Such a service ensures that the latest signature files are present, without using the execution of a service to trigger an external check.

In summary, the concept of instance security for cloud services does not so much rely upon new technology, but more a rethink in terms of how existing solutions are deployed.

19.3.5 Application Architecture

A common approach to application architecture is that of separating the architecture into tiers, whereby communication access between tiers is tightly controlled. This serves to constrain any problems in one tier without adversely affecting the other. For instance, a Web application tier might be kept separate from the back-office system tier. In between the tiers would be a firewall that restricts the network traffic between the two tiers.

As mentioned earlier in the chapter, the multitenant environment of a public cloud prevents physical network infrastructure from being inserted. At best, an enterprise could create VLANs, albeit that the network traffic is still physically intermingled. One approach is to seek out a cloud provider who is willing to allow the user to define subnets as part of the rented infrastructure. This would permit an enterprise to replicate some of its more traditional architecture within a virtualized environment and achieve its desired network topology. Another approach would be to persist with instance-level management and implement tier separation at the instance firewalls. This increases complexity somewhat, but it could be argued that it fits more cohesively with the *instance approach* to managing security in a robust way by treating each appliance as a discrete service provider.

An alternative option is to adopt the services of a cloud provider to help in the management of security. Some cloud providers are now offering security layers that mimic instance-level firewalls. Such layers are convenient to use, though they have the potential to increase vendor lock-in until more open security standards are developed.

19.3.6 Patch Management

Patch management refers to the constant checking and maintenance of software during its use. As bugs and vulnerabilities are discovered, the corrections may result in software patches that need to be applied. The resourcing and management of software patching is perhaps one of the motivators for considering cloud adoption, since anything relating to the services that are provided will be managed by the cloud provider. SaaS is the best example of fully delegating the responsibility to the provider; for IaaS, it only applies to the infrastructure itself, not the systems or services that run on them.

However, one aspect that enterprises should consider is the way in which a cloud provider manages the installation of patches, particularly those cloud providers who originate from a history of being a data center. Traditionally, images of the system would be created by the data center, which was a snapshot of a particular instance. This could then be deployed rapidly, without having to build an installation from scratch every time. In the era of rapid provisioning, this means that new images have to be created each time a patch is installed, resulting in lots of images being created. As the cloud computing industry matures, more and more cloud providers will reject this practice and utilize automation to dynamically build instances on top of very basic images. This ensures that the latest software updates are incorporated into the instance but also permits extra instances to be dynamically provisioned to enable service elasticity. Thus, software patches can be kept in one repository (and therefore managed) and be called upon only when they are required.

This is an important issue to consider for an enterprise. The IT department does not want to be investing significant resources into protecting instances, only to find that the instance is built upon an image with a known vulnerability.

19.4 Cloud Security Concerns

The top security concerns of cloud users are as follows:

1. *Availability*: Cloud requirements for availability are concerned with denying illegitimate access to computing resources and preventing external attacks such as denial-of-service attacks. Additional issues to address include attempts by malicious entities to control, destroy, or damage computing resources and deny legitimate access to systems. While availability is being preserved, confidentiality and integrity have to be maintained. Requirements for this category should address how to ensure that computing resources are available to authorized users when needed.

2. *Authentication*: Cloud requirements for authentication specify the means of authenticating a user when the user is requesting service on a cloud resource and presenting his or her identity. The authentication must be performed in a secure manner. Strong authentication

using a public key certificate should be employed to bind a user to an identity. Exchanged information should not be alterable. This safeguard can be accomplished using a certificate-based digital signature. Some corresponding requirements include the following:

a. Mechanisms for determining identity

b. Binding of a resource to an identity

c. Identification of communication origins

d. Management of out-of-band authentication means

e. Reaffirmations of identities

3. *Authorization*: Subsequent to authentication, cloud requirements for authorization address authorization to allow access to resources, including the following:

a. A user requesting that specified services not be applied to his or her message traffic

b. Bases for negative or positive responses

c. Specifying responses to requests for services in a simple and clear manner

d. Including the type of service and the identity of the user in an authorization to access services

e. Identification of entities that have the authority to set authorization rules between users and services

f. Means for the provider of services to identify the user and associated traffic

g. Means for the user to acquire information concerning the service profile kept by the service provider on the user

Consequent to the authorization, the system must address the following:

a. Specific mechanisms to provide for access control

b. Privileges assigned to subjects during the system's life

c. Management of access control subsystems

4. *Integrity*: Cloud requirements for integrity ensure the integrity of data both in transit and in storage. It should also specify means to recover from detectable errors, such as deletions, insertions, and modifications. The means to protect the integrity of information include access control policies and decisions regarding who can transmit and receive data, and which information can be exchanged. Derived requirements for integrity should address the following:

a. Validating the data origin

b. Detecting the alteration of data

c. Determining whether the data origin has changed

5. *Confidentiality*: Cloud requirements for confidentiality are concerned with protecting data during transfers between entities. A policy defines the requirements for ensuring the confidentiality of data by preventing unauthorized disclosure of information being sent between two end points. The policy should specify who can exchange information and what type of data can be exchanged. Related issues include intellectual property rights, access control, encryption, inference, anonymity, and covert channels. These policy statements should translate into requirements that address the following:

    a. Mechanisms that should be applied to enforce authorization

    b. What form of information is provided to the user and what the user can view

    c. The means of identity establishment

    d. What other types of confidentiality utilities should be used

6. *Auditing*: Cloud requirements for auditing include the following:

    a. Determination of the audit's scope

    b. Determination of the audit's objectives

    c. Validation of the audit plan

    d. Identification of necessary resources

    e. Conduct of the audit

    f. Documentation of the audit

    g. Validation of the audit results

    h. Report of final results

The audit should also consider organizational characteristics such as supervisory issues, institutional ethics, compensation policies, organizational history, and the business environment. In particular, the following elements of the cloud system management should be considered:

1. Organizational roles and responsibilities

    a. Separation of duties

2. IS management

    a. Qualifications of IS staff

    b. IS training

3. Third party–provided services

    a. Managing of contracts

    b. Service-level agreements (SLAs)

4. Infrastructure Management
   a. Capacity management
   b. Database administration
   c. Information system security management
   d. Business continuity management
5. Quality management and assurance standards
6. Change management
7. Problem management
8. Project management
   a. Performance management and indicators
9. Economic performance
10. Expense management and monitoring

The cloud policy decomposition for the audit component is recursive in that the audit has to address the cloud system security policy, standards, guidelines, and procedures. It should also delineate the three basic types of controls, which are preventive, detective, and corrective; and it should provide the basis for a qualitative audit risk assessment that includes the following:

- Identification of all relevant assets
- Valuation of the assets
- Identification of threats
- Identification of regulatory requirements
- Identification of organizational risk requirements
- Identification of the likelihood of threat occurrence
- Definition of organizational entities or subgroupings
- Review of previous audits
- Determination of audit budget constraints

Users are greatly concerned about the legal framework for enforcing cloud computing security. The cloud technology has moved much faster than cloud security and privacy legislation, so users have legitimate concerns regarding the ability to defend their rights. Because the data centers of a cloud system may be located in several countries, it is difficult to understand which laws apply—the laws of the country where information is stored and processed, the laws of the countries where the information crossed from the user to the data center, or the laws of the country where the user is located.

19.5 Cloud Security Solutions

Here is an overview of what cloud users can and should do to minimize security risks regarding data handling by the Cloud Service Provider (CSP). First, users should evaluate the security policies and the mechanism the CSP has in place to enforce these policies. Then users should analyze the information that would be stored and processed on the cloud. Finally, the contractual obligations should be clearly spelled out. The contract between the user and the CSP should do the following:

- State explicitly the CSP's obligations to securely handle sensitive information and its obligation to comply with privacy laws
- Spell out CSP liabilities for mishandling sensitive information
- Spell out CSP liabilities for data loss
- Spell out the rules governing the ownership of the data
- Specify the geographical regions where information and backups can be stored

To minimize security risks, a user may try to avoid processing sensitive data on a cloud.

19.5.1 Aspects of Cloud Security Solutions

19.5.1.1 Operating System Security

Operating System (OS) is a complex software system consisting of millions of lines of code, and it is vulnerable to a wide range of malicious attacks. An OS does not insulate completely one application from another, and once an application is compromised, the entire physical platform and all applications running on it can be affected. The platform security level is thus reduced to the security level of the most vulnerable application running on the platform. Operating systems provide only weak mechanisms for applications to authenticate to one another and do not have a trusted path between users and applications. These shortcomings add to the challenges of providing security in a distributed computing environment.

An OS allows multiple applications to share the hardware resources of a physical system, subject to a set of policies. A critical function of an OS is to protect applications against a wide range of malicious attacks such as unauthorized access to privileged information, tampering with executable code, and spoofing. Such attacks can now target even single-user systems such as personal computers, tablets, or smartphones. Data brought into the system may contain malicious code; this could occur via a Java applet, or data imported by a browser from a malicious Website. The existence of trusted paths, mechanisms supporting user interactions with trusted software, is critical to system security. If such mechanisms do not exist, malicious software can impersonate trusted software. Some systems provide trust paths for a few functions such as log-in authentication and password changing and allow servers to authenticate their clients. A trusted-path mechanism is required to prevent malicious software invoked by an authorized application to tamper with the attributes of the object and/or with the policy rules.

A highly secure operating system is necessary but not sufficient unto itself; application-specific security is also necessary. Sometimes security implemented above the operating system is better. This is the case for electronic commerce that requires a digital signature on each transaction. Applications with special privileges that perform security-related functions are called trusted applications. Such applications should only be allowed the lowest level of privileges required to perform their functions. For example, type enforcement is a mandatory security mechanism that can be used to restrict a trusted application to the lowest level of privileges.

19.5.1.2 *Virtual Machine (VM) Security*

VM technology provides a stricter isolation of virtual machines from one another than the isolation of processes in a traditional operating system. Indeed, a VMM controls the execution of privileged operations and can thus enforce memory isolation as well as disk and network access. The VMMs are considerably less complex and better structured than traditional operating systems; thus, they are in a better position to respond to security attacks. A major challenge is that a VMM sees only raw data regarding the state of a guest operating system, whereas security services typically operate at a higher logical level, for example, at the level of a file rather than a disk block. Virtual security services are typically provided by the VMM or through a dedicated security services VM. A secure trusted computing base (TCB) is a necessary condition for security in a virtual machine environment; if the TCB is compromised, the security of the entire system is affected.

A guest OS runs on simulated hardware, and the VMM has access to the state of all virtual machines operating on the same hardware. The state of a guest virtual machine can be saved, restored, cloned, and encrypted by the VMM. Not only can replication ensure reliability, it can also support security, whereas cloning could be used to recognize a malicious application by testing it on a cloned system and observing whether it behaves normally.

One of the most significant aspects of virtualization is that the complete state of an operating system running under a virtual machine is captured by the VM. This state can be saved in a file and then the file can be copied and shared. Thus, creating a VM reduces ultimately to copying a file; therefore there will be a natural explosion in the number of VMs, and the only limitation for the number of VMs is the amount of storage space available. While traditional organizations install and maintain the same version of system software, in a virtual environment the number of different operating systems, their versions, and the patch status of each version will be very diverse, taxing the support team. A side effect of the ability to record in a file the complete state of a VM is the possibility to roll back a VM. This opens wide the door for a new type of vulnerability caused by events recorded in the memory of an attacker.

In case of an infection, in nonvirtual environments, once it is detected, the infected systems are quarantined and then cleaned up. The systems will then behave normally until the next episode of infection occurs. However, in case of virtual environments, the infected VMs may be dormant at the time when the measures to clean up the systems are taken and then, at a later time, they could wake up and infect other systems. This scenario can repeat itself indefinitely.

Another undesirable effect of the virtual environment affects the trust. *Trust is conditioned by the ability to guarantee the identity of entities involved.* Each computer system in a network has a unique physical, or *MAC*, address; the uniqueness of this address guarantees that an infected or malicious system can be identified and then cleaned, shut down, or denied network access. This process breaks down for virtual systems when VMs are created dynamically. Often, to avoid name collision, a random *MAC* address is assigned to a new VM.

There is price to be paid for the better security provided by virtualization. This price includes: higher hardware costs, because a virtual system requires more resources, such as CPU cycles, memory, disk, and network bandwidth; the cost of developing VMMs and modifying the host operating systems in case of paravirtualization; and the overhead of virtualization because the VMM is involved in privileged operations.

19.5.1.3 Security Threats from Shared VM Images

One of the major security risks, especially associated with the IaaS cloud delivery model, is the sharing of VM images like Amazon Machine Images (AMIs).

19.6 Cloudware Security, Governance, Risk, and Compliance

Cloud computing is a combination of virtualization, process automation, and dynamic response to changing application conditions. None of these, on its own, are anything more than a logical extension of existing IT; the combination, however, changes the way IT operates.

Some of the changes are as follows:

- *Dynamism*: The rapid provisioning of computing resources, as well as rapid change in application topologies as computing resources are dynamically added and subtracted in response to changing application load.

First of all, cloud computing leverages virtualization, which breaks the association between application and physical server. Consequently, assuming security can be tied to physical resources is no longer practical. Cloud computing extends virtualization to add dynamism—the ability for applications to rapidly change deployment topology. Moreover, user self-service (NIST Cloud Computing characteristic Number One) means that the assumptions of extended deployment timelines, with sufficient opportunity for security review and implementation prior to moving an application into production, are no longer valid.

- Pooled resources cloud computing abstracts use from assets, where use (i.e., application operation) is not associated with a particular set of computing resources, but instead is hosted in a general pool of computing resources. This means that the location of specific application components may change from time to time as loads are rebalanced within the resource pool. It also means that security measures must not be associated with specific hardware, but must instead migrate dynamically along with the application as it moves from one set of computing resources to another.

  In a pooled resource environment, no user controls the infrastructure, so the common appliance solution is not possible—after all, one user's traffic examination appliance is another user's intrusion threat. Consequently, the shared environment of Cloud Computing negates many traditional security practices.

- Security deperimeterization: Because cloud applications operate in a dynamic, shared resource pool, traditional security solutions are often unusable. Relying on a network-attached appliance to examine all network traffic is unworkable, due to restrictions imposed by cloud providers. Moreover, the ongoing opening up of applications to external parties like partners and customers also means that the traditional model of imposing strong security at the data center perimeter (i.e., relying on a restrictive firewall to prevent traffic from reaching internal resources) is unsustainable as well. The new model of security requires that each endpoint implements security measures to protect itself as appropriate.

These changes, combined, mean that the traditional models of security, governance, and compliance all change in the world of cloud computing. In the areas of security and compliance, the cloud user and cloud provider both hold some of the responsibility. The interface between where one party's responsibility ends and the other begins may be referred to as the trust boundary. In its basic form, the trust boundary represents a demarcation line: on one side of the line, the cloud provider possesses responsibility for security measures; on the other, the cloud user possesses responsibility (see Figure 13.2). On the cloud provider's side of the trust boundary, the user is a passive assessor of what the cloud provider implements in terms of security practices. On the cloud user's side of the trust boundary, the user is an active implementer of security practices.

The location of the trust boundary varies according to what model of cloud computing is being used: IaaS, PaaS, or SaaS. Each model has the cloud provider taking on differing levels of responsibility for the total application, and thereby affects where the trust boundary is located. The thick black zigzag line indicates where the trust boundary lies for each Cloud delivery model. As an example, in a PaaS environment, the Cloud provider is responsible for the security of the infrastructure and the middleware, while the Cloud user retains responsibility for the security of the application itself. As you can see from the descriptions, this means that the Cloud user would need to audit and evaluate whether the security measures of the provider in its areas of responsibility are sufficient.

Every Cloud provider offers a security framework into which users integrate their application. Naturally, every provider has a somewhat different framework, so it is incumbent upon users to understand the framework and ensure that they integrate with it properly. In fact, it is more than crucial. Without understanding the security framework presented by the Cloud provider, it is likely that the Cloud user will fail to configure its usage properly, and thereby leave security vulnerabilities that may be exploited by attackers.

Figure 13.2 is a chart of security responsibilities of each of the three Cloud delivery models, along three key areas of responsibility: infrastructure, operating system (and middleware), and application. Here are brief descriptions of each area of responsibility:

1. Application: This area refers to the software used to provide the actual functionality of the application itself. Falling under this area are topics like: software component version verification, patch installation practices, application identity management, and the like.
2. Operating system and middleware: This area refers to software components that provide the operating environment within which the application runs. Key security issues for this area of responsibility include whether appropriate security software is installed within the OS, patch installation practices, administrative access to manage these components, and so on.

3. Infrastructure: This area addresses both physical security and software security. Physical security refers both to the physical infrastructure of the cloud computing environment (i.e., the data center itself) and the security practices surrounding the physical infrastructure. For example, this area covers whether the data center has redundant Internet access methods, as well as what practices are in place regarding access to the physical facility (e.g., requiring both identification documents and biometric scanning as prerequisites for entering the data center facility). Physical infrastructure would also refer to the hardware within the data center like back-up generators and so on. Finally, infrastructure also refers to the software infrastructure used to implement the cloud computing environment. Most cloud computing (though not all) uses virtualization as a foundation for the cloud computing environment, so this security area would cover the virtualization hypervisor, including security practices related to controlling access to logging into administer the virtualization and ensuring proper security patches are installed.

19.7 Assessing a Cloud Service Provider

A report by the European Network and Information Security Agency suggests the following security risks as being priorities for cloud-specific architectures:

1. *Loss of governance*: Governance is complicated by the fact that some responsibilities are delegated to the cloud provider, but the lines of responsibility do not fall across traditional boundaries. These boundaries are less well established and may need to be debated with the cloud provider in terms of what responsibility they will wholly adopt and what responsibility needs to be shared. The traditional approach is to draw up service-level agreements (SLAs), though these, without sufficient scrutiny, may leave security gaps.

2. *Lock-in*: At the time of writing, there is a lack of consensus with regard to tools, procedures, or standard data formats to enable data, application, and service portability. Cloud consumers who wish to migrate to other cloud platforms in the future may find the costs too prohibitive, therefore increasing the dependency between a cloud provider and consumer.

3. *Isolation failure*: The pooling of computing resources among multiple tenants is a classic cloud environment. It is therefore necessary to have mechanisms in place to isolate any failures to the minimum number of instances possible. This can also be the source of

a security vulnerability exploit (*guest hopping* attacks), although it is more challenging to successfully attack a hypervisor (where the resource is isolated from the infrastructure) rather than a traditional operating system architecture.

4. *Compliance risks*: Enterprises that have invested heavily in the attainment of industry standard or regulatory certification may risk noncompliance if the cloud provider cannot evidence their compliance or if the cloud provider does not permit audit of its own facilities. In practice, this means that certain certifications cannot be achieved or maintained when using public clouds.

5. *Management interface compromise*: The web-based customer management interfaces that cloud providers supply are also an extra risk with regard to extra opportunities to compromise a cloud system.

6. *Data protection*: An enterprise may not be able to verify how a cloud provider handles its data and therefore cannot establish whether the practices employed are lawful. In the case of federated clouds, where different clouds are linked by a trusted network, this challenge is more complicated. Some cloud providers have achieved certified status with regard to data handling.

7. *Insecure or incomplete data deletion*: The request to delete a cloud may not result in the data being completely expunged, and there may be instances when immediate wiping is not desirable.

> The main purpose of any interaction with a potential cloud provider is to establish whether their security strategy is harmonious with a particular enterprise. This becomes more difficult as an enterprise moves up the cloud computing stack. For instance, if an enterprise is seeking IaaS, then all of the OS security upward is the responsibility of the enterprise, not the cloud provider. If PaaS is required, then there is some responsibility for the provider to ensure that the OS and platform layers are secure, but again, role-based permissions that are part of the application layer are ultimately the responsibility of the consuming enterprise. This becomes more complex if access control is associated with OS-level security. For SaaS, the provider maintains the access to the application, but access within the application may be managed by the customer.

Again, this becomes ever more complicated as external services are consumed, thus reinforcing the need to understand security in the context of an enterprise, before a cloud migration takes place. A solid implementation of security in a service-based environment is much easier to transfer to the cloud, irrespective of the level in the stack that is required. The key concept

TABLE 19.1

Security Risk Areas When Selecting a Potential Public Cloud Provider

Issue	Questions for Potential Public Cloud Providers
Architecture	Is the provider's security architecture available for scrutiny? What is the architecture for access management?
Risk assessment	Do you utilize an independent authority to assess and monitor security risks?
Legislation, compliance, and governance	What controls do you have in place to ensure that domain-specific legislation is complied with?
Information location	Where will the information reside?
Segregation	Will the applications/tools be shared with other tenants? Which application/tools will be shared?
Service level	What service level is to be guaranteed and what measures are in place for access to data during downtime? What is the scope of any penalties for downtime/loss of access?
Portability	What standards are employed to guarantee data/application/tools/process portability?
Physical security	To what standard is physical security provided?
Management tools	How are software updates and patches managed to minimize service disruption? What monitoring tools are provided?
Perimeter security	What controls are in place for firewalls and the management of Virtual Private Network (VPN) access?
Encryption	What standards for encryption are in place? How are public keys managed? How is single sign-on (SSO) implemented?

here is to establish where the boundary of trust exists between the consumer and the provider; this needs to be established up front to prevent costly confusion in the future. Table 19.1 illustrates some pertinent issues to raise with a potential cloud provider.

19.7.1 Requisite Certifications

It is evident that to stand any chance of successfully evaluating a cloud offering requires considerable expertise. Arguably, the expertise can only be acquired by undertaking the provision of cloud services itself. The nature of IT is that it is a domain that is constantly faced with newly emerging technologies, approaches, and models, and therefore it is not atypical to be faced with the business case driving a fundamental change, without fully comprehending the impact that this change is likely to make.

The normal response from the IT industry is to create a body of industrial partners, many of whom will have a vested interest in the products that are on offer (it's usually a technology for sale). Industrial parties are joined by representatives from government or trade bodies and sometimes from regulatory agencies. Once formed, the body works toward a standard that can be used to harmonize the approaches taken toward the adoption of the new technology, including the specification of new products that adhere to the standard. The implementation of a standard makes the process of evaluation, and ultimately comparison with other technologies, much more straightforward and also enables comparisons to be made more effectively with technologies that are clearly different to the standard.

Unfortunately, the standard itself does not mean that an end user is proficient enough to evaluate against it correctly, and there is always the risk of bias affecting any qualitative judgments that have favorable market consequences for the enterprise concerned. This is normally addressed by the use of an impartial third party, who can conduct the evaluation without any vested interest and, through the process of auditing, can certify that a given standard or standards have been complied with. This process is more rigorous in terms of quality assurance and more expedient in that end users are not repeatedly conducting evaluations, an activity that they are not practiced at.

The certification process consists of an approved auditor inspecting the system or infrastructure to be scrutinized, and then making an assessment against a set of formal criteria. Satisfaction must be achieved in the criteria assessed in order to be awarded a certificate of compliance. Having such a scheme in place is invaluable when faced with the prospect of selecting a provider of services; a simple filter is whether the candidate service provider has the relevant certification. From then on, there is the assurance that the standards for that particular domain have been certified.

Table 19.2 summarizes some of the more prevalent standards that are applicable to cloud computing. The multiplicity of standards does imply overlap, and this can create complications for organizations engaged in a variety of industrial domains (public cloud providers, for instance). Having said this, should an enterprise who already has certified compliance with a standard decide to select a cloud provider, the details of the compliance will make comparison with the cloud provider's offering more straightforward.

The Cloud Security Alliance (CSA) is a not-for-profit organization that provides recommendations for the planning and implementation of security in cloud systems. Its mission is as follows:

To promote the use of best practices for providing security assurance within Cloud Computing, and provide education on the uses of Cloud Computing to help secure all other forms of computing.

https://cloudsecurityalliance.org/

It comprises technology vendors, users, security experts, and service providers, who collaborate to establish industrial standards for the execution of secure cloud environments. The CSA has an international remit and organizes conferences and local meetings to exchange ideas with regard to cloud security. This focus upon security has led to the publication of research into cloud computing, including user's experiences of compliance with the myriad standards.

TABLE 19.2

Cloud Computing Provisioning IT Certification Standards

Standard	Remit
Control Objectives for Information and Related Technology (COBIT)	A set of process declarations that describe how IT should be managed by an organization.
National Institute of Standards and Testing (NIST) SP 800-53	The quality assurance of secure information provision to US government agencies, being audited against the Federal Information Security Management Act (FIMSA).
Federal Risk and Authorization Management Program (FedRAMP)	Quality assurance is achieved by collectively achieving multiple certifications that are compliant with FIMSA. This is intended for large IT infrastructures where compliance can be a largely repetitive process.
ISO/IEC 27001:2005—Information Technology, security techniques, information security management systems—requirements	Security controls to assure the quality of information service provision.
Statement on Standards for Attestation Engagements (SSAE) No. 16, Reporting on Controls at a Service Organization	This standard supersedes the Statement on Auditing Standards (SAS) No. 70. Service Organizations. SSAE 16 describes controls for organizations that provide services to users, including an assessment of the reliability and consistency of process execution.
Generally Accepted Privacy Principles (GAPP)	This standard is primarily concerned with information privacy policies and practices.

19.8 Summary

This chapter dealt with the security characteristics and challenges of cloud computing environments. The chapter started with an introduction to the requirements of governance and security for an enterprise. It described the various dimensions of security essential for an enterprise followed by a priority list of security concerns related to the operations of cloudware applications. The later part of the chapter briefly sketched aspects of security solutions at the Operating System (OS) and Virtual Machine (VM) levels. In the end, it described issues related to the assessment and selection of a cloud service provider (CSP).

20

Migrating to Cloudware

Cloud computing is an on-demand access to a shared pool of computing resources. It helps consumers to reduce costs, reduce management responsibilities, and increase business agility. For this reason, it is becoming a popular paradigm, and increasingly more companies are shifting toward IT cloud computing solutions. Advantages are many but, being a new paradigm, there are also challenges and inherent issues.

20.1 Cloud Computing

Cloud computing is an important development in the landscape of computing options. Although cloud technologies have been around for some time, it has only been until very recently that the cloud as a business model has gathered momentum. Cloud is viewed as a business–IT innovation that is expected to accelerate the pace of some of the changes mentioned earlier as well as an opportunity to foster new business models and innovative solutions. Cloud technologies have come a long way and are maturing fast along the evolving business models, and companies need to look beyond the hype to fully exploit these emerging opportunities. Most of the conversations revolve around placing the strategy focus on the technology often at the expense of real business benefits. The cloud hype proposes that cloud adoption is a sure path to business benefits, often not articulating in balance the cloud option against the trade-offs and interdependencies.

There is a need for a business-focused strategy that takes account of cloud and its capabilities, and accords within the enterprises business and technology portfolio. IT decision makers need to determine where and how cloud could offer business value for their enterprises by establishing a strong foundation for a long-term evolution toward cloud and other emerging options as they mature over time. Just like any (information) technology solution, cloud computing offers tools and techniques that may or may not fit business needs. Hence, it is imperative to consider the impact of cloud adoption on short-, medium-, and long-term business strategy and operational delivery.

Companies of all sizes have increasingly been affected by unpredictable and uncertain changing business environments in recent years:

1. Externally, current business landscape is driven by new hurdles and opportunities such as challenging economy, emerging technologies, increasing commoditization of technology, shorter information life cycle, increased transparency along the supply chains, changing customer demands and preferences, evolving issues around transparency, privacy, security, as well as associated regulations, compliance, and standards.

2. Internally, enterprises are challenged by tighter budgets and emphasis on managing investments effectively, reducing costs, managing the evolving complexity of markets, and managing new relationships including developing new business models to differentiate and stay competitive.

3. Technologically, enterprises are challenged for harnessing the technology capabilities for efficient and effective business delivery, process, and operational efficiency, strategically aligning IT resources to business needs for optimized business benefit realization, managing risk, and many more.

Addressing these concerns and maintaining growth and delivering business value has more than ever driven enterprises to focus on their core competencies by carefully designing and leveraging the unique and differentiated processes, architectures, skills and competencies, and relationships across their business ecosystems. IT as a valuable strategic asset has also been the center of attention due to its significant business impact and also the investment requirements and the need to lower operations cost, deriving business benefits, improved productivity and performance, and streamlined processes and services, just to name a few.

Strategic value of cloud is demonstrated when it plays a key role in an enterprise's achievement of overall business strategy. As an ingredient of the IT portfolio, cloud strategy needs to be focused on business outcomes and plays a significant role in optimizing and improving core value chain processes and drives innovation that enables new technology-enabled product and service revenue streams. The results can be translated into and measured by improved customer satisfaction and market share.

Therefore, the underpinning premises of the planning for cloud include the following:

- Going into the cloud requires strategic planning that leverages the business effectiveness.
- A strategic planning framework can help ensure that cloud investments are aligned with and support the strategic business objectives.

- Cloud can contribute to developing a sustainable competitive advantage.
- Cloud and other emerging technologies should be recognized as strategic opportunities.
- Cloud and other emerging technologies need to be managed in a wider context of a sustainable IT portfolio.
- Business and enterprise architectures should be made cloud compliant to link the multiple strategic business and technology priorities and operational delivery.
- Internal and external business environments will have bearing on the cloud options and provision configuration.
- Cloud adoption will imply business and organizational change and can in turn drive or support it.
- Perhaps most importantly, it is imperative to strike a balance between business and information technology.

20.2 Planning for Migration

As enterprises take the strategic information era further into the innovation era, they need to be able to develop a fitting strategic view of the nexus between their Information Technology and business performance and in the way it impacts their competitive position. Cloud computing reduced barriers to entry; accelerates innovation cycles, as well as time to value and markets; and commoditizes the once unique technical capabilities companies invested in to design and develop. The dynamics are creating a new paradigm: commoditize to differentiate. Enterprises of all sizes now have incredible opportunity to review their current business processes and technology portfolios and identify scope for simplification and new efficiencies and focus on their core business where they can develop differentiating capacities. This exercise requires smart planning not only of technological order but rather a very important business activity. Enterprises need to appreciate the relationship between technology in general and cloud fit, more specifically, to their business and develop a holistic view of impacts of various business models and scenarios in the context of their industry and their business. It is important to realize the need to look beyond the boundaries of their enterprises and consider the value chains and complex ecosystems that make up the future successful cloud adoptions. Although it is stated that switching cost from one cloud provider to another is significantly lower than the equivalent traditional IT, the industry is still taking shape and the risks need to be fully accounted for as well. The outcome would be developing integrative business–IT strategies that leverage cloud for business benefits realization.

It is important to consider cloud in its totality in relation to as-is and the to-be context of the business. This means that companies should see cloud as a mix of many options in terms of business models, deployment models, and technology architecture. Business benefits will often be achieved as a result of careful and smart integration and configuration of various cloud options.

Developing a cloud strategy is not easy and can be a complex task even more so if these strategies need to support business strategies. How these strategies are formulated and managed will depend on the organizational structure, culture, and organizational pressures both external and internal. The environment may be stable and allow for structured planning to take place, or the enterprise may be operating in a volatile changing environment where there is a requirement to respond to immediate issues and develop emergent strategies fast.

Hence, a cloud computing strategy needs to aim to

- Articulate the value, benefits, risks, and trade-offs of cloud computing
- Provide a decision framework to evaluate the cloud opportunity in relation to strategic business and technology direction
- Support with development of an appropriate target architecture
- Support with adopting an appropriate cloud choice and infrastructure to deliver
- Provide a plan and methodology for successful adoption
- Identify activities, roles, and responsibilities for catalyzing cloud adoption
- Identify measures, targets, and initiatives that support monitoring, optimization, and continuous improvement

Strategic planning for cloud is the effective and sustainable management of cloud provision and its optimal technology and business impact. It also includes the inter- and intraorganizational elements such as change management, process management, governance, managing policies, and service-level agreements. Broadly speaking, there are three areas that should be considered in any such strategic planning exercise:

1. Managing the long-term and optimal impact of cloud provision in the context of enterprise including management of the cloud service and the inter- and intrarelationships with the stakeholders including the user community, service providers and other stakeholders, policies, and process management.

   This level of strategy is predominantly concerned with management controls for cloud and broader IT systems, management roles and responsibilities, cloud performance measurement, optimization, and continuous improvement.

2. Managing the long-term and optimal impact of cloud in the broader context of the information systems strategy.

This level of strategy will be primarily concerned with aligning cloud investment with business requirements and seeking strategic advantage from cloud provision. The typical approach is to formulate these strategies following formal enterprise's strategic planning frameworks to align with the enterprise's business strategy. The resulting action plan should consist of a mix of short-term and medium-term application requirements following a thorough review of current technology and systems use and emerging technology issues.

3. Managing the long-run and optimal impact of cloud in the broader context of underlying IT infrastructure and application portfolio.

The cloud strategy will be primarily concerned with the technical policies relating to cloud architecture (and its fit within the broader IT architectures), including risks and service-level agreements (SLAs). It seeks to provide a framework within which the cloud provision can deliver services required by the users. It is heavily influenced by the CIO and IT specialists and is likely to be a single corporate strategy. Separate federated cloud strategies may be required if there are very different needs in particular business units.

In practice, all three of these are interlinked and need to be managed accordingly to ensure a coherent and strategic cloud provision.

The planning process must address the following aspects:

- What is the current state of the business in terms of the management and utilization of current business and supporting information system and technologies: *as-is* (baseline/current state) analysis
- What is the desired state of the enterprise in terms of the management and utilization of business assets and IT resources: *to-be* (target/future state) analysis
- How can the enterprise close any strategic gaps identified in its management and utilization of cloud as an IT resource: realization plan
- How can the resulting information systems plans be implemented successfully: go live
- How to review progress through effective business metrics and control structures and procedures: maintenance

What is clearly evident is that the whole process of strategy development relies on different groups of stakeholders in the enterprise working together to assess business and process requirements and the information needs and the steps toward developing implementable strategy. The caveat to this is that there are wide-ranging issues that need to be mitigated to ensure that adoption of the cloud strategy can be successful. The issues encompass various areas including internal competencies and experience of the strategy development as well as cloud delivery teams, organizational change management, project management, technical solution management, and data governance, just to name a few. It is therefore imperative that the cloud business case and plan consider a holistic view that cuts across people, systems, architectures and business processes for successful adoption and optimizing business benefits.

> The method we have described in the following is a generic approach toward adoption of most systems including cloud and respective organizational requirements. In practice, it is of course necessary to modify or extend the plan to suit specific needs. One of the key tasks before applying the model is to produce an organizations-specific adoption plan. To do so, you need to review various steps and building blocks for relevance and fit and adapt them appropriately to the circumstances of the enterprise. Depending on the maturity of the enterprise in terms of their business and enterprise architectures and alignment of current systems adoption processes with their business model, the cloud adoption plan may involve the following changes:
>
> - Organizations may opt for either an as-is or to-be as the first approach depending on their level of understanding of cloud and business and technology maturity. In as-is, first, an assessment of the current state landscape informs the gaps and opportunities for cloud adoption, whereas in the latter, a target solution is elaborated and mapped back to current state; gaps and change requirements are identified and assessed to inform the realization plans.
> - In case of SMEs, it might be more appropriate to use a *cutdown* model adjusted to their typically lower resource levels and system complexity. The plan adaptation is a function of practical assessment of resource and competency availability and the value that can realistically be delivered.

20.3 Deployment Model Scenarios

In the early days of service-oriented architecture (SOA), there was great debate in the industry about where to begin with SOA and Web Services—internally within the four walls of your enterprise or externally with customer- and partner-facing Web Services. Central to the debate were the challenges around Web Services security.

Cloud has surfaced the very same challenges, especially in the data security and privacy arena. Many of the cloud security challenges are those that have been by and large addressed through SOA and Web Services security standards and solutions, fundamentally because cloud is so heavily dependent on SOA and Web Services as the means of exposing prepackaged cloud-enabled resources and capabilities via interfaces that can be discovered, bound to, and leveraged via an service-level agreement (SLA).

The question of where to begin with cloud revolves around the same internal versus external debate we had with SOA. In cloud vernacular, this relates to the various cloud deployment models available—internal private clouds, external public clouds, and lastly hybrid clouds that blend private and public cloud capabilities. There are a variety of advantages for starting with each of the three: public clouds, private clouds, and hybrid clouds. The respective reasons for beginning your cloud initiative with private, public, or hybrid cloud deployment models will vary by industry and business need. Every organization must justify its decision based on tolerance for risk, stance toward emerging technology adoption, and other factors.

20.3.1 Public Cloud

Public cloud deployments, in which an organization migrates an application, its data, or a business process onto a third-party cloud service provider's platform via the Internet, are excellent ways to begin exploring cloud computing in a cost-effective and agile rapid time-to-market fashion. Leveraging various cloud offerings from Amazon, Google, Salesforce, and others is an excellent way to explore what cloud can offer to your enterprise.

The following is a list of reasons an organization would choose to begin its cloud computing initiative with a public cloud service:

- Low cost. Public clouds offer a very low cost of entry into cloud computing, which supports a POC or pilot project with limited research and development (R&D) funding.
- Cloud solution variety. There is a wide variety of cloud-enabled resources to assemble into complete cloud solutions, from virtualization and cloud operating system (OS) or platform technologies to Platforms-as-a-Service (PaaS) and Software-as-a-Service (SaaS) offerings.
- Low risk. An organization can quickly experiment with cloud computing solutions with minimal risk exposure.
- Pay for what you need/use. Public clouds are based on a completely variable, utility cost model, whereby once the initial project has

completed, or if you no longer need the cloud services, you can stop paying the fees.

- Rapid accumulation of knowledge, skills, and experience. Public clouds offer a way to quickly gain experience, knowledge, and skills on the emerging technology trend of cloud computing. Leveraging public clouds enables your organization to tap into the knowledge and experience of your third-party cloud service provider. This is a tremendous competitive advantage for any organization seeking first-mover advantage for its cloud computing strategy.

20.3.2 Private Cloud

Private cloud deployment scenarios, in which an organization implements cloud technologies on its internal network, or its Intranet, behind its security firewalls, enable the organization to explore cloud capabilities internally without the risk exposure of moving its data or applications outside of its own internal and corporate security controls.

The following is a list of reasons an organization would choose to begin its cloud computing initiative with a private cloud deployment model:

- Security and privacy. Mitigates privacy and security concerns by maintaining data behind your own firewalls.
- Strategic opacity. Maintains strategic opacity, so your competitors cannot ascertain your intentions.
- Focus on internal optimization first. Internally optimize internal utilization of infrastructure assets.
- Become an internal cloud service provider. Beginning your cloud strategy with a private cloud focus will accelerate your ability to become an internal cloud service provider to the enterprise. This is a key benefit of beginning your cloud initiative internally with a private cloud deployment model.

20.3.3 Hybrid Cloud

Hybrid clouds leverage aspects of both public and private clouds to address a broader set of operational use cases and business scenarios. For example, an organization may use private cloud capabilities to federate two data centers and optimize utilization and availability of computer, storage, and network resources and may also in parallel leverage public cloud capabilities from Amazon to offer a new application or service accessible via Amazon's e-commerce storefronts. This hybrid cloud mixes multiple cloud patterns to satisfy this requirement example.

The following is a list of reasons an organization would choose to begin its cloud computing initiative with a hybrid cloud deployment model:

- Begin with the end game. A hybrid cloud deployment as your cloud start point supports the ultimate end state of cloud computing. Most industry analysts feel that in a short time, there will be only hybrid clouds, and the separation into public and private clouds is an artificial distinction given the infancy of the cloud industry.
- Cloud solution range. Hybrid clouds offer a great magnitude of solution variations that address business models and solutions that we can barely imagine now. Why not begin with hybrid clouds early to better understand what the true potential of cloud is in the bigger picture? There is no reason why you should constrain your learning process out of the gate. Hybrid clouds offer that to you.
- Explore cloud-based business models. Hybrid clouds allow you to explore and create new business models that exploit the combination of private and public cloud use cases. In this manner, you can actually explore business model innovation through new channels to market and new distribution models of internal processes across your extended value chain. Hybrid clouds offer this unique experience to your organization.
- Extra-enterprise thinking. Hybrid clouds encourage extra-enterprise thinking with respect to business processes, cloud solutions, and capabilities. If you begin your cloud initiative with an extended enterprise frame of reference, you will be in a better position to innovate your business model, operations model, and business processes by leveraging cloud solutions.
- End-state knowledge acceleration. Beginning with hybrid clouds allows your organization to practice the cloud end state sooner by learning, gaining expertise, and accelerating the knowledge accumulation your team will benefit from in the short and long term. The more you understand about how cloud will evolve and the sooner you develop that understanding, the sooner you can exploit first-mover advantage.

20.4 Cloud Adoption Plan

The cloud adoption plan should ensure close alignment of the target architecture goals, measures, and initiatives with the strategic business objectives that address the operational requirements of the business.

20.4.1 As-Is (Baseline/Current State) Analysis

The assessment of the current broader information systems and supporting technology landscape indicates the existing capabilities and resources in support of the current business strategy. There will almost certainly be gaps between the current resources and competencies and those needed to satisfy the cloud-enabled information system. The main deliverable of the as-is analysis is establishing business direction and needs stemming from its strategy and business architecture and the collection of demands from the current business operations in its internal and external context.

20.4.1.1 Analyzing the Business Context and Technology Requirements and Opportunities

This stage of analysis considers the macrocontext of the business. It is important to indentify the trends that can affect the business positively or negatively. This has strategic significance and cloud affects the choice of cloud model an organization opts for or the pace of change required and potential risks that might affect the adoption plans.

It is also important to consider the level at which the framework is applied. It is useful to consider the local as well as global context of the business in the cloud. Similarly, it may be helpful to narrow down to a particular part of the business and analyze the impact of cloud at that level as it will provide an opportunity to focus on more relevant and specific influences and issues.

Some of the other concepts, tools, and techniques that help with internal and external analysis include the following:

- Porter's five forces analysis to identify possible supply chain linkages with.
- Customers and suppliers and other actors such as service providers or service brokers should also assist in evaluating stakeholder activity and their interests in the cloud ecosystem.
- Value chain analysis to identify if cloud could improve efficiency or effectiveness or provide competitive advantage through linkages within the internal and external value chains.
- SWOT analysis for each business.

20.4.1.2 Analyzing the As-Is Business Architecture

This stage of analysis requires critical evaluation of the business environment, business and technology drivers, business processes, and organizational change program. In essence, it is an audit of the enterprise's current context, direction, and capabilities. This exercise is similar to an enterprise architecture development, and its purpose is to establish the business case for cloud by modeling a clear understanding of the vision and existing level of business and technical maturity to determine what factors need to be introduced or improved in successful path toward the cloud adoption.

Analyzing the cloud in isolation of the business architecture and narrowly focusing the plan on the technical adoption of cloud is unlikely to deliver a holistic picture of existing concerns nor will it realistically address the potential business benefits and the wider impact that it has on the entire organization.

Some examples of useful concepts and tools that can contribute to this stage of analysis are

- Use-case modeling and analysis
- Business process modeling and analysis

20.4.1.3 Analyzing the Current IT and IS Architecture and Systems

At this stage, we want to establish how information systems (IS) contribute to the business and what underlying IT infrastructure and architectures support the IS provision. The key activities involve compilation of the current systems portfolio followed by critical evaluation of effectiveness and efficiencies and any issues or shortcomings that can be addressed. In this context, we need to critically evaluate the nexus between IT and business, that is, IT contribution to business and the extent to which business is able to leverage the existing IT capabilities to achieve strategic objectives and goals. Establishing the strengths and weaknesses of the current provision and identifying potential opportunities and risks will also be an important deliverable at this stage.

Some of the useful concepts, tools, and techniques that can help with this analysis would be the following:

- McFarlan's IT portfolio grid analysis: that is, strategic, high-potential, key operational, support systems.
- Evaluate each system employing the business value add versus delivered technical quality assessments.

20.4.2 To-Be (Target/Future State) Analysis

The key task here is to develop appropriate target/desired cloud solutions. This encompasses both business and technical aspects of cloud deployment. The core focus is to identify, define, and configure cloud-enabled information system architecture and application portfolio that support the business architecture.

At this stage, we also consider what aspects of business operation and what business processes in what part of the business will be affected and how. A detailed definition of target business and technology infrastructure, application, and data architectures in relation to a suitable cloud deployment model will be a key deliverable.

The scope and level of granularity and rigor of decomposition in this exercise will depend on the relevance of the technical elements to attaining the target cloud model, maturity of existing infrastructure, existing architecture models and descriptions, and the extent it can sufficiently enable identification of gaps and the scope of candidate/target cloud provision.

20.4.2.1 Data for the Cloud

The purpose of the to-be architecture for data for cloud is to define the data entities (and their characteristics) relevant to the enterprise that will be processed by the service and cloud-enabled applications. Moving to the cloud may critically require data transformation and migration; hence, one concern will be adopting suitable standards and processes that may affect all data entities and processing services and applications whether existing or in the target applications and architecture.

Undertaking transformation toward cloud requires careful understanding and addressing information life-cycle management and governance issues. Some of these issues relate to risk and security concerns that will constrain the choice of platforms, cloud deployment architecture, service-level agreements, etc. Some of the finer-grain considerations in data management for cloud include access and identification, authentication, confidentiality, availability, privacy, ownership and distribution, retention, and auditability.

Data engineering requirements for the cloud include the following:

- Data architecture (and structure)
- Data management (processes)
- Data migration
- Data integration
- Data governance

20.4.2.2 Applications for the Cloud

The core objective here is to define the target platform and application architecture including cloud services, referenced across the technical and business functions, data requirements, and business processes. The exercise will highlight the extent and nature of interoperability and integration, migration, and operational concerns that need to be addressed. Cloud will not only consolidate resources and change the way the services are consumed and managed but also open new requirements and opportunities due to its properties and limitations.

The objective is to define the relevant types of applications and enable technologies necessary to process the data and information effectively and efficiently to support the business. In this sense, we need to logically group applications in terms of capabilities that manage the data objects defined earlier in the data architecture to support the business functions in the business architecture. Although generally these application building blocks are stable and remain relatively unchanged over time, we are seeking fresh opportunities to improve, simplify, augment, or replace components for the target cloud-enabled systems landscape. The business architecture and the application portfolio analysis in the as-is assessment earlier will inform the list of affected or required logical and physical components in the application stack.

20.4.2.3 Technology and Infrastructure for the Cloud

This aspect of to-be analysis is concerned with defining target cloud technology portfolio components. In this exercise, we map underlying technology and respective application components defined in the application architecture depending on the cloud model of choice. Depending on the business vision, and business architecture and business and technology maturity, the technology deployment may be a hybrid configuration, which means that cloud will coexist within a modified current technology architecture. In this manner, we are making the technology platform cloud enabled to make necessary exchanges across a heterogeneous environment not only possible but optimized for business efficiency and effectiveness. In case of full migration to a new cloud platform again, we need to assess the technology and business readiness.

This element of the to-be analysis is essential for the realization phase.

20.4.3 Realization

Realization plan consists of the activities that the organization needs to maintain to close any strategic and delivery gaps identified in its adoption, utilization, and management of cloud as an IT resource. The key components of this aspect of the cloud plan primarily draw on the iterative assessment of fit–gap analysis in as-is and to-be analyses. Any gaps and dependencies and appropriate solutions need to be identified, analyzed, and consolidated ensuring that any constraints on the adoption or migration plans are addressed.

An important activity here involves assessment, validation, and consolidation of business, functional, and systems requirements through reviewing cloud technology requirements and gaps and solutions proposed. Once consolidated, it is important to identify functional integration points. As cloud services are delivered based on shared services and shared resources, it is important to assess the need and scope for reuse and interoperability. It is likely that business is not fully migrating to cloud and there is still need for utilizing bespoke, COTS (commercial off the shelf), or third-party services; hence, verifying service integration and interoperability is an important task.

The realization plan will be articulated as an overall cloud adoption strategy. The merits of different strategic approaches characterized by pace and scope of change/adoption need to be balanced against business constraints such as funding and resource concerns, technology maturity, existing business capabilities, and opportunity costs. The organization may opt for phased and incremental introduction of cloud or a big bang approach. Cloud computing requires a different strategy to the upgrade of existing IT infrastructure. Formulation of an appropriate strategic approach for this will rely on careful analysis of risks and change requirements.

20.4.3.1 Fit–Gap Analysis

The fit–gap analysis involves linking the as-is and to-be models for business, information system, and IT architectures and compares, contrasts, and rationalizes the trade-offs and interdependencies of the activities to validate the cloud requirement specification (within broader IS/IT portfolio) for organization, business, and technical perspectives. Logically grouping and organizing activities into program and project work packages will feed into the realization and adoption plans and requirements.

Fit–gap analysis is widely used in many problem-solving-based methods and is concerned with validating the issues, solutions, plans, etc. The primary goal at this stage is to highlight any overlaps or shortfalls between the as-is and to-be states. It concentrates on the evaluation of gaps, including further fine-tuning of priorities to consolidate, integrate, and analyze the information to determine the best way to contribute to cloud implementation and transition plans.

There are different categories of gaps to consider relating to business, systems, technologies, and data. Various techniques could be used in this exercise such as creating a simple gap matrix that lists *as-is* and *to-be* components in columns and rows with respective gaps in the intersection. Similarly, to consolidate gaps and assess potential solutions and dependencies, a gaps-solution dependency table will be an easy technique to use. This can serve as a planning tool toward creating project work packages and initiatives.

20.4.3.2 Change Management

Adopting cloud computing will inevitably require change that cuts across the enterprise both in terms of understanding the cloud and its value proposition, and opportunities and different ways it affects the business from new ways to engage with the old and new business applications, possible changes to the current processes, new data related policies, the new operating model that uses "services" as the key organizational construct, the new approach to accounting for cloud and management of IT resources, new roles, skills and competencies.

Change will be part of the business alignment exercise or *organizational readiness* building block in the plan and important work package in the cloud program or project management. The plan should outline the required change, goals and objectives and critical success factors, impact of change, and strategies for implementation. The key activities involve change planning (change plan), change enablement, change management, and embedding the change.

A change and capability maturity assessment early on will help determine readiness factors, qualified and quantified, as well as potential constraints and risks, identified and planned. These form important deliverables of the change/transformation plan.

Change management is a structured approach to transitioning the enterprise, teams, individuals, and culture from the as-is state to the desired target (*to-be*) state. It aims to align the enterprise and strategies with the target architecture through identifying, characterizing, and addressing issues relating to communications, process and system design, workforce management, performance management, and overall governance features that can help to support the organizational and other changes as a result of adopting cloud.

20.4.3.3 Risk Analysis

Cloud adoption comes with its bag of issues, risks, and security concerns. It is important to perform risk assessment to identify and validate potential issues and consolidate appropriate mitigation strategies. This will form an integral part of the transition plans that will feed into the cloud project management. The business and technology drivers for cloud have to be assessed against the risk factors at the earlier stages of the process. The risks can be business, organizational, systems, data, or people related.

Risk mitigation strategy may leverage existing enterprise's governance and capabilities or may require extending or developing capacity and competencies.

As far as security of the cloud is concerned, there is a requirement to implement updated security strategies and governance framework needed to govern cloud operations and services to ensure their long-term stability. Some examples of technical consideration may include policies in relation to service availability, service scalability, performance management, service security policies, acceptable usage, auditing, and data retention.

There are two broad categories of risk:

1. Delivery risk: entails risk of not delivering the required or expected capabilities, for instance, unproven technology provision, reliance of vendors, lack of clarity of scope and deliverables, provider or

technology compliance and standards issues, level of integration/interface required of cloud solution with existing systems, and quality of project management.

2. Benefits risk: entails risk of not achieving expected business benefits such as lack of business–IT alignment, lack of alignment with technical standards and architectures, lack of appropriate security compliance, lack of credibility and measurability of business outcomes, change management requirements, and senior management involvement.

There are various tools and techniques to help with risk assessment. A risk classification matrix, followed by a tiered probability–impact matrix analysis, that is, identify risk elements in terms of their impact and probability being low, medium, or high, serves as a useful technique. Similarly, a risk scorecard and risk matrix can illustrate the risk, quantified impact, and mitigation measures in a structured and traceable manner.

20.4.4 Go Live

The main objective of go-live plan is supporting cloud business value delivery through creation of a viable cloud adoption plan and recommendations. The key activities include assessing the dependencies, costs, and benefits of the various realization projects. The prioritized list of projects will form the basis of the detailed go-live plan.

Some of the key deliverables of this phase include the following:

1. Addressing/establishing management framework capability for cloud transformation aligned with enterprise's strategic planning, operations management, performance management, program and project management, and governance frameworks.

2. Value planning: Identifying business value drivers and assigning them to all relevant activities and projects. This will serve as a benchmark for performance management and monitoring. Analysis and assignment of strategic fit, critical success factors, business key performance indicators, cloud performance indicator, process performance indicators, return on investment, etc., will form a set of criteria and benchmarks to assess business value realization of the cloud project.

3. Cost–benefit analysis and validation using TCO, ROI, and risk models assigned to each relevant project.

4. Resource allocation, coordination and scheduling, and capability planning to establish the resource requirements and operational delivery implications of the project.

5. Aggregating, categorizing, and prioritizing projects to establish project sequence, timescales, timeline, and key milestones in line with value delivery plans.

6. Best practice recommendations for successful implementation.

7. Plan governance process that entails the process of managing and evaluating the cloud plan and its realization success.

8. Integrating and documenting the plan.

20.5 Summary

It is important to prioritize the cloud initiative and the adoption model according to business drivers and priorities. These may be affected by high-level strategic business objectives and the overall business model of the organization. In any case, articulating the organization-specific guiding principles for the plan development at the planning stage will serve an important milestone toward successful adoption of the cloud initiative.

Section IV

Cloudware Applications

This section presents an overview of the areas of cloudware applications of significance for the future, namely, big data, mobile (i.e., enterprise mobilization), and context-aware applications. Chapter 21 defines and identifies the common characteristics of big data applications along with corresponding tools, techniques, and technologies. Chapter 22 deals with enterprise agility and its realization through mobilization applications. Chapter 23 introduces the concept of context-aware applications and explains how they can significantly enhance the efficiency and effectiveness of even routinely occurring transactions.

An end-user application's effectiveness and performance can be enhanced by transforming it from a *bare* transaction to a transaction *clothed* by a surrounding context formed as an aggregate of all relevant decision patterns in the past.

21

Big Data Computing Applications

The rapid growth of the Internet and World Wide Web has led to vast amounts of information available online. In addition, business and government organizations create large amounts of both structured and unstructured information that needs to be processed, analyzed, and linked. It is estimated the amount of information currently stored in a digital form in 2007 at 281 exabytes and the overall compound growth rate at 57% with information in organizations growing at even a faster rate. It is also estimated that 95% of all current information exists in unstructured form with increased data processing requirements compared to structured information. The storing, managing, accessing, and processing of this vast amount of data represent a fundamental need and an immense challenge in order to satisfy needs to search, analyze, mine, and visualize these data as information.

The Web is believed to have well over a trillion Web pages, of which at least 50 billion have been catalogued and indexed by search engines such as Google, making them searchable by all of us. This massive Web content spans well over 100 million domains (i.e., locations where we point our browsers, such as http://www.wikipedia.org). These are themselves growing at a rate of more than 20,000 net domain additions daily. Facebook and Twitter each have over 900 million users, who between them generate over 300 million posts a day (roughly 250 million tweets and over 60 million Facebook updates). Added to this are the over 10,000 credit-card payments made per second, the well over 30 billion point-of-sale transactions per year (via dial-up POS devices), and finally the over 6 billion mobile phones, of which almost 1 billion are smartphones, many of which are GPS-enabled, and which access the Internet for e-commerce, tweets, and post updates on Facebook. Finally, and last but not least, there are the images and videos on YouTube and other sites, which by themselves outstrip all these put together in terms of the sheer volume of data they represent.

21.1 Big Data

This deluge of data, along with emerging techniques and technologies used to handle it, is commonly referred to today as *big data*. Such big data are both valuable and challenging, because of their sheer volume. So much so that the volume of data being created in the current 5 years from 2010 to 2015 will far exceed all the data generated in human history. The Web, where all these data are being produced and reside, consists of millions of servers, with data storage soon to be measured in zettabytes.

Cloud computing provides the opportunity for organizations with limited internal resources to implement large-scale big data computing applications in a cost-effective manner. The fundamental challenges of big data computing are managing and processing exponentially growing data volumes, significantly reducing associated data analysis cycles to support practical, timely applications, and developing new algorithms that can scale to search and process massive amounts of data. The answer to these challenges is a scalable, integrated computer systems hardware and software architecture designed for parallel processing of big data computing applications. This chapter explores the challenges of big data computing.

21.1.1 What Is Big Data?

Big data can be defined as volumes of data available in varying degrees of complexity, generated at different velocities and varying degrees of ambiguity, which cannot be processed using traditional technologies, processing methods, algorithms, or any commercial off-the-shelf solutions.

Data defined as big data include weather; geospatial and GIS data; consumer-driven data from social media; enterprise-generated data from legal, sales, marketing, procurement, finance, and human-resources department; and device-generated data from sensor networks, nuclear plants, x-ray and scanning devices, and airplane engines.

21.1.1.1 Data Volume

The most interesting data for any organization to tap into today are social media data. The amount of data generated by consumers every minute provides extremely important insights into choices, opinions, influences, connections, brand loyalty, brand management, and much more. Social media sites provide not only consumer perspectives but also competitive positioning, trends, and access to communities formed by common interest. Organizations today leverage the social media pages to personalize marketing of products and services to each customer.

Every enterprise has massive amounts of e-mails that are generated by its employees, customers, and executives on a daily basis. These e-mails are all considered an asset of the corporation and need to be managed as such. After Enron and the collapse of many audits in enterprises, the US government mandated that all enterprises should have a clear life-cycle management of e-mails and that e-mails should be available and auditable on a case-by-case basis. There are several examples that come to mind like insider trading, intellectual property, competitive analysis, and much more, to justify governance and management of e-mails.

TABLE 21.1

Scale of Data

Size of Data	Scale of Data
1000 megabytes	1 gigabyte (GB)
1000 gigabytes	1 terabyte (TB)
1000 terabytes	1 petabyte (PB)
1000 petabytes	1 exabyte (EB)
1000 exabytes	1 zettabyte (ZB)
1000 zettabytes	1 yottabyte (YB)

If companies can analyze petabytes of data (equivalent to 20 million four drawer file cabinets filled with text files or 13.3 years of HDTV content) with acceptable performance to discern patterns and anomalies, businesses can begin to make sense of data in new ways. Table 21.1 indicates the escalating scale of data.

The list of features for handling data volume included the following:

- Nontraditional and unorthodox data processing techniques need to be innovated for processing this data type.
- Metadata is essential for processing these data successfully.
- Metrics and KPIs are key to provide visualization.
- Raw data do not need to be stored online for access.
- Processed output needs to be integrated into an enterprise level analytical ecosystem to provide better insights and visibility into the trends and outcomes of business exercises including CRM, optimization of inventory, and clickstream analysis.
- The enterprise data warehouse (EDW) is needed for analytics and reporting.

21.1.1.2 Data Velocity

The business models adopted by Amazon, Facebook, Yahoo, and Google, which became the de facto business models for most web-based companies, operate on the fact that by tracking customer clicks and navigations on the website, you can deliver personalized browsing and shopping experiences. In this process of clickstreams, there are millions of clicks gathered from users at every second, amounting to large volumes of data. These data can be processed, segmented, and modeled to study population behaviors based on time of day, geography, advertisement effectiveness, click behavior, and guided navigation response. The result sets of these models can be stored to create a better experience for the next set of clicks exhibiting similar behaviors. The velocity of data produced by user clicks on any website today is a prime example for big data velocity.

The most popular way to share pictures, music, and data today is via mobile devices. The sheer volume of data that is transmitted by mobile networks provides insights to the providers on the performance of their network, the amount of data processed at each tower, the time of day, the associated geographies, user demographics, location, latencies, and much more. The velocity of data movement is unpredictable and sometimes can cause a network to crash. The data movement and its study have enabled mobile service providers to improve the QoS (quality of service), and associating these data with social media inputs has enabled insights into competitive intelligence.

The list of features for handling data velocity included the following:

- System must be elastic for handling data velocity along with volume.
- System must scale up and scale down as needed without increasing costs.
- System must be able to process data across the infrastructure in the least processing time.
- System throughput should remain stable independent of data velocity.
- System should be able to process data on a distributed platform.

21.1.1.3 Data Variety

Data come in multiple formats as it ranges from e-mails to tweets to social media and sensor data. There is no control over the input data format or the structure of the data. The processing complexity associated with a variety of formats is the availability of appropriate metadata for identifying what is contained in the actual data. This is critical when we process images, audio, video, and large chunks of text. The absence of metadata or partial metadata means processing delays from the ingestion of data to producing the final metrics and, more importantly, in integrating the results with the data warehouse (Tables 21.2 and 21.3).

The list of features for handling data variety included the following:

- Scalability
- Distributed processing capabilities
- Image processing capabilities
- Graph processing capabilities
- Video and audio processing capabilities

21.1.2　Common Characteristics of Big Data Computing Systems

There are several important common characteristics of big data computing systems that distinguish them from other forms of computing:

TABLE 21.2

Value of Big Data across Industries

	Volume of Data	Velocity of Data	Variety of Data	Underutilized Data (*Dark Data*)	Big Data Value Potential
Banking and securities	High	High	Low	Medium	High
Communications and media services	High	High	High	Medium	High
Education	Very low	Very low	Very low	High	Medium
Government	High	Medium	High	High	High
Health-care providers	Medium	High	Medium	Medium	High
Insurance	Medium	Medium	Medium	Medium	Medium
Manufacturing	High	High	High	High	High
Chemicals and natural resources	High	High	High	High	Medium
Retail	High	High	High	Low	High
Transportation	Medium	Medium	Medium	High	Medium
Utilities	Medium	Medium	Medium	Medium	Medium

1. *Principle of colocation of the data and programs or algorithms to perform the computation*: To achieve high performance in big data computing, it is important to minimize the movement of data. This principle—*Move the code to the data*—which was designed into the data-parallel processing architecture implemented by Seisint in 2003, is extremely effective since program size is usually small in comparison to the large data sets processed by big data systems and results in much less network traffic since data can be read locally instead of across the network. In direct contrast to other types of computing and super-computing that utilize data stored in a separate repository or servers and transfer the data to the processing system for computation, big data computing uses distributed data and distributed file systems in which data are located across a cluster of processing nodes and, instead of moving the data, the program or algorithm is transferred to the nodes with the data that need to be processed. This character-istic allows processing algorithms to execute on the nodes where the data reside reducing system overhead and increasing performance.

2. *Programming model utilized*: Big data computing systems utilize a machine-independent approach in which applications are expressed in terms of high-level operations on data and the runtime system transparently controls the scheduling, execution, load balancing, com-munications, and movement of programs and data across the distrib-uted computing cluster. The programming abstraction and language

TABLE 21.3

Industry Use Cases for Big Data

Manufacturing	*Retail*
Product research	Customer relationship management
Engineering analysis	Store location and layout
Predictive maintenance	Fraud detection and prevention
Process and quality metrics	Supply-chain optimization
Distribution optimization	Dynamic pricing
Media and telecommunications	*Financial services*
Network optimization	Algorithmic trading
Customer scoring	Risk analysis
Churn prevention	Fraud detection
Fraud prevention	Portfolio analysis
Energy	*Advertising and public relations*
Smart grid	Demand signaling
Exploration	Targeted advertising
Operational modeling	Sentiment analysis
Power-line sensors	Customer acquisition
Health care and life sciences	*Government*
Pharmacogenomics	Market governance
Bioinformatics	Weapon systems and counter terrorism
Pharmaceutical research	Econometrics
Clinical outcomes research	Health informatics

tools allow the processing to be expressed in terms of dataflows and transformations incorporating new dataflow programming languages and shared libraries of common data manipulation algorithms such as sorting. Conventional supercomputing and distributed computing systems typically utilize machine-dependent programming models that can require low-level programmer control of processing and node communications using conventional imperative programming languages and specialized software packages, which adds complexity to the parallel programming task and reduces programmer productivity. A machine-dependent programming model also requires significant tuning and is more susceptible to single points of failure.

3. *Focus on reliability and availability*: Large-scale systems with hundreds or thousands of processing nodes are inherently more susceptible to hardware failures, communications errors, and software bugs. Big data computing systems are designed to be fault resilient. This includes redundant copies of all data files on disk, storage of intermediate processing results on disk, automatic detection of node or processing failures, and selective recomputation of results.

A processing cluster configured for big data computing is typically able to continue operation with a reduced number of nodes following a node failure with automatic and transparent recovery of incomplete processing.

A final important characteristic of big data computing systems is the inherent scalability of the underlying hardware and software architecture. Big data computing systems can typically be scaled in a linear fashion to accommodate virtually any amount of data or to meet time-critical performance requirements by simply adding additional processing nodes to a system configuration in order to achieve billions of records per second processing rates (BORPS). The number of nodes and processing tasks assigned for a specific application can be variable or fixed depending on the hardware, software, communications, and distributed file system architecture. This scalability allows computing problems once considered to be intractable due to the amount of data required or amount of processing time required to now be feasible and affords opportunities for new breakthroughs in data analysis and information processing.

One of the key characteristics of the cloud is elastic scalability: users can add or subtract resources in almost real time based on changing requirements. The cloud plays an important role within the big data world. Dramatic changes happen when these infrastructure components are combined with the advances in data management. Horizontally expandable and optimized infrastructure supports the practical implementation of big data. Cloudware technologies like virtualization increases the efficiency of the cloud that makes many complex systems easier to optimize. As a result, organizations have the performance and optimization to be able to access data that were previously either unavailable or very hard to collect. Big data platforms are increasingly used as sources of enormous amounts of data about customer preferences, sentiment, and behaviors. Companies can integrate this information with internal sales and product data to gain insight into customer preferences to make more targeted and personalized offers.

21.1.3 Big Data Appliances

Big data analytics applications combine the means for developing and implementing algorithms that must access, consume, and manage data. In essence, the framework relies on a technology ecosystem of components that must be combined in a variety of ways to address each application's requirements, which can range from general information technology (IT) performance scalability to detailed performance improvement objectives associated with specific algorithmic demands. For example, some algorithms expect that massive amounts of data are immediately available quickly, necessitating large amounts of core memory. Other applications may need numerous iterative exchanges of data between different computing nodes, which would require high-speed networks.

The big data technology ecosystem stack may include the following:

1. Scalable storage systems that are used for capturing, manipulating, and analyzing massive data sets.

2. A computing platform, sometimes configured specifically for large-scale analytics, often composed of multiple (typically multicore) processing nodes connected via a high-speed network to memory and disk storage subsystems. These are often referred to as appliances.

3. A data management environment, whose configurations may range from a traditional database management system scaled to massive parallelism to databases configured with alternative distributions and layouts, to newer graph-based or other NoSQL data management schemes.

4. An application development framework to simplify the process of developing, executing, testing, and debugging new application code. This framework should include programming models, development tools, program execution and scheduling, and system configuration and management capabilities.

5. Methods of scalable analytics (including statistical and data mining models) that can be configured by the analysts and other business consumers to help improve the ability to design and build analytical and predictive models.

6. Management processes and tools that are necessary to ensure alignment with the enterprise analytics infrastructure and collaboration among the developers, analysts, and other business users.

21.2 Tools, Techniques, and Technologies of Big Data

21.2.1 Big Data Architecture

Analytical environments are deployed in different architectural models. Even on parallel platforms, many databases are built on a shared everything approach in which the persistent storage and memory components are all shared by the different processing units.

Parallel architectures are classified by what shared resources each processor can directly access. One typically distinguishes shared memory, shared disk, and shared nothing architectures (as depicted in Figure 21.1).

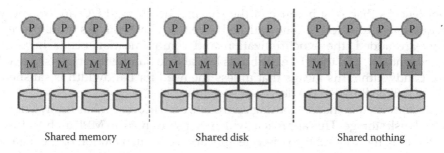

Shared memory   Shared disk   Shared nothing

FIGURE 21.1
Parallel architectures.

1. In a shared memory system, all processors have direct access to all memory via a shared bus. Typical examples are the common symmetric multiprocessor systems, where each processor core can access the complete memory via the shared memory bus. To preserve the abstraction, processor caches, buffering a subset of the data closer to the processor for fast access, have to be kept consistent with specialized protocols. Because disks are typically accessed via the memory, all processes also have access to all disks.

2. In a shared disk architecture, all processes have their own private memory, but all disks are shared. A cluster of computers connected to a SAN is a representative for this architecture.

3. In a shared nothing architecture, each processor has its private memory and private disk. The data is distributed across all disks, and each processor is responsible only for the data on its own connected memory and disks. To operate on data that spans the different memories or disks, the processors have to explicitly send data to other processors. If a processor fails, data held by its memory and disks is unavailable. Therefore, the shared nothing architecture requires special considerations to prevent data loss.

When scaling out the system, the two main bottlenecks are typically the bandwidth of the shared medium and the overhead of maintaining a consistent view of the shared data in the presence of cache hierarchies. For that reason, the shared nothing architecture is considered the most scalable one, because it has no shared medium and no shared data. While it is often argued that shared disk architectures have certain advantages for transaction processing, the shared nothing is the undisputed architecture of choice for analytical queries.

A shared-disk approach may have isolated processors, each with its own memory, but the persistent storage on disk is still shared across the system. These types of architectures are layered on top of SMP machines. While there may be applications that are suited to this approach, there are bottlenecks that exist because of the sharing, because all I/O and memory requests are transferred (and satisfied) over the same bus. As more processors are added, the synchronization and communication needs increase exponentially, and therefore the bus is less able to handle the increased need for bandwidth. This means that unless the need for bandwidth is satisfied, there will be limits to the degree of scalability.

In contrast, in a shared-nothing approach, each processor has its own dedicated disk storage. This approach, which maps nicely to an MPP architecture, is not only more suitable to discrete allocation and distribution of the data, it enables more effective parallelization and consequently does not introduce the same kind of bus bottlenecks from which the SMP/shared-memory and shared-disk approaches suffer. Most big data appliances use a collection of computing resources, typically a combination of processing nodes and storage nodes.

21.2.2 Row versus Column-Oriented Data Layouts

Most traditional database systems employ a row-oriented layout, in which all the values associated with a specific row are laid out consecutively in memory. That layout may work well for transaction processing applications that focus on updating specific records associated with a limited number of transactions (or transaction steps) at a time; these are manifested as algorithmic scans that are performed using multiway joins. Accessing whole rows at a time when only the values of a smaller set of columns are needed may flood the network with extraneous data that are not immediately needed and ultimately will increase the execution time.

Big data analytics applications scan, aggregate, and summarize over massive data sets. Analytical applications and queries will only need to access the data elements needed to satisfy join conditions. With row-oriented layouts, the entire record must be read in order to access the required attributes, with significantly more data read than is needed to satisfy the request. Also, the row-oriented layout is often misaligned with the characteristics of the different types of memory systems (core, cache, disk, etc.), leading to increased access latencies. Subsequently, row-oriented data layouts will not enable the types of joins or aggregations typical of analytic queries to execute with the anticipated level of performance.

Hence, a number of appliances for big data use a database management system that uses an alternate, columnar layout for data that can help to reduce the negative performance impacts of data latency that plague databases with a row-oriented data layout. The values for each column can be stored separately, and because of this, for any query, the system is able to selectively access the specific column values requested to evaluate the join conditions. Instead of requiring separate indexes to tune queries, the data values themselves within each column form the index. This speeds up data access while reducing the overall database footprint, while dramatically improving query performance. The simplicity of the columnar approach provides many benefits, especially for those seeking a high-performance environment to meet the growing needs of extremely large analytic data sets.

21.2.3 NoSQL Data Management

NoSQL suggests environments that combine traditional SQL (or SQL-like query languages) with alternative means of querying and access. NoSQL data systems hold out the promise of greater flexibility in database management while reducing the dependence on more formal database administration. NoSQL databases have more relaxed modeling constraints, which may benefit both the application developer and the end-user analysts when their interactive analyses are not throttled by the need to cast each query in terms of a relational table-based environment.

Different NoSQL frameworks are optimized for different types of analyses. For example, some are implemented as key–value stores, which nicely align to certain big data programming models, while another emerging model is a graph database, in which a graph abstraction is implemented to embed both semantics and connectivity within its structure. In fact, the general concepts for NoSQL include schema-less modeling in which the semantics of the data are embedded within a flexible connectivity and storage model; this provides for automatic distribution of data and elasticity with respect to the use of computing, storage, and network bandwidth in ways that don't force specific binding of data to be persistently stored in particular physical locations. NoSQL databases also provide for integrated data caching that helps reduce data access latency and speed performance.

A relatively simple type of NoSQL data store is a key–value store, a schema-less model in which distinct character strings called keys are associated with values (or sets of values, or even more complex entity objects)—not unlike hash table data structure. If you want to associate multiple values with a single key, you need to consider the representations of the objects and how they are associated with the key. For example, you may want to associate a list of attributes with a single key, which may suggest that the value stored with the key is yet another key–value store object itself.

> The key–value store does not impose any constraints about data typing or data structure—the value associated with the key is the value, and it is up to the consuming business applications to assert expectations about the data values and their semantics and interpretation. This demonstrates the schema-less property of the model.

Key–value stores are essentially very long and presumably thin tables (in that there are not many columns associated with each row). The table's rows can be sorted by the key–value to simplify finding the key during a query. Alternatively, the keys can be hashed using a hash function that maps the key to a particular location (sometimes called a *bucket*) in the table. The representation can grow indefinitely, which makes it good for storing large amounts of data that can be accessed relatively quickly, as well as allows massive amounts of indexed data values to be appended to the same key–value table, which can then be sharded or distributed across the storage nodes. Under the right conditions, the table is distributed in a way that is aligned with the way the keys are organized, so that the hashing function that is used to determine where any specific key exists in the table can also be used to determine which node holds that key's bucket (i.e., the portion of the table holding that key).

NoSQL data management environments are engineered for two key criteria:

1. Fast accessibility, whether that means inserting data into the model or pulling it out via some query or access method
2. Scalability for volume, so as to support the accumulation and management of massive amounts of data

The different approaches are amenable to extensibility, scalability, and distribution, and these characteristics blend nicely with programming models (like MapReduce) with straightforward creation and execution of many parallel processing threads. Distributing a tabular data store or a key–value store allows many queries/accesses to be performed simultaneously, especially when the hashing of the keys maps to different data storage nodes. Employing different data allocation strategies will allow the tables to grow indefinitely without requiring significant rebalancing. In other words, these data organizations are designed for high-performance computing for reporting and analysis.

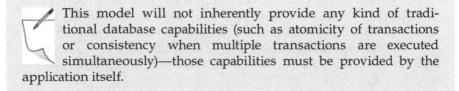 This model will not inherently provide any kind of traditional database capabilities (such as atomicity of transactions or consistency when multiple transactions are executed simultaneously)—those capabilities must be provided by the application itself.

21.2.4 In-Memory Computing

The idea of running databases in memory was used by business intelligence (BI) product company QlikView. In-memory allows the processing of massive quantities of data in main memory to provide immediate results from analysis and transaction. The data to be processed is ideally real-time data or as close to real time as is technically possible. Data in main memory (RAM) can be accessed 100,000 times faster than data on a hard disk; this can dramatically decrease access time to retrieve data and make it available for the purpose of reporting, analytics solutions, or other applications.

The medium used by a database to store data, that is, RAM, is divided into pages. In-memory databases saves changed pages in savepoints, which are asynchronously written to persistent storage in regular intervals. Each committed transaction generates a log entry that is written to nonvolatile storage—this log is written synchronously. In other words, a transaction does not return before the corresponding log entry has been written to persistent storage—in order to meet the durability requirement that was described earlier—thus ensuring that in-memory databases meet (and pass) the ACID test (see Section 5.7, "Transaction Processing Monitors" for a Note on ACID). After a power failure, the database pages are restored from the savepoints; the database logs are applied to restore the changes that were not captured in the savepoints. This ensures that the database can be restored in memory to exactly the same state as before the power failure.

21.2.5 Developing Big Data Applications

For most big data appliances, the ability to achieve scalability to accommodate growing data volumes is predicated on multiprocessing—distributing the computation across the collection of computing nodes in ways that are aligned with the distribution of data across the storage nodes. One of the key objectives of using a multiprocessing node environment is to speed application execution by breaking up large *chunks* of work into much smaller ones that can be farmed out to a pool of available processing nodes. In the best of all possible worlds, the data sets to be consumed and analyzed are also distributed across a pool of storage nodes. As long as there are no dependencies forcing any one specific task to wait to begin until another specific one ends, these smaller tasks can be executed at the same time, that is, *task parallelism*. More than just scalability, it is the concept of *automated scalability* that has generated the present surge of interest in big data analytics (with corresponding optimization of costs).

A good development framework will simplify the process of developing, executing, testing, and debugging new application code, and this framework should include

1. A programming model and development tools
2. Facility for program loading, execution, and for process and thread scheduling
3. System configuration and management tools

The context for all of these framework components is tightly coupled with the key characteristics of a big data application—algorithms that take advantage of running lots of tasks in parallel on many computing nodes to analyze lots of data distributed among many storage nodes. Typically, a big data platform will consist of a collection (or a *pool*) of processing nodes; the optimal performances can be achieved when all the processing nodes are kept busy, and that means maintaining a healthy allocation of tasks to idle nodes within the pool. Any big application that is to be developed must map to this context, and that is where the programming model comes in. The programming model essentially describes two aspects of application execution within a parallel environment:

1. How an application is coded
2. How that code maps to the parallel environment

MapReduce programming model is a combination of the familiar procedural/imperative approaches used by Java or C++ programmers embedded within what is effectively a functional language programming model such as the one used within languages like Lisp and APL. The similarity is based on MapReduce's dependence on two basic operations that are applied to sets or lists of data value pairs:

1. Map, which describes the computation or analysis applied to a set of input key–value pairs to produce a set of intermediate key–value pairs
2. Reduce, in which the set of values associated with the intermediate key–value pairs output by the map operation are combined to provide the results

A MapReduce application is envisioned as a series of basic operations applied in a sequence to small sets of many (millions, billions, or even more) data items. These data items are logically organized in a way that enables the MapReduce execution model to allocate tasks that can be executed in parallel.

Combining both data and computational independence means that both the data and the computations can be distributed across multiple storage and processing units and automatically parallelized. This parallelizability allows the programmer to exploit scalable massively parallel processing resources for increased processing speed and performance.

21.3 Additional Details on Big Data Technologies

21.3.1 Processing Approach

Current big data computing platforms use a *divide and conquer* parallel processing approach combining multiple processors and disks in large computing clusters connected using high-speed communications switches and networks that allows the data to be partitioned among the available computing resources and processed independently to achieve performance and scalability based on the amount of data (Figure 5.1). We define a cluster as "a type of parallel and distributed system, which consists of a collection of inter-connected stand-alone computers working together as a single integrated computing resource."

This approach to parallel processing is often referred to as a *shared-nothing* approach since each node consisting of processor, local memory, and disk resources shares nothing with other nodes in the cluster. In parallel computing, this approach is considered suitable for data processing problems that are *embarrassingly parallel*, that is, where it is relatively easy to separate the problem into a number of parallel tasks and there is no dependency or communication required between the tasks other than overall management of the tasks. These types of data processing problems are inherently adaptable to various forms of distributed computing including clusters and data grids and cloud computing.

21.3.2 Big Data System Architecture

A variety of system architectures have been implemented for big data and large-scale data analysis applications including parallel and distributed relational database management systems that have been available to run on shared-nothing clusters of processing nodes for more than two decades. These include database systems from Teradata, Netezza, Vertica, and Exadata/Oracle, and others, which provide high-performance parallel database platforms. Although these systems have the ability to run parallel applications and queries expressed in the SQL, they are typically not general-purpose processing platforms and usually run as a back-end to a separate front-end application processing system.

Although this approach offers benefits when the data utilized are primarily structured in nature and fits easily into the constraints of a relational database, and often excels for transaction processing applications, most data growth is with data in unstructured form and new processing paradigms with more flexible data models were needed. Internet companies such as Google, Yahoo, Microsoft, Facebook, and others required a new processing approach to effectively deal with the enormous amount of Web data for applications such as search engines and social networking. In addition, many government and business organizations were overwhelmed with data that could not be effectively processed, linked, and analyzed with traditional computing approaches.

Several solutions have emerged including the MapReduce architecture pioneered by Google and now available in an open-source implementation called Hadoop used by Yahoo, Facebook, and others (see Section 17.3, "Hadoop").

21.3.2.1 Brewer's CAP Theorem and the BASE Principle

In the previous section, we briefly discussed techniques for achieving ACID properties in a database system. However, applying these techniques in large-scale scenarios such as data services in the cloud leads to scalability problems: the amount of data to be stored and processed and the transaction and query load to be managed are usually too large to run the database services on a single machine. To overcome this data storage bottleneck, the database must be stored on multiple nodes, for which horizontal scaling is the typically chosen approach.

The database is partitioned across the different nodes: either tablewise or by sharding (see Section 21.3.3 below). Both cases result in a distributed system for which Eric Brewer has formulated the famous CAP theorem, which characterizes three of the main properties of such a system:

1. *Consistency*: All clients have the same view, even in the case of updates. For multisite transactions, this requires all-or-nothing semantics. For replicated data, this implies that all replicas have always consistent states.
2. *Availability*: Availability implies that all clients always find a replica of data even in the presence of failures.
3. *Partition tolerance*: In the case of network failures that split the nodes into groups (partitions), the system is still able to continue the processing.

The CAP theorem further states that in a distributed, shared-data system, these three properties cannot be achieved simultaneously in the presence of failures. In order to understand the implications, we have to consider possible failures. For scalability reasons, the database is running on two sites S1 and S2 sharing a data object o, for example, a flight booking record. This data sharing should be transparent to client applications, that is, an application AS1 connected to site A and AS2 accessing the database via site S2. Both clients should always see the same state of o even in the presence of an update. Hence, in order to ensure a consistent view, any update performed for instance by AS1 and changing o to a new state o has to be propagated by sending a message m to update o at S2 so AS2 reads o. To understand why the CAP theorem holds, we consider the scenario where the network connecting S1 and S2 fails, resulting in a network partitioning and whether all three properties can be simultaneously fulfilled. In this situation, m cannot be delivered resulting in an inconsistent (outdated) value of o at site S2. If we want to avoid this to ensure consistency, m has to be sent synchronously, that is, in an atomic operation with the updates. However, this procedure sacrifices the availability property: if m cannot be delivered, the update on node S1 cannot be performed. However, sending m asynchronously does not solve the problem because then S1 does not know when S2 receives the message. Hence, any approach trying to achieve a strong consistent view such as locking and centralized management would either violate availability or partition tolerance.

In order to address these restrictions imposed by CAP, the system designer has to choose to relax or give up one of these three properties:

- *Consistency*: If we want to preserve availability and partition tolerance, the only choice is to give up or relax consistency: the data can be updated on both sites, and both sites will converge to the same state when the connection between them is re-established and a certain time has elapsed.
- *Availability*: Availability is given up by simply waiting when a partition event occurs until the nodes come back and the data are consistent again. The service is unavailable during the waiting time. Particularly, for large settings with many nodes, this could result in long downtimes.
- *Partition tolerance*: Basically, this means avoiding network partitioning in the case of link failures. Partition tolerance can be achieved by ensuring that each node is connected to each other or making a single atomically failing unit, but obviously, this limits scalability.

The CAP theorem implies that consistency guarantees in large-scale distributed systems cannot be as strict as those in centralized systems. Specifically, it suggests that distributed systems may need to provide BASE guarantees instead of the ACID guarantees provided by traditional database systems (see note on ACID in Chapter 5, Section 5.7, "Transaction Processing Monitors"). The CAP theorem states that no distributed system can provide more than two of the following three guarantees: consistency, availability, and partitioning tolerance. Here, consistency is defined as in databases; that is, if multiple operations are performed on the same object (which is actually stored in a distributed system), the results of the operations appear as if the operations were carried out in some definite order on a single system. Availability is defined to be satisfied if each operation on the system (e.g., a query) returns some result. The system provides partitioning tolerance if the system is operational even when the network between two components of the system is down.

Since distributed systems can satisfy only two of the three properties due to the CAP theorem, there are three types of distributed systems. CA (consistent, available) systems provide consistency and availability, but cannot tolerate network partitions. An example of a CA system is a clustered database, where each node stores a subset of the data. Such a database cannot provide availability in the case of network partitioning, since queries to data in the partitioned nodes must fail. CA systems may not be useful for cloud computing, since partitions are likely to occur in medium to large networks (including the case where the latency is very high). If there is no network partitioning, all servers are consistent, and the value seen by both clients is the correct value.

However, if the network is partitioned, it is no longer possible to keep all the servers consistent in the face of updates. There are then two choices. One choice is to keep both servers up and ignore the inconsistency. This leads to AP (available, partition-tolerant) systems where the system is always available, but may not return consistent results. The other possible choice is to bring one of the servers down, to avoid inconsistent values. This leads to CP (consistent, partition-tolerant) systems where the system always returns consistent results but may be unavailable under partitioning—including the case where the latency is very high. AP systems provide weak consistency. An important subclass of weakly consistent systems is those that provide eventual consistency. A system is defined as being eventually consistent if the system is guaranteed to reach a consistent state in a finite amount of time if there are no failures (e.g., network partitions) and no updates are made. The inconsistency window for such systems is the maximum amount of time that can elapse between the time that the update is made and the time that the update is guaranteed to be visible to all clients. If the inconsistency window is small compared to the update rate, then one method of dealing with stale data is to wait for a period greater than the inconsistency window and then retry the query.

Classic database systems focus on guaranteeing the ACID properties and, therefore, favor consistency over partition tolerance and availability. This is achieved by employing techniques like distributed locking and two-phase commit protocols. In certain circumstances, data needs are not transactionally focused, and at such times, the relational model is not the most appropriate one for what we need to do with the data we are storing. However, giving up availability is often not an option in Web business where users expect a 24×7 or always-on operation.

Most traditional RDBMS would guarantee that all the values in all our nodes are identical before it allows another user to read the values. But as we have seen, that is at a significant cost in terms of performance. Relational databases, with their large processing overhead in terms of maintaining the ACID attributes of the data they store and their reliance on potentially processor hungry joins, are not the right tool for the task they have before them: quickly finding relevant data from terabytes of unstructured data (Web content) that may be stored across thousands of geographically desperate nodes. In other words, relational model does not scale well for this type of data. Thus, techniques for guaranteeing strong consistency in large distributed systems limit scalability and results in latency issues. To cope with these problems, BASE was proposed as an alternative to ACID.

21.3.2.2 BASE (Basically Available, Soft State, Eventual Consistency)

BASE follows an optimistic approach accepting stale data and approximate answers while favoring availability. Some ways to achieve this are by supporting partial failures without total system failures, decoupling updates on different tables (i.e., relaxing consistency), and item potent operations that can be applied multiple times with the same result. In this sense, BASE describes more a spectrum of architectural styles than a single model. The eventual state of consistency can be provided as a result of a read repair, where any outdated data are refreshed with the latest version of the data as a result of the system detecting stale data during a read operation. Another approach is that of weak consistency. In this case, the read operation will return the first value found, not checking for staleness. Any stale nodes discovered are simply marked for updating at some stage in the future. This is a performance-focused approach but has the associated risk that data retrieved may not be the most current. In the following sections, we will discuss several techniques for implementing services following the BASE principle.

Conventional storage techniques may not be adequate for big data and, hence, the cloud applications. To scale storage systems to cloud scale, the basic technique is to partition and replicate the data over multiple independent storage systems. The word independent is emphasized, since it is well-known that databases can be partitioned into mutually dependent sub-databases that are automatically synchronized for reasons of performance and availability. Partitioning and replication increases the overall throughput of the system, since the total throughput of the combined system is the aggregate of the individual storage systems. To scale both the throughput and the maximum size of the data that can be stored beyond the limits of traditional database deployments, it is possible to partition the data, and store each partition in its own database. For scaling the throughput only, it is possible to use replication. Partitioning and replication also increase the storage capacity of a storage system by reducing the amount of data that needs to be stored in each partition. However, this creates synchronization and consistency problems, and discussion of this aspect is out of scope for this book.

The other technology for scaling storage described in this section is known by the name Not only SQL (NoSQL). NoSQL was developed as a reaction to the perception that conventional databases, focused on the need to ensure data integrity for enterprise applications, were too rigid to scale to cloud levels. As an example, conventional databases enforce a schema on the data being stored, and changing the schema is not easy. However, changing the schema may be a necessity in a rapidly changing environment like the cloud. NoSQL storage systems provide more flexibility and simplicity compared to relational databases. The disadvantage, however, is greater application complexity. NoSQL systems, for example, do not enforce a rigid schema. The trade-off is that applications have to be written to deal with data records of varying formats (schema). BASE is the NoSQL operating premise, in the same way that traditional transactionally focused databases use ACID: one moves from a world of certainty in terms of data consistency to a world where all we are promised is that all copies of the data will, at some point, be the same.

Partitioning and replication techniques used for scaling are as follows:

1. The first possible method is to store different tables in different databases (as in multidatabase systems).
2. The second approach is to partition the data within a single table onto different databases. There are two natural ways to partition the data from within a table: to store different rows in different databases and to store different columns in different databases (more common for NoSQL databases)

21.3.2.3 Functional Decomposition

As stated previously, one technique for partitioning the data to be stored is to store different tables in different databases, leading to the storage of the data in a multi-database system (MDBS).

21.3.2.4 Master–Slave Replication

To increase the throughput of transactions from the database, it is possible to have multiple copies of the database. A common replication method is master–slave replication. The master and slave databases are replicas of each other. All writes go to the master and the master keeps the slaves in sync. However, reads can be distributed to any database. Since this configuration distributes the reads among multiple databases, it is a good technology for read-intensive workloads. For write-intensive workloads, it is possible to have multiple masters, but then ensuring consistency if multiple processes update different replicas simultaneously is a complex problem. Additionally, time to write increases, due to the necessity of writing to all masters and the synchronization overhead between the masters rapidly becomes a limiting overhead.

21.3.3 Row Partitioning or Sharding

In cloud technology, sharding is used to refer to the technique of partitioning a table among multiple independent databases by row. However, partitioning of data by row in relational databases is not new and is referred to as horizontal partitioning in parallel database technology. The distinction between sharding and horizontal partitioning is that horizontal partitioning is done transparently to the application by the database, whereas sharding is explicit partitioning done by the application. However, the two techniques have started converging, since traditional database vendors have started offering support for more sophisticated partitioning strategies. Since sharding is similar to horizontal partitioning, we first discuss different horizontal partitioning techniques. It can be seen that a good sharding technique depends upon both the organization of the data and the type of queries expected.

The different techniques of sharding are as follows:

1. *Round-robin partitioning*: The round-robin method distributes the rows in a round-robin fashion over different databases. In the example, we could partition the transaction table into multiple databases so that the first transaction is stored in the first database, the second in the second database, and so on. The advantage of round-robin partitioning is its simplicity. However, it also suffers from the disadvantage of losing associations (say) during a query, unless all databases are queried. Hash partitioning and range partitioning do not suffer from the disadvantage of losing record associations.

2. *Hash partitioning method*: In this method, the value of a selected attribute is hashed to find the database into which the tuple should be stored. If queries are frequently made on an attribute (say Customer_Id), then associations can be preserved by using this attribute as the attribute that is hashed, so that records with the same value of this attribute can be found in the same database.

3. *Range partitioning*: The range partitioning technique stores records with *similar* attributes in the same database. For example, the range of Customer_Id could be partitioned between different databases. Again, if the attributes chosen for grouping are those on which queries are frequently made, record association is preserved and it is not necessary to merge results from different databases. Range partitioning can be susceptible to load imbalance, unless the partitioning is chosen carefully. It is possible to choose the partitions so that there is an imbalance in the amount of data stored in the partitions (data skew) or in the execution of queries across partitions (execution skew). These problems are less likely in round-robin and hash partitioning, since they tend to uniformly distribute the data over the partitions.

Thus, hash partitioning is particularly well suited to large-scale systems. Round-robin simplifies a uniform distribution of records but does not facilitate the restriction of operations to single partitions. While range partitioning does support this, it requires knowledge about the data distribution in order to properly adjust the ranges.

21.4 NoSQL Databases

NoSQL databases have been classified into four subcategories:

1. *Column family stores*: An extension of the key–value architecture with columns and column families; the overall goal was to process distributed data over a pool of infrastructure, for example, HBase and Cassandra.
2. *Key–value pairs*: This model is implemented using a hash table where there is a unique key and a pointer to a particular item of data creating a key–value pair, for example, Voldemort.
3. *Document databases*: This class of databases is modeled after Lotus Notes and similar to key–value stores. The data are stored as a document and is represented in JSON or XML formats. The biggest design feature is the flexibility to list multiple levels of key–value pairs, for example, Riak and CouchDB.
4. *Graph databases*: Based on the graph theory, this class of database supports the scalability across a cluster of machines. The complexity of representation for extremely complex sets of documents is evolving, for example, Neo4J.

21.4.1 Column-Oriented Stores or Databases

Hadoop HBase is the distributed database that supports the storage needs of the Hadoop distributed programming platform. HBase is designed by taking inspiration from Google BigTable; its main goal is to offer real-time read/write operations for tables with billions of rows and millions of columns by leveraging clusters of commodity hardware. The internal architecture and logic model of HBase is very similar to Google BigTable, and the entire system is backed by the Hadoop Distributed File System (HDFS), which mimics the structure and services of GFS.

21.4.2 Key–Value Stores (K–V Store) or Databases

Apache Cassandra is a distributed object store from an aging large amounts of structured data spread across many commodity servers. The system is designed to avoid a single point of failure and offer a highly reliable service. Cassandra was initially developed by Facebook; now, it is part of the Apache incubator initiative. Facebook in the initial years had used a leading commercial database solution for their internal architecture in conjunction with some Hadoop. Eventually, the tsunami of users led the company to start thinking in terms of unlimited scalability and focus on availability and distribution. The nature of the data and its producers and consumers did not mandate consistency but needed unlimited availability and scalable performance. The team at Facebook built an architecture that combines the data model approaches of BigTable and the infrastructure approaches of Dynamo with scalability and performance capabilities, named Cassandra. Cassandra is often referred to as hybrid architecture since it combines the column-oriented data model from BigTable with Hadoop MapReduce jobs, and it implements the patterns from Dynamo like eventually consistent, gossip protocols, a master–master way of serving both read and write requests. Cassandra supports a full replication model based on NoSQL architectures.

The Cassandra team had a few design goals to meet, considering the architecture at the time of first development and deployment was primarily being done at Facebook. The goals included

- High availability
- Eventual consistency
- Incremental scalability
- Optimistic replication
- Tunable trade-offs between consistency, durability, and latency
- Low cost of ownership
- Minimal administration

Amazon Dynamo is the distributed key–value store that supports the management of information of several of the business services offered by Amazon Inc. The main goal of Dynamo is to provide an incrementally scalable and highly available storage system. This goal helps in achieving reliability at a massive scale, where thousands of servers and network components build an infrastructure serving 10 million requests per day. Dynamo provides a simplified interface based on get/put semantics, where objects are stored and retrieved with a unique identifier (key). The main goal of achieving an extremely reliable infrastructure has imposed some constraints on the properties of these systems. For example, ACID properties on data have been sacrificed in favor of a more reliable and efficient infrastructure. This creates what it is called an eventually consistent model (i.e., in the long term, all the users will see the same data).

21.4.3 Document-Oriented Databases

Document-oriented databases or document databases can be defined as a schema-less and flexible model of storing data as documents, rather than relational structures. The document will contain all the data it needs to answer specific query questions. Benefits of this model include

- Ability to store dynamic data in unstructured, semistructured, or structured formats
- Ability to create persisted views from a base document and store the same for analysis
- Ability to store and process large data sets

The design features of document-oriented databases include

- Schema-free—There is no restriction on the structure and format of how the data need to be stored. This flexibility allows an evolving system to add more data and allows the existing data to be retained in the current structure.
- Document store—Objects can be serialized and stored in a document, and there is no relational integrity to enforce and follow.
- Ease of creation and maintenance—A simple creation of the document allows complex objects to be created once and there is minimal maintenance once the document is created.
- No relationship enforcement—Documents are independent of each other and there is no foreign key relationship to worry about when executing queries. The effects of concurrency and performance issues related to the same are not a bother here.
- Open formats—Documents are described using JSON, XML, or some derivative, making the process standard and clean from the start.
- Built-in versioning—Documents can get large and messy with versions. To avoid conflicts and keep processing efficiencies, versioning is implemented by most solutions available today.

Document databases express the data as files in JSON or XML formats. This allows the same document to be parsed for multiple contexts and the results scrapped and added to the next iteration of the database data.

Apache CouchDB and MongoDB are two examples of document stores. Both provide a schema-less store whereby the primary objects are documents organized into a collection of key–value fields. The value of each field can be of type string, integer, float, date, or an array of values. The databases expose a RESTful interface and represent data in JSON format. Both allow querying and indexing data by using the MapReduce programming model, expose JavaScript as a base language for data querying and manipulation rather than SQL, and support large files as documents. From an infrastructure point of view, the two systems support data replication and high availability. CouchDB ensures ACID properties on data. MongoDB supports sharding, which is the ability to distribute the content of a collection among different nodes.

21.4.4 Graph Stores or Databases

Social media and the emergence of Facebook, LinkedIn, and Twitter have accelerated the emergence of the most complex NoSQL database, the graph database. The graph database is oriented toward modeling and deploying data that is graphical by construct. For example, to represent a person and their friends in a social network, we can either write code to convert the social graph into key–value pairs on a Dynamo or Cassandra or simply convert them into a node-edge model in a graph database, where managing the relationship representation is much more simplified.

A graph database represents each object as a node and the relationships as an edge. This means person is a node and household is a node and the relationship between them is an edge. Like the classic ER model for RDBMS, we need to create an attribute model for a graph database. We can start by taking the highest level in a hierarchy as a root node (similar to an entity) and connect each attribute as its subnode. To represent different levels of the hierarchy, we can add a subcategory or subreference and create another list of attributes at that level. This creates a natural traversal model like a tree traversal, which is similar to traversing a graph. Depending on the cyclic property of the graph, we can have a balanced or skewed model. Some of the most evolved graph databases include Neo4J, InfiniteGraph, GraphDB, and AllegroGraph.

21.4.5 Comparison of NoSQL Databases

1. Column-based databases allow for rapid location and return of data from one particular attribute. They are potentially very slow with writing, however, since data may need to be shuffled around to allow a new data item to be inserted. As a rough guide then, traditional transactionally oriented databases will probably fair better in an RDBMS. Column based will probably thrive in areas where speed of access to nonvolatile data is important, for example, in some decision support applications. You only need to review marketing material from commercial contenders, like Ingres Vectorwise, to see that business analytics is seen as the key market and speed of data access the main product differentiator.

2. If you do not need large and complex data structures and can always access your data using a known key, then key–value stores have a performance advantage over most RDBMS. Oracle has a feature within their RDBMS that allows you to define a table at an index-organized table (IOT), and this works in a similar way. However, you do still have the overhead of consistency checking, and these IOTs are often just a small part of a larger schema. RDBMS have a reputation for poor scaling in distributed systems, and this is where key–value stores can be a distinct advantage.

3. Document-centric databases are good where the data are difficult to structure. Web pages and blog entries are two oft-quoted examples. Unlike RDBMS, which impose structure by their very nature, document-centric databases allow free-form data to be stored. The onus is then on the data retriever to make sense of the data that are stored.

21.5 Summary

This chapter introduces big data systems that are associated with big volume, variety, and velocity. It describes the characteristic features of such systems including big data architecture, row versus column-oriented data layouts, NoSQL data management, in-memory computing, and developing big data applications. In the later part of the chapter, it provides a brief on the various types of NoSQL databases including column-oriented stores, key–value stores, document-oriented databases, and graph stores.

22

Mobile Applications

Mobile computing represents a fundamentally new paradigm in enterprise computing. Mobile computing enables operating a job- and role-specific application loaded on a handheld or tablet device that passes only relevant data between a field worker and the relevant back-end enterprise systems regardless of connectivity availability.

22.1 Agile Enterprises

The difficult challenges facing businesses today require organizations to be transitioned into flexible, agile structures that can respond to new market opportunities quickly with a minimum of new investment and risk. As enterprises have experienced the need to be simultaneously efficient, flexible, responsive, and adaptive, they have transitioned themselves into agile enterprises with small, autonomous teams that work concurrently and reconfigure quickly and adopt highly decentralized management that recognizes its knowledge base and manages it effectively.

Enterprise agility is the ability to be

1. Responsive—Adaptability is enabled by the concept of loosely coupled interacting components reconfigurable within a unified framework. This is essential for ensuring opportunity management to sustain viability.

   The ability to be responsive involves the following aspects:

   a. An organizational structure that enables change is based on reusable elements that are reconfigurable in a scalable framework. Reusability and reconfigurability are generic concepts that are applicable to work procedures, manufacturing cells, production teams, and information automation systems.

   b. An organizational culture that facilitates change and focuses on change proficiency.

2. Intelligence intensive or ability to manage and apply knowledge effectively whether it is knowledge of a customer, a market opportunity, a competitor's threat, a production process, a business practice,

a product technology, or an individual's competency. This is essential for ensuring innovation management to sustain leadership.

The ability to be intelligence intensive involves the following aspects:

a. Enterprise knowledge management

b. Enterprise collaborative learning

> When confronted with a competitive opportunity, a smaller company is able to act more quickly, whereas a larger company has access to more comprehensive knowledge (options, resources, etc.) and can decide to act sooner and more thoroughly.

Agility is the ability to respond to (and ideally benefit from) unexpected change. Agility is unplanned and unscheduled adaption to unforeseen and unexpected external circumstances. However, we must differentiate between agility and flexibility. Flexibility is scheduled or planned adaptation to unforeseen yet expected external circumstances.

One of the foremost abilities of an agile enterprise is its ability to quickly react to change and adapt to new opportunities. This ability to change works along two dimensions:

1. The number or *types of change* an organization is able to undergo

2. The *degree of change* an organization is able to undergo

The former is termed as range, and the latter is termed as response ability. The more response able an enterprise is, the more radical a change it can gracefully address. Range refers to how large a domain is covered by the agile response system; in other words, how far from the expected set of events one can go and still have the system respond well. However, given a specific range, how well the system responds is a measure of response or change ability.

> Enterprises primarily aim progressively for efficiency, flexibility, and innovation in that order. The Model Builder's kit, Erector Set kit, and LEGO kit are illustrations of enterprises targeting for efficiency, flexibility, and innovation (i.e., agility), respectively.

Construction toys offer a useful metaphor because the enterprise systems we are concerned with must be configured and reconfigured constantly, precisely the objective of most construction toys. An enterprise system architecture and structure consisting of reusable components reconfigurable in a scalable framework can be an effective base model for creating variable (or built-for-change) systems. For achieving this, the nature of the framework appears to be a critical factor. We can introduce the framework/component concept, by looking at three types of construction toys and observing how they are used in practice, namely, Erector Set kit, LEGO kit, and Model Builder's kit.

You can build virtually anything over and over again with either of these toys; but fundamental differences in their architectures give each system unique dynamic characteristics. All consist of a basic set of core construction components and also have an architectural and structural framework that enables connecting the components into an unbounded variety of configurations. Nevertheless, the Model Builder is not as reusable in practice as Erector Set; and, the Erector Set is not as reusable or reconfigurable or scalable in practice as LEGO, but LEGO is more reusable, reconfigurable, and scalable than either of them. LEGO is the dominant construction toy of choice among preteen builders—who appear to value experimentation and innovation.

The Model Builder's kit can be used to construct one object like airplane of one intended size. A highly integrated system, this construction kit offers maximum esthetic appeal for one-time construction use; but the parts are not reusable, the construction cannot be reconfigured, and one intended size precludes any scalability. But it will remain what it is for all time—there is zero variability here.

Erector Set kits can be purchased for constructing specific models, such as a small airplane that can be assembled in many different configurations. With the Erector Set kit, the first built model is likely to remain as originally configured in any particular play session. Erector Set, for all its modular structure, is just not as reconfigurable in practice as LEGO. The Erector Set connectivity framework employs a special-purpose intermediate subsystem used solely to attach one part to another—a nut-and-bolt pair and a 90° elbow. The components in the system all have holes through which the bolts may pass to connect one component with another. When a nut is lost, a bolt is useless, and vice versa; when all the nuts and bolts remaining in a set have been used, any remaining construction components are useless, and vice versa. All the parts in a LEGO set can always be used and reused, but the Erector Set, for all its modularity, is not as reusable in practice as LEGO.

LEGO offers similar kits, and both toys include a few necessary special parts, like wheels and cowlings, to augment the core construction components. Watch a child work with either and you will see the LEGO construction undergoes constant metamorphosis; the child may start with one of the pictured configurations, but then reconfigures the pieces into all manner of other imagined styles. LEGO components are plug compatible with each other, containing the connectivity framework as an integral feature of the component. A standard grid of bumps and cavities on component surfaces allow them to snap together into a larger configuration—without limit.

The Model Builder's kit has a tight framework: a precise construction sequence, no part interchangeability, and high integration. Erector Set has a loose framework that does not encourage interaction among parts and insufficiently discriminates among compatible parts. In contrast, each component in the LEGO system carries all it needs to interact with other components (the interaction framework rejects most unintended parts) and can grow without end.

22.1.1 Stability versus Agility

Most large-scale change efforts in established enterprises fail to meet expectations because nearly all models of organization design, effectiveness, and change assume stability is not only desirable but also attainable. The theory and practice in organization design explicitly encourage organizations to seek alignment, stability, and equilibrium. The predominant logic of organizational effectiveness has been that an organization's fit with its environment, its execution, and its predictability are the keys to its success. Organizations are encouraged to institutionalize best practices, freeze them into place, focus on execution, stick to their knitting, increase predictability, and get processes under control. These ideas establish stability as the key to performance.

Stability of a distinctive competitive advantage is a strong driver for organization design because of its expected link to excellence and effectiveness. Leveraging an advantage requires commitments that focus attention, resources, and investments to the chosen alternatives. In other words, competitive advantage results when enterprises finely hone their operations to perform in a particular way. This leads to large investments in operating technologies, structures, and ways of doing things. If such commitments are successful, they lead to a period of high performance and a considerable amount of positive reinforcement. Financial markets reward stable competitive advantages and predictable streams of earnings: a commitment to alignment reflects a commitment to stability.

Consequently, enterprises are built to support stable strategies, organizational structures, and enduring value creations, not to vary. For example, the often-used strengths, weaknesses, opportunities, and threats (SWOT) analysis encourages the firm to leverage opportunities while avoiding weaknesses and threats. This alignment among positive and negative forces is implicitly assumed to remain constant, and there is no built-in assumption of agility. When environments are stable or at least predictable, enterprises are characterized by rules, norms, and systems that limit experimentation, control variation, and rewarded consistent performance. They have many checks and balances in place to ensure that the organization operates in the prescribed manner. Thus, to get the high performance they want, enterprises put in place practices they see as a good fit, without considering whether they can be changed and whether they will support changes in future, that is, by aligning themselves to achieve high performance today, enterprises often make it difficult to vary, so that they can have high performance tomorrow.

When the environment is changing slowly or predictably, these models are adequate. However, as the rate of change increases with increasing globalization, technological breakthroughs, associative alliances, and regulatory changes, enterprises have to look for greater agility, flexibility, and innovation from their companies. Instead of pursuing strategies, structures, and cultures that are designed to create long-term competitive advantages, companies must seek a string of temporary competitive advantages through an approach to organization design that assumes change is normal. With the advent of the Internet and the accompanying extended *virtual* market spaces, enterprises are now competing based on intangible assets like identity, intellectual property, ability to attract and stick to customers, and their ability to organize, reorganize frequently, or organize differently in different areas depending on the need. Thus, the need for changes in management and organization is much more frequent, and excellence is much more a function of possessing the ability for changes. Enterprises need to be built around practices that encourage change, not thwart it. Instead of having to create change efforts, disrupt the status quo, or adapt to change, enterprises should be built for change.

To meet the conflicting objectives of performing well against current set of environmental demands and changing themselves to face future business environments, enterprises must engender two types of changes: the natural process of evolution, or what we will call strategic adjustments, and strategic reorientations:

1. Strategic adjustments involve the day-to-day tactical changes required to bring in new customers, make incremental improvements in products and services, and comply with regulatory requirements. This type of change helps fine-tune current strategies and structures to achieve short-term results; it is steady, incremental, and natural. This basic capability to evolve is essential if an enterprise is to survive to thrive.

2. Strategic reorientation involves altering an existing strategy and, in some cases, adopting a new strategy. When the environment evolves or changes sufficiently, an organization must significantly adjust some elements of its strategy and the way it executes that strategy. More often than not, enterprises have to face a transformational change that involves not just a new strategy but a transformation of the business model that leads to new products, services, and customers and requires markedly new competencies and capabilities. However, operationally, all these changes can be seen as manifestations of the basic changes only differing in degrees and multiple dimensions.

Maintaining an agile enterprise is not a matter of searching for the strategy but continuously strategizing, not a matter of specifying an organization design but committing to a process of organizing, and not generating value but continuously improving the efficiency and effectiveness of the value generation process. It is a search for a series of temporary configurations that create short-term advantages. In turbulent environments, enterprises that string together a series of temporary but adequate competitive advantages will outperform enterprises that stick with one advantage for an extended period of time. The key issue for the built-for-change enterprise is orchestration or coordinating the multiple changing subsystems to produce high levels of current enterprise performance.

22.1.2 Aspects of Agility

This section addresses the analytical side of agility or change proficiency of the enterprise. It highlights the fundamental principles that underlie an enterprise's ability to change, and by indicating how to apply these principles in real situations, it illustrates what it is that makes a business and any of its constituting systems easy to change.

Agility or change proficiency enables both efficiency programs (e.g., lean production) and transformation programs; if the enterprise is proficient at change, it can adapt to take advantage of an unpredictable opportunity and can also counter the unpredictable threat. Agility can embrace semantics across the whole spectrum: it can capture cycle-time reduction, with everything happening faster; it can build on lean production, with high resource productivity; it can encompass mass customization, with customer-responsive product variation; it can embrace virtual enterprise, with streamlined supplier networks and opportunistic partnerships; it can echo reengineering, with a process and transformation focus; and it can demand a learning organization, with systemic training and education. Being agile means being proficient at change. Agility allows an enterprise to do anything it wants to do whenever it wants to—or has to—do it. Thus, an agile enterprise can employ business process reengineering as a core competency when transformation is called for; it can hasten its conversion to lean production when greater efficiencies are useful; and it can continue to succeed when constant innovation becomes the dominant competitive strategy. Agility can be wielded overtly as a business strategy as well as inherently as a sustainable-existence competency.

Agility derives from both the physical ability to act (change ability) and the intellectual ability to find appropriate things to act on (knowledge management). Agility can be expressed as the ability to manage and apply knowledge effectively, so that enterprise has the potential to thrive in a continuously changing and unpredictable business environment. Agility derives from two sources: an enterprise architecture that enables variation and an organizational culture that also facilitates required change or variation. The enterprise architecture that enables variation is based on reusable elements that are reconfigurable in a scalable framework.

Agility is a core fundamental requirement of all enterprises. It was not an area of interest when environmental variation was relatively slow and predictable. Now, there is virtually no choice; enterprises must develop a conscious competency. Practically, all enterprises now need some method to assess their agility and determine whether it is sufficient or needs improvement. This section introduces techniques for characterizing, measuring, and comparing variability in all aspects of business and among different businesses.

22.1.3 Principles of Built-for-Change Systems

Christopher Alexander introduced the concept of patterns in the late 1970s in the field of architecture. A pattern describes a commonly occurring solution that generates decidedly successful outcomes.

A list of success patterns for agile enterprises (and systems) in terms of their constituting elements or functions or components are as follows.

22.1.3.1 Reusable

Agility Pattern 1 Self-contained units (components): *The components of agile enterprises are autonomous units cooperating toward a shared goal.*

Agility Pattern 2 Plug compatibility: *The components of agile enterprises are reusable and multiply replicable, that is, depending on requirements multiple instances of the same component can be invoked concurrently.*

Agility Pattern 3 Facilitated reuse: *The components of agile enterprises share well-defined interaction and interface standards and can be inserted, removed, and replaced easily and noninvasively.*

22.1.3.2 Reconfigurable

Agility Pattern 4 Flat interaction: *The components of agile enterprises communicate, coordinate, and cooperate with other components concurrently and in real-term sharing of current, complete, and consistent information on interactions with individual customers.*

Agility Pattern 5 Deferred commitment: *The components of agile enterprises establish relationships with other components in the real term to enable deferment of customer commitment to as late a stage as possible within the sales cycle, coupled with the corresponding ability to postpone the point of product differentiation as close as possible to the point of purchase by the customer.*

Agility Pattern 6 Distributed control and information: *The components of agile enterprises are defined declaratively rather than procedurally;*

the network of components displays the defining characteristics of any "small worlds" network, namely, local robustness and global accessibility.

Agility Pattern 7 Self-organization: *The components of agile enterprises are self-aware, and they interact with other components via on-the-fly integration, adjustment, or negotiation.*

22.1.3.3 Scalable

Agility Pattern 8 Evolving standards (framework): *The components of agile enterprises operate within predefined frameworks that standardize inter-component communication and interaction, determine component compatibility, and evolve to accommodate old, current, and new components.*

Agility Pattern 9 Redundancy and diversity: *The components of agile enterprises replicate components to provide the desired capacity, load balancing and performance, fault tolerance, as well as variations on the basic component functionality and behavior.*

Agility Pattern 10 Elastic capacity: *The components of agile enterprises enable dynamic utilization of additional or a reduced number of resources depending on the requirements.*

Chapter 3, Section 3.1.2 "Enterprise Component Architecture" presents contemporary view to enterprise agility and architecture.

22.1.4 Framework for Change Proficiency

How do we measure enterprise agility? This section establishes a metric framework for proficiency at variation i.e. change; an enterprise's change proficiency may exist in one or more of dimensions of variations. And, these dimensions of changes can form a structural framework for understanding current capabilities and setting strategic priorities for improvement: how does the agile enterprise know when it is improving its changeability or losing ground; how does it know if it is less changeable than its competition; and how does it set improvement targets? Thus, a practical measure of variation proficiency is needed before we can talk meaningfully about getting more of it or even getting some of it.

It must be highlighted that measuring competency is generally not unidimensional nor likely to result in an absolute and unequivocal comparative metric. Change proficiency has both reactive and proactive modes. Reactive change is opportunistic and responds to a situation that threatens viability. Proactive change is innovative and responds to a possibility for leadership. An organization sufficiently proficient at reactive change, when prodded, should be able to use that competency proactively and let others do the reacting.

Would it be proficient if a short-notice variation was completed in the time required, but at a cost that eventually bankrupted the company? Or if the changed environment, thereafter, required the special wizardry and constant attention of a specific employee to keep it operational? Is it proficient if the change is virtually free and painless, but out of synch with market opportunity timing? Is it proficient if it can readily accommodate a broad latitude of change that is no longer needed, or too narrow for the latest challenges thrown at it by the business environment? Are we change proficient if we can accommodate any change that comes our way as long as it is within a narrow 10% of where we already are?

Thus, change proficiency can be understood to be codetermined by four parameters:

1. *Time*: a measure of elapsed time to complete a change (fairly objective)

2. *Cost*: a measure of monetary cost incurred in a change (somewhat objective)

3. *Quality*: a measure of prediction quality in meeting change time, cost, and specification targets robustly (somewhat subjective)

4. *Range*: a measure of the latitude of possible change, typically defined and determined by mission or charter (fairly subjective)

22.1.5 Enhancing Enterprise Agility

22.1.5.1 E-Business Strategy

E-business refers to an enterprise that has reengineered itself to conduct its business via the Internet and Web. Successful enterprises need to reconceptualize the very nature of their business.

As customers begin to buy via Internet and enterprises rush to use the Internet to create new operational efficiencies, most enterprises seek to update their business strategies. Enterprises survey the changing environment and then modify their company strategies to accommodate these changes. This involves major changes in the way companies do business, including changes in marketing, sales, service, product delivery, and even manufacturing and inventory. Changed strategies will entail changed business processes that, in turn, imply changed software systems or, better still, software systems that are changeable!

22.1.5.2 Business Process Reengineering (BPR)

Although, BPR has its roots in information technology (IT) management, it is basically a business initiative that has a major impact on the satisfaction of both the internal and external customers. Michael Hammer, who triggered the BPR revolution in 1990, considers BPR as a *radical change* for which IT is the key enabler. BPR can be broadly termed as *the rethinking and change of business processes to achieve dramatic improvements in the measures of performances such as cost, quality, service, and speed.*

Some of the principles advocated by Hammer are as follows:

- Organize around outputs, not tasks.
- Put the decisions and control, and hence all relevant information, into the hands of the performer.
- Have those who use the outputs of a process to perform the process, including the creation and processing of the relevant information.
- The location of user, data, and process information should be immaterial; it should function as if all were in a centralized place.

When perusing the aforementioned points, it will become evident that the implementation of SAP CRM possesses most of the characteristics mentioned.

The most important outcome of BPR has been viewing business activities as more than a collection of individual or even functional tasks; it has engendered the process-oriented view of business. However, BPR is different from quality management efforts like TQM and ISO 9000, which refer to programs and initiatives that emphasize bottom-up incremental improvements in existing work processes and outputs on a continuous basis. In contrast, BPR usually refers to top-down dramatic improvements through redesigned or completely new processes on a discrete basis. In the continuum of methodologies ranging from ISO 9000, TQM, ABM, and so on, at one end and BPR on the other, SAP CRM implementation definitely lies on the BPR side of the spectrum when it comes to corporate change management efforts.

22.1.5.3 Mobilizing Enterprise Processes

This strategy entails replacing the process or process segment under consideration by a mobile-enabled link. In the next subsection, we discuss an overview of business processes before discussing the characteristics of mobilized processes.

Mobility offers new opportunities to dramatically improve business models and processes and will ultimately provide new, streamlined business processes that never would have existed if not for this new phenomenon.

22.1.5.3.1 Extending Web to Wireless

The first step in the evolution of mobility is to extend the Web to wireless; this is also known as *webifying*. For the most part, business processes are minimally affected in this phase. The goal is to provide value-added services through mobility with minimal disruption to existing processes. An example might be creating a new company website accessible through WAP (Wireless Application Protocol) phones or Palm OS–based personal digital assistants (PDAs). Firms attain immediate value through realizing additional exposure and market presence, and customers realize value through additional services.

22.1.5.3.2 *Extending Business Processes with Mobility*

The next step in the evolution of mobility is to extend existing business processes. New opportunities to streamline company business processes emerge and evolve to produce new revenue opportunities. One example is the way that mobility extends business processes through a supply chain optimization model. New business processes emerge through these new mechanisms that ultimately shorten the supply chain cycle, thus minimizing error and maximizing efficiency and realizing the utmost customer satisfaction. Real-time tracking and alert mechanisms provide supply chain monitors with the capability to monitor shipments and product line quality in ways that traditional business models were not capable of doing.

22.1.5.3.3 *Enabling a Dynamic Business Model*

The final phase in the evolution of mobility is the one that has only been touched upon in today's world. The unique attributes of mobility will provide new and exciting ways of managing processes and allow for efficiencies never before attainable. The convergence of wireless technologies with existing business models will result in fully dynamic business processes.

22.1.6 Network Enterprises

Agile companies produce the right product, at the right place, at the right time, at the right price for the right customer. As pointed out by Jagdish Sheth in these times of market change and turbulence, the *half-life* (i.e., the time within which it loses currency by 50%) of customer knowledge is getting shorter and shorter. The difficult challenges facing businesses today require organizations to transition into flexible, agile structures that can respond to new market opportunities quickly with a minimum of new investment and risk.

As enterprises have experienced the need to be simultaneously efficient, flexible, responsive, and adaptive, they have turned increasingly to the network form of organization with the following characteristics:

- Networks rely more on market mechanisms rather than on administrative processes to manage resource flows. These mechanisms are not simple *arms-length* relationships usually associated with independently owned economic entities. Instead, to maintain the position within the network, members recognize their interdependence and are willing to share information, cooperate with each other, and customize their product or service.

- While a network of sub-contractors have been common for many years, recently formed networks expect members to play a much more proactive role in improving the final product or service.

- Instead of holding all assets required to produce a given product or service in-house, networks use the collective assets of several firms located along the value chain.

The agile organization is composed of small, autonomous teams or subcontractors who work concurrently and reconfigure quickly to thrive in an unpredictable and rapidly changing customer environment. Each constituent has the full resources of the company or the value chain at its disposal and has a seamless information exchange between the lead organization and the virtual partners.

Thus, a network enterprise is a coalition of enterprises that work collectively and collaboratively to create value for the customers of a focal enterprise. Sometimes, the coalition is loosely connected; at other times, it is tightly defined, as in the relationship between Dell and its component suppliers. An enterprise network consists of a wide range of companies—suppliers, joint venture (JV) partners, contractors, distributors, franchisees, licensees, and so on—that contribute to the focal enterprise's creation and delivery of value to its customers. Each of these enterprises in turn will have their own enterprise networks focused around themselves. Thus, relationships between enterprises in the network both enable and constrain focal companies in the achievement of their goals. Therefore, liberating the potential value in customer relationships hinges on enterprises effectively managing their non-customer-network relationships.

Though they appear similar, there are fundamental differences between the *agile* and *lean* approaches for running a business. Lean production is at heart simply an enhancement of mass production methods, whereas agility implies breaking out of the mass production mold and into mass customization. Agility focuses on economies of scope rather than economies of scale, ideally serving ever-smaller niche markets—even quantities of one—without the high cost traditionally associated with customization. A key element of agility is an enterprise-wide view, whereas lean production is usually associated with the efficient use of resources on the operations floor.

22.2 Process-Oriented Enterprise

Enterprise systems (ESs) enable an organization to truly function as an integrated organization, integration across all functions or segments of the traditional value chain—sales order, production, inventory, purchasing, finance and accounting, personnel and administration, and so on. They do this by modeling primarily the business processes as the basic business entities of the enterprise rather than by modeling data handled by the enterprise (as done by the traditional IT systems). However, every ES might not be completely successful in doing this. In a break with the legacy enterprise-wide solutions, modern ES treats business processes as more fundamental than data items.

Collaborations or relationships manifest themselves through the various organizational and interorganizational processes. A process may be generally defined as the set of resources and activities necessary and sufficient to convert some form of input into some form of output. Processes are internal, external, or a combination of both; they cross functional boundaries; they have starting and ending points; and they exist at all levels within the organization.

The significance of a process to the success of the enterprise's business is dependent on the value, with reference to the customer, of the collaboration that it addresses and represents. In other words, the nature and extent of the value addition by a process to a product or services delivered to a customer is the best index of the contribution of that process to the company's overall customer satisfaction or customer collaboration. Customer knowledge by itself is not adequate; it is only when the organization has effective processes for sharing this information and integrating the activities of frontline workers and has the ability to coordinate the assignment and tracking of work that organizations can become effective.

Thus, Management by Collaboration (MBC) not only recognizes inherently the significance of various process-related techniques and methodologies such as process innovation (PI), business process improvement (BPI), business process redesign (BPRD), business process re-engineering (BPR), and business process management (BPM) but also treats them as fundamental, continuous, and integral functions of the management of a company itself. A collaborative enterprise enabled by the implementation of an ES is inherently amenable to business process involvement, which is also the essence of any total quality management (TQM)-oriented effort undertaken within an enterprise.

22.2.1 Value-Add Driven Enterprise

Business processes can be seen as the very basis of the value addition within an organization that was traditionally attributed to various functions or divisions in an organization. As organizational and environmental conditions become more complex, global, and competitive, processes provide a framework for dealing effectively with the issues of performance improvement, capability development, and adaptation to the changing environment.

Along a value stream (i.e., a business process), analysis of the absence or creation of added value or (worse) destruction of value critically determines the necessity and effectiveness of a process step. The understanding of value-adding and non-value-adding processes (or process steps) is a significant factor in the analysis, design, benchmarking, and optimization of business processes leading to BPM in the companies. As discussed in "Implementing SAP CRM", Chapter 7, "SAP and Business Process Re-engineering," SAP provides an environment for analyzing and optimizing business processes.

Values are characterized by both value determinants such as

- Time (cycle time and so on)
- Flexibility (options, customization, composition, and so on)
- Responsiveness (lead time, number of hand-offs, and so on)
- Quality (rework, rejects, yield, and so on)
- Price (discounts, rebates, coupons, incentives, and so on)

We must hasten to add that we are not disregarding cost (materials, labor, overhead, and so forth) as a value determinant. However, the effect of cost is truly a result of a host of value determinants such as time, flexibility, and responsiveness.

The nature and extent of a value addition to a product or service is the best measure of that addition's contribution to the company's overall goal for competitiveness. Such value expectations are dependent on the following:

- The customer's experience of similar product(s) and/or service(s)
- The value delivered by the competitors
- The capabilities and limitations of locking into the base technological platform

However, value as originally defined by Michael Porter in the context of introducing the concept of the value chain is meant more in the nature of the cost at various stages. Rather than a value chain, it is more of a cost chain! Porter's value chain is also a structure-oriented and hence a static concept. Here, we mean value as the satisfaction of not only external but also internal customers' requirements, as defined and continuously redefined, as the least total cost of acquisition, ownership, and use.

Consequently, in this formulation, one can understand the company's competitive gap in the market in terms of such process-based, customer-expected levels of value and the value delivered by the company's process for the concerned products or services. Customer responsiveness focuses on costs in terms of the yield. Therefore, we can perform market segmentation for a particular product or services in terms of the most significant customer values and the corresponding value determinants or what we term as critical value determinants (CVDs).

Strategic planning exercises can then be understood readily in terms of devising strategies for improving on these process-based CVDs based on the competitive benchmarking of these *collaborative* values and processes between the enterprise and customers. These strategies and the tactics resulting from analysis, design, and optimization of the process will in turn focus on the restrategizing of all relevant business process at all levels. This can result in the modification or deletion of the process or creation of a new one.

22.2.2 Business Process Management (BPM)

Business Process Management (BPM) addresses the following two important issues for an enterprise:

1. The strategic long-term *positioning* of the business with respect to the current and envisaged customers, which will ensure that the enterprise would be competitively and financially successful, locally and globally

2. The enterprise's *capability/capacity* that is the totality of all the internal processes that dynamically realize this positioning of the business

Traditionally, positioning has been considered as an independent set of functional tasks split within the marketing, finance, and strategic planning functions. Similarly, capability/capacity has usually been considered the preserve of the individual operational departments that may have mutually conflicting priorities and measures of performances (see book "Implementing SAP CRM", Chapter 2, Section "CRM System Reflects and Mimics the Integrated Nature of an Enterprise").

The problem for many enterprises lies in the fact that there is a fundamental flaw in the organizational structure—organizational structures are *hierarchical*, while the transactions and workflows that deliver the solutions (i.e., products and services) to the customers are *horizontal*. Quite simply, the *structure* determines who the customer really is. The traditional management structures condition managers to put functional needs above those of the multifunctional processes to which their functions contribute. This results in

- Various departments competing for resources
- Collective failure in meeting or exceeding the customers' expectations
- Inability to coordinate and collaborate on multifunctional customer-centric processes that would truly provide the competitive differentiation in future markets

The traditional mass marketing type of organization works well for researching market opportunities, planning the offering, and scheduling all of the steps required to produce and distribute the offering to the marketplace (where it is selected or rejected by the customer). It takes a very different kind of organization, namely, the customized marketing type organization to build long-term relationships with customers so that they call such organizations first when they have a need because they trust that such enterprises will be able to respond with an effective solution. This is customer-responsive management, which we will discuss in the section that follows.

BPM is the process that manages and optimizes the inextricable linkage between the positioning and the capability/capacity of an enterprise. A company cannot position the organization to meet a customer need that it cannot fulfill without an unprofitable level of resources nor can it allocate enhanced resources to provide a cost-effective service that no customer wants!

Positioning leads to higher levels of revenue through increasing the size of the market, retaining the first-time customers, increasing the size of the wallet share, and so on. Positioning has to do with factors such as

- Understanding customer needs
- Understanding competitor initiatives
- Determining the businesses' financial needs
- Conforming with legal and regulatory requirements
- Conforming with environmental constraints

The capability/capacity has to be aligned with the positioning or else it has to be changed to deliver the positioning. Capability/capacity has to do with internal factors such as

- Key business processes
- Procedures and systems
- Competencies, skills, training, and education

The key is to have a perceived differentiation of being better than the competition in whatever terms the customers choose to evaluate or measure and to deliver this at the lowest unit cost.

In practice, BPM has developed a focus on changing capability/capacity in the short term to address current issues. This short-term change in capability/capacity is usually driven by the need to

- Reduce the cycle time to process customer orders
- Improve quotation times
- Lower variable overhead costs
- Increase product range to meet an immediate competitor threat
- Rebalance resources to meet current market needs
- Reduce work-in-progress stocks
- Meet changed legislation requirements
- Introduce short-term measures to increase market share (e.g., increased credit limit from customers hit by recessionary trends)

22.2.3 Business Process Reengineering (BPR) Methodology

An overview of a seven-step methodology is as follows:

1. Develop the context for undertaking the BPR and in particular reengineer the enterprise's business processes. Then identify the reason behind redesigning the process to represent the value perceived by the customer.
2. Select the business processes for the reengineering effort.
3. Map the selected processes.
4. Analyze the process maps to discover opportunities for reengineering.
5. Redesign the selected processes for increased performance.
6. Implement the reengineered processes.
7. Measure the implementation of the reengineered processes.

 Outsourcing is distancing the company from noncore but critical functions; as against this, reengineering is exclusively about the core.

The BPR effort within an enterprise is not a one-time exercise but an ongoing one. One could also have multiple BPR projects in operation simultaneously in different areas within the enterprise. The BPR effort involves business visioning, identifying the value gaps, and hence, selection of the corresponding business processes for the BPR effort. The reengineering of the business processes might open newer opportunities and challenges, which in turn triggers another cycle of business visioning followed by BPR of the concerned business processes.

The competitive gap can be defined as the gap between the customer's minimum acceptance value (MAV) and the customer value delivered by the enterprise. Companies that consistently surpass MAVs are destined to thrive, those that only meet the MAVs will survive, and those that fall short of the MAVs may fail. Customers will take their business to the company that can deliver the most value for their money. Hence, the MAVs have to be charted in detail. MAV is dependent on several factors, such as

- The customer's prior general and particular experience base with an industry, product, and/or service
- What competition is doing in the concerned industry, product, or service
- What effect technological limitations have on setting the upper limit

As mentioned earlier, MAVs can be characterized in terms of the CVDs; only four to six value determinants may be necessary to profile a market segment. CVDs can be defined by obtaining data through

1. The customer value survey
2. Leaders in noncompeting areas
3. The best-in-class performance levels
4. Internal customers

A detailed customer value analysis analyzes the value gaps and helps in further refining the goals of the process reengineering exercise. The value gaps are as follows:

- Gaps that result from different value perceptions in different customer groups
- Gaps between what the company provides and what the customer has established as the minimum performance level
- Gaps between what the company provides and what the competition provides
- Gaps between what the organization perceives as the MAV for the identified customer groups and what the customer says are the corresponding MAVs

It must be noted that analyzing the value gaps is not a one-time exercise; neither is it confined to the duration of a cycle of the breakthrough improvement exercise. Like the BPR exercise itself, it is an activity that must be done on an ongoing basis. Above all, selecting the right processes for an innovative process reengineering effort is critical. The processes should be selected for their high visibility, relative ease of accomplishing goals, and at the same time, their potential for great impact on the value determinants.

22.3 Mobile-Enabling Business Processes

This section explores the motivations behind mobile enabling business processes. The key motivation for mobile enabling business processes is the need to serve customers faster and reduce costs. Globalization and intense competition demand that businesses respond faster to changing market and customer needs, by reducing time-to-market for products and services and serving their customers in a near instantaneous fashion. Businesses are continuing basic drivers for addressing process mobility:

1. *Process efficiency*: A set of factors that demand cost reduction and improved customer service and response time. An example of mobile enabling a process for efficiency is sales staff being able to create online quotations and orders at the customer site using their mobile devices.

2. *Increased personal productivity of employees*: Time and travel management are some of the common processes for mobile enabling and achieving significant productivity improvements.

In some areas of business, the benefits of enterprise systems cannot be fully realized, as a large number of mobile workers cannot access wire-bound systems. As a result, many organizations cannot realize the expected return on investments in expensive enterprise systems. In addition to that, most business processes and supporting systems are designed around office-based employees and are not friendly to mobile workers. A good example is an expense claim form, which is typically made available on company intranets and cannot be accessed by sales staff when they are away from the office. Mobility of processes supported by systems can help achieve additional return on investments in enterprise systems such as ERP, CRM, and SCM. Business managers do not introduce new technologies into the business processes to become technology leaders; they are interested in the added business value and how technology contributes to it; in how they can achieve the goals of streamlining their processes. From the point of view of mobility, managers are interested in what value mobile enabling can add to the business processes.

22.4 Mobile Enterprise

Businesses that aim to support mobile workers and enhance process effectiveness will need to consider extending their process and systems beyond the workplace. In order to achieve this, they will have to change their processes and systems in line with the objectives of process mobility. An enterprise that can transform its processes to make its mobile workers and processes more effective can be considered a mobile enterprise.

Smart organizations will aim to leverage process mobility for strategic advantages. Such businesses derive tactical and strategic value from mobile enabling processes. In order to gain maximum benefits from mobility, organizations should have a mobility strategy defined—aligned to its business strategy—and it should complement the IT strategy.

22.4.1 Mobile Business Processes

Processes can be decomposed to smaller subprocesses and viewed at lower levels of detail. The lowest level subprocess is an activity with a well-defined input and output. Mobile business process, according to which mobility is given for a business process, when at least for one of the process activities there is externally determined uncertainty of location and the process needs cooperation with external resources for its execution. I will discuss this in more detail in one of the later sections of this chapter. To simplify the afore-mentioned definition, a mobile business process is one that consists of one or more activities being performed at an uncertain location and requiring the worker to be mobile. Such a process can be supported by mobile systems to increase process efficiency. For processes that are supported by mobile systems, the term mobile-enabled business process is more appropriate to differentiate from a mobile business process.

22.4.2 Mobile Enterprise Systems

Mobile systems can be beneficial across a number of processes in most business areas. Large corporations are embracing mobile systems in almost all major areas of business such as sales, procurement, warehouse management, and so on. A number of mobile solutions available in the market are truly enterprise encompassing in nature as they fully integrate with the existing enterprise systems and bridge across major processes of the enterprise. Companies like SAP and Oracle offer mobile solutions that build on their existing enterprise systems offerings, primarily ERP, CRM, and SCM applications. These solutions address mobility in areas such as sales, field service, procurement, supply chain, and asset management.

Mobile applications can be used to redesign or improve processes at a specific activity level (e.g., e-mail) or can be used to mobile enable large end-to-end processes that cut across functions (e.g., procure-to-pay processes). Similarly, mobile applications can be utilized for most processes in sales, supply chain, asset management, and plant maintenance. Systems that mobile enable core processes and key activities across multiple functions in an organization can be seen as mobile enterprise systems. Mobile enterprise systems can either be enterprise systems extended to support process mobility or separate mobile systems integrated with existing enterprise systems. There are two key aspects to a mobile enterprise system. First, the mobile system should support one or more core business processes, and second, it works on the existing enterprise data.

As discussed earlier, the drivers for process mobility are location uncertainty and user mobility.

As most sales and service staffs are highly mobile with activities carried out at external (uncertain) locations, these areas make a strong case for mobile systems. With mobile systems, the sales and field service staff can access business information wherever they like and capture data wherever it is generated. For example, sales personnel can view order status from a customer site and create new orders online using their mobile devices; service engineers can input job completion details on handheld computers that update centralized databases in real time. With the effective use of mobile systems, sales teams can spend more time with customers and prospects.

Other high-potential areas are asset management, plant maintenance, and materials management. Mobility of workers in these processes is usually within a limited area but requires movement around that area—for example, capturing technical measurements of equipment around the plant and updating equipment repair history on a handheld device. Mobile technologies, when combined with other technologies such as bar code and more recently RFID (radio frequency identification), can offer more appealing solutions and bring about substantial efficiency improvements. For instance, radio frequency (RF) tags can be used to store maintenance and service data pertaining to equipment. With the help of mobile devices, users can instantly view the equipment maintenance information and repair history stored on the RF tag attached on the equipment. Supply chain management and procurement management processes can also benefit from mobile enabling. Mobile enabling employee-oriented processes such as filling and approving time sheets, travel expense forms, and leave requests can increase employee efficiency by making effective use of unproductive time such as travel by train, taxi, or air, as well as waiting periods.

With the growing use of mobile systems, mobile technologies are finding their place in enterprises systems' architecture and technology strategy. Organizations considering mobile systems for their core processes are viewing mobility as a strategic and not just technological element. From a systems perspective, mobile systems can be seen as a virtual mobile layer around the enterprise architecture. Similarly, the business process architecture can be understood to have a mobility layer that represents the mobility of business processes and supporting systems. Business organizations striving to be competitive will have to address mobility requirements and capture the opportunities arising from mobile enabling business processes. Such organizations will require not only a mobile layer in their technology stack but also a corporate-level strategy for mobility. Figure 22.1 shows an architectural view of the enterprise systems of an engineering business. The figure shows typical business areas using mobile systems as an extension of the enterprise systems. The mobile systems in the diagram are shown as a virtual layer around the enterprise systems.

22.4.3 Redesigning for Mobility

In order to leverage the capabilities of mobile systems, the design of business processes needs to be assessed from a mobility perspective and, if required,

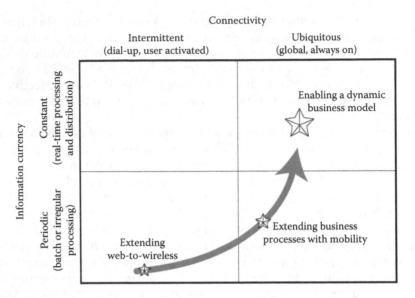

FIGURE 22.1
The evolution of mobile business models.

to be redesigned to maximize the advantage of mobile enabling. Mobile applications should not just be utilized to extend business processes onto handheld or mobile devices. Instead, mobile systems should be tightly integrated with enterprise systems to streamline processes. Thus, technology innovation can be a stepping stone to achieve business process innovation. The ultimate objective of mobility in business can be seen as providing the power of a desktop computer in the hands of the mobile worker with the ability to connect to a network or the Internet from virtually anywhere. However, this may not be required or even feasible in all situations—technically or economically. Thus, it is important to identify and assess processes and aspects of processes that would benefit from mobile enabling to make a sound business case.

22.5 Mobile Web Services

Web Services are the cornerstone toward building a global distributed information system, in which many individual applications will take part; building a powerful application whose capability is not limited to local resources will unavoidably require interacting with other partner applications through Web Services across the Internet. The strengths of Web

Services come from the fact that Web Services use XML and related technologies connecting business applications based on various computers and locations with various languages and platforms. The counterpart of the WSs in the context of mobile business processes would be mobile Web Services (MWS).

The proposed MWS are to be the base of the communications between the Internet network and wireless devices such as mobile phones, PDAs, and so forth. The integration between wireless device applications with other applications would be a very important step toward global enterprise systems. Similar to WS, MWS is also based on the industry-standard language XML and related technologies such as SOAP, WSDL, and UDDI.

Many constraints make the implementation of WSs in a mobile environment very challenging. The challenge comes from the fact that mobile devices have smaller power and capacities as follows:

- Small power limited to a few hours
- Small memory capacity
- Small processors not big enough to run larger applications
- Small screen size, especially in mobile phones, which requires developing specific websites with suitable size
- Small keypad that makes it harder to enter data
- Small hard disk
- The speed of the data communication between the device and the network, and that varies

The most popular MWS is a proxy-based system where the mobile device connects to the Internet through a proxy server. Most of the processing of the business logic of the mobile application will be performed on the proxy server that transfers the results to the mobile device that is mainly equipped with a user interface to display output on its screen. The other important advantage a proxy server provides in MWS is, instead of connecting the client application residing on the mobile device to many service providers and consuming most of the mobile processor and the bandwidth, the proxy will communicate with service providers, do some processing, and send back only the final result to the mobile device. In the realistic case where the number of mobile devices becomes in the range of tens of millions, the proxy server would be on the cloud and the service providers would be the cloud service providers.

Mobile Web Services use existing industry-standard XML-based Web Services architecture to expose mobile network services to the broadest audience of developers. Developers will be able to access and integrate mobile network services such as messaging, location-based content delivery, syndication, personalization, identification, authentication, and billing services into their applications. This will ultimately enable solutions that work seamlessly across stationary networks and mobile environments. Customers will be able to use mobile Web Services from multiple devices on both wired and wireless networks.

The aim of the mobile Web Services effort is twofold:

1. To create a new environment that enables the IT industry and the mobile industry to create products and services that meet customer needs in a way not currently possible within the existing Web Services practices. With Web Services being widely deployed as the SOA of choice for internal processes in organizations, there is also an emerging demand for using Web Services enabling mobile working and e-business. By integrating Web Services and mobile computing technologies, consistent business models can be enabled on a broad array of endpoints: not just on mobile devices operating over mobile networks but also on servers and computing infrastructure operating over the Internet. To make this integration happen at a technical level, mechanisms are required to expose and leverage existing mobile network services. Also, practices for how to integrate the various business needs of the mobile network world and their associated enablers such as security must be developed. The result is a framework, such as the Open Mobile Alliance, that demonstrates how the Web Service specifications can be used and combined with mobile computing technology and protocols to realize practical and interoperable solutions.

   Successful mobile solutions that help architect customers' service infrastructures need to address security availability and scalability concerns both at the functional level and at the end-to-end solution level, rather than just offering fixed-point products. What is required is a standard specification and an architecture that tie together service discovery, invocation, authentication, and other necessary components—thereby adding context and value to Web Services. In this way, operators and enterprises will be able to leverage the unique capabilities of each component of the end-to-end network and shift the emphasis of service delivery from devices to the human user. Using a combination of wireless, broadband, and wireline devices, users can then access any service on demand, with a single identity and single set of service profiles, personalizing service delivery as dictated by the situation. There are three important requirements to accomplish user (mobile-subscriber)-focused delivery of mobile services: federated identity, policy, and federated context.

   Integrating identity, policy, and context into the overall mobile services architecture enables service providers to differentiate the user from the device and deliver the right service to the right user on virtually any device:

a. *Federated identity*: In a mobile environment, users are not seen as individuals (e.g., mobile subscribers) to software applications and processes who are tied to a particular domain, but rather as entities that are free to traverse multiple service networks. This requirement demands a complete federated network identity model to tie the various *personas* of an individual without compromising privacy or loss of ownership of the associated data. The federated network identity model allows the implementation of seamless single sign-on for users interacting with applications (Nokia 2004). It also ensures that user identity, including transactional information and other personal information, is not tied to a particular device or service, but rather is free to move with the user between service providers. Furthermore, it guarantees that only appropriately authorized parties are able to access protected information.

b. *Policy*: User policy, including roles and access rights, is an important requirement for allowing users not only to have service access within their home network but also to move outside it and still receive the same access to services. Knowing who the user is and what role they fulfill at the moment they are using a particular service is essential to providing the right service at the right instance. The combination of federated identity and policy enables service providers and users to strike a balance between access rights and user privacy

c. *Federated context*: Understanding what the user is doing, what they ask, why it is being requested, where they are, and what device they are using is an essential requirement. The notion of federated context means accessing and acting upon a user's current location, availability, presence, and role, for example, at home, at work, on holiday, and other situational attributes. This requires the intelligent synthesis of information available from all parts of the end-to-end network and allows service providers and enterprises to deliver relevant and timely applications and services to end users in a personalized manner. For example, information about the location and availability of a user's device may reside on the wireless network, the user's calendar may be on the enterprise intranet, and preferences may be stored in a portal.

2. To help create Web Services standards that will enable new business opportunities by delivering integrated services across stationary (fixed) and wireless networks. Mobile Web Services use existing industry-standard XML-based Web Services architecture to expose mobile network services to the broadest audience of developers.

> Developers will be able to access and integrate mobile network services such as messaging, location-based content delivery, syndication, personalization, identification, authentication, and billing services into their applications. This will ultimately enable solutions that work seamlessly across stationary networks and mobile environments. Customers will be able to use mobile Web Services from multiple devices on both wired and wireless networks.

Delivering appealing, low-cost mobile data services, including ones that are based on mobile Internet browsing and mobile commerce, is proving increasingly difficult to achieve. The existing infrastructure and tools as well as the interfaces between Internet/Web applications and mobile network services remain largely fragmented, characterized by tightly coupled, costly, and close alliances between value-added service providers and a complex mixture of disparate and sometimes overlapping standards (WAP, MMS, Presence, Identity, etc.) and proprietary models (e.g., propriety interfaces). This hinders interoperability solutions for the mobile sector and at the same time drives up the cost of application development and ultimately the cost of services offered to mobile users. Such problems have given rise to initiatives for standardizing mobile Web Services. The most important of these initiatives are the Open Mobile Alliance and the mobile Web Services frameworks that are examined below.

The Open Mobile Alliance (www.openmobilealliance.org). The OMA is a group of wireless vendors, IT companies, mobile operators, and application and content providers, who have come together to drive the growth of the mobile industry. The objective of OMA is to deliver open technical specifications, based on market requirements, for the mobile industry, that enable interoperable solutions across different devices, geographies, service providers, operators, and networks. OMA includes all key elements of the wireless value chain and contributes to the timely availability of mobile service enablers. For enterprises already using a multitiered network architecture based on open technologies, such as Web Services, which implement wireless services, OMA is a straightforward extension of existing wireline processes and infrastructures. In this way, wireless services become simply another delivery channel for communication, transactions, and other value-added services. Currently, the OMA is defining core services such as location, digital rights, and presence services and using cases involving mobile subscribers, mobile operators, and service providers; an architecture for the access and deployment of core services; and a Web Services framework for using secure SOAP.

The technical working groups within OMA address the need to support standardized interactions. To achieve this, the OMA is currently addressing how mobile operators can leverage Web Services and defines a set of common protocols, schemas, and processing rules using Web Services technologies that are the elements that can be used to create or interact with a number of different services. The OMA Web Services Enabler (OWSER) specification capitalizes on all the benefits of Web Services technologies to simplify the task of integrators, developers, and implementers of service enablers by providing them with common mechanisms and protocols for interoperability of service enablers. Examples of functionality common across service enablers range from transport and message encoding definitions to security concerns, service discovery, charging, definition, and management of SLAs, as well as management, monitoring, and provisioning of the service enablers that exist within a service provider's network.

The OMA Web Service interfaces are intended to enhance a service provider's data for a particular mobile subscriber. A common scenario starts with a data request from some application (perhaps a mobile browser) to a service provider. The service provider then uses Web Services to interact with a subscriber's mobile operator to retrieve some relevant data about the subscriber such as location or presence. These data can be used to enhance the service provider's response to the initial request. Mobile Web Services are envisioned to support server-to-server, server-to-mobile terminal, mobile terminal-to-server, and mobile terminal-to-mobile terminal (or peer-to-peer) interactions.

Similarly, the objective of the mobile Web Services framework is to meet the requirements for bridging stationary enterprise infrastructure and the mobile world, and it enables the application of Web Services specifications, SOA implementations, and tools to the problem of exposing mobile network services in a commercially viable way to the mass market of developers. The focus of the work concentrates on mechanisms to orchestrate the calls to mobile Web Services.

The mobile Web Services framework places particular emphasis on core mechanisms such as security, authentication, and payment. Core security mechanisms are offered that apply WS-Security to mobile network security services, such as the use of a GSM-style SIM security device within a Web Services end point to provide a means for authentication. In addition, a set of core payment mechanisms within the WSs architecture have been proposed that understand how to interact with the participating WS end points. It is expected that a number of services dependent on the mobile Web Services framework and that rely on its core mechanisms will be developed. SMS services, MMS services, and location-based services have been identified as common services that are candidates for specification activity. Specification work will include profiling and optimization of the core Web Services protocols so that they can easily be realized over any bearer, on any device, or both. This addresses the inefficiencies that current Web Services specifications exhibit when used over a narrowband and possibly intermittent bearer or when being processed by a low-performance mobile device.

22.6 Mobile Field Cloud Services

Companies that can outfit their employees with devices like PDAs, laptops, multifunction smartphones, or pagers will begin to bridge the costly chasm between the field and the back office. For example, transportation costs for remote employees can be significantly reduced, and productivity can be significantly improved by eliminating needless journeys back to the office to file reports, collect parts, or simply deliver purchase orders.

Wireless services are evolving toward the goal of delivering the right cloud service to whoever needs it, for example, employees, suppliers, partners, and customers, at the right place, at the right time, and on any device of their choice. The combination of wireless handheld devices and cloud service delivery technologies poses the opportunity for an entirely new paradigm of information access that in the enterprise context can substantially reduce delays in the transaction and fulfillment process and lead to improved cash flow and profitability.

A field cloud services solution automates, standardizes, and streamlines manual processes in an enterprise and helps centralize disparate systems associated with customer service life-cycle management including customer contact, scheduling and dispatching, mobile workforce communications, resource optimization, work order management, time, labor, material tracking, billing, and payroll. A field Web Services solution links seamlessly all elements of an enterprise's field service operation—customers, service engineers, suppliers, and the office—to the enterprise's stationary infrastructure, wireless communications, and mobile devices. Field Web Services provide real-time visibility and control of all calls and commitments, resources, and operations. They effectively manage business activities such as call taking and escalation, scheduling and dispatching, customer entitlements and SLAs, work orders, service contracts, time sheets, labor and equipment tracking, preinvoicing, resource utilization, reporting, and analytics.

Cloud service optimization solutions try to automatically match the most cost-effective resource with each service order based on prioritized weightings assigned to every possible schedule constraint. To accommodate evolving business priorities, most optimization solutions allow operators to reorder these weightings and to execute ad hoc *what-if* scenario analyses to test the financial and performance impacts of scheduling alternatives. In this way, they help enhance supply chain management by enabling real-time response to changing business conditions.

Of particular interest to field services are location-based services, notification services, and service disambiguation as these mechanisms enable developers to build more sophisticated cloud service applications by providing accessible interfaces to advanced features and *intelligent* mobile features:

1. Location-based services provide information specific to a location using the latest positioning technologies and are a key part of the mobile Web Services suite. Dispatchers can use GPS or network-based positioning information to determine the location of field workers and optimally assign tasks (push model) based on geographic proximity. Location-based services and applications enable enterprises to improve operational efficiencies by locating, tracking, and communicating with their field workforce in real time. For example, location-based services can be used to keep track of vehicles and employees, whether they are conducting service calls or delivering products. Trucks could be pulling in or out of a terminal, visiting a customer site, or picking up supplies from a manufacturing or distribution facility. With location-based services, applications can get such things such as real-time status alerts, for example, estimated time of approach, arrival, departure, duration of stop, current information on traffic, weather, and road conditions for both home-office and en route employees.

2. Notification services allow critical business to proceed uninterrupted when employees are away from their desks, by delivering notifications to their preferred mobile device. Employees can thus receive real-time notification when critical events occur, such as when incident reports are completed. The combination of location-based and notification services provides added value by enabling such services as proximity-based notification and proximity-based actuation. Proximity-based notification is a push or pull interaction model that includes targeted advertising, automatic airport check-in, and sightseeing information. Proximity-based actuation is a push–pull interaction model, whose most typical example is payment based on proximity, for example, toll watch.

3. Service instance disambiguation helps distinguish between many similar candidate service instances, which may be available inside close perimeters. For instance, there may be many on-device payment services in proximity of a single point of sale. Convenient and natural ways for identifying appropriate service instances are then required, for example, relying on closeness or pointing rather than identification by cumbersome unique names.

22.7 Summary

This chapter described mobile-based cloudware applications. It discusses the need and characteristics of agile enterprises. It presents various strategies for enabling enterprise agility ranging from eBusiness transformations to mobilizing the business processes. It provides an introduction to the mobilization of business processes and consequently the enterprises. It describes mobile Web Services and how they can be provisioned via cloud services.

23

Context-Aware Applications

Most mobile applications are location-aware systems. Specifically, tourist guides are based on users' locations in order to supply more information on the city attraction closer to them or the museum exhibit they are seeing. Nevertheless, recent years have seen many mobile applications trying to exploit information that characterizes the current situation of users, places, and objects in order to improve the services provided.

The principle of context-aware applications (CAA) can be explained using the metaphor of the Global Positioning System (GPS). In aircraft navigation, for example, a GPS receiver derives the speed and direction of an aircraft by recording over time the coordinates of longitude, latitude, and altitude. This contextual data is then used to derive the distance to the destination, communicate progress to date, and calculate the optimum flight path.

For a GPS-based application to be used successfully, the following activities are a prerequisite:

1. The region in focus must have been GPS-mapped accurately.
2. The GPS map must be superimposed with the relevant information regarding existing landmarks, points of civic and official significance, and, facilities and service points of interest in the past to the people—this is the context in this metaphor.
3. There must be a system available to ascertain the latest position per the GPS system.
4. The latest reported position must be mapped and transcribed on to the GPS-based map of the region.
5. This latest position relative to the context (described in the 2nd activity) is used as the point of reference for future recommendation(s) and action(s).

It should be noted that the initial baseline of the context (described in the 2nd activity) is compiled and collated separately and then uploaded into the system to be accessible by the CAA. However, with passage of time, this baseline gets added further with the details of each subsequent transaction.

We can also imagine an equivalent of the Global Positioning System for calibrating the performance of enterprises. The coordinates of longitude, latitude, and altitude might be replaced by ones of resource used, process performed, and product produced. If we designed a GPS for an enterprise, we could measure its performance (e.g., cost or quality) in the context of the resource used, the process performed, and the product delivered as compared with its own performance in the past (last month or one year back) or in particular cases that of another target organization. Such an approach could help us specify our targets, communicate our performance, and signal our strategy.

Most of the current context-aware systems have been built in an ad hoc approach and are deeply influenced by the underlying technology infrastructure utilized to capture the context. To ease the development of context-aware ubicomp (ubiquitous computing) and mobile applications, it is necessary to provide universal models and mechanisms to manage context. Even though significant efforts have been devoted to research methods and models for capturing, representing, interpreting, and exploiting context information, we are still not close to enabling an implicit and intuitive awareness of context nor efficient adaptation to behavior at the standards of human communication practice.

Context information can be a decisive factor in mobile applications in terms of selecting the appropriate interaction technique. Designing interactions among users and devices, as well as among devices themselves, is critical in mobile applications. Multiplicity of devices and services calls for systems that can provide various interaction techniques and the ability to switch to the most suitable one according to the user's needs and desires. Current mobile systems are not efficiently adaptable to the user's needs. The majority of ubicomp and mobile applications try to incorporate the users' profile and desires into the system's infrastructure either manually or automatically observing their habits and history. According to our perspective, the key point is to give them the ability to create their own mobile applications instead of just customizing the ones provided.

Thus, mobile applications can be used not only for locating users and providing them with suitable information but also for

- Providing them with a tool necessary for composing and creating their own mobile applications
- Supporting the system's selection of appropriate interaction techniques

- A selection of recommendation(s) and consequent action(s) conforming with the situational constraints judged via the business logic and other constraints sensed via the context
- Enabling successful closure of the interaction (answering to a query, qualifying an objection, closing an order, etc.)

23.1 Decision Patterns as Context

This chapter discusses location-based services applications as a particular example of context-aware applications. But, context-aware applications can significantly enhance the efficiency and effectiveness of even routinely occurring transactions. This is because most end-user applications' effectiveness and performance can be enhanced by transforming them from a *bare* transaction to a transaction *clothed* by a surround of a context formed as an aggregate of all relevant decision patterns utilized in the past.

The decision patterns contributing to a transaction's context include the following:

- Characteristic and sundry details associated with the transaction under consideration
- Profiles of similar or proximate transactions in the immediately prior week or month or 6 months or last year or last season
- Profiles of similar or proximate transactions in same or adjacent or other geographical regions
- Profiles of similar or proximate transactions in same or adjacent or other product groups or customer groups

To generate the context, the relevant decision patterns can either be discerned or discovered by mining the relevant pools or streams of primarily the transaction data. Or they could be augmented or substituted by conjecturing or formulating decision patterns that explain the existence of these characteristic pattern(s) (in the pools or streams of primarily the transaction data). In the next subsection we look at function-specific decision patterns with particular focus on financial decision patterns.

 Thus, generation of context itself is critically dependent on employing big data and mobilized applications, which in turn needs cloud computing as a prerequisite.

23.1.1 Concept of Patterns

The concept of patterns used in this book originated from the area of real architecture. Alexander gathered architectural knowledge and best practices regarding building structures in a pattern format. This knowledge was obtained from years of practical experience. A pattern according to Alexander is structured text that follows a well-defined format and captures nuggets of advice on how to deal with recurring problems in a specific domain. It advises the architect on how to create building architectures, defines the important design decisions, and covers limitations to consider. Patterns can be very generic documents, but may also include concrete measurements and plans. Their application to a certain problem is, however, always a manual task that is performed by the architect. Therefore, each application of a pattern will result in a differently looking building, but all applications of the pattern will share a common set of desired properties. For instance, there are patterns describing how eating tables should be sized so that people can move around the table freely, get seated comfortably, find enough room for plates and food, while still being able to communicate and talk during meals without feeling too distant from people seated across the table. While the properties of the table are easy to enforce once concrete distances and sizes are specified, they are extremely hard to determine theoretically or by pure computation using a building's blueprint.

In building architecture, pattern-based descriptions of best practices and design decisions proved especially useful, because many desirable properties of houses, public environments, cities, streets, etc., are not formally measurable. They are perceived by humans and, thus, cannot be computed or predicted in a formal way. Therefore, best practices and well-perceived architectural styles capture a lot of implicit knowledge how people using and living in buildings perceive their structure, functionality, and general feel. Especially, the indifferent emotion that buildings trigger, such as awe, comfort, coziness, power, cleanness, etc., are hard to measure or explain and are also referred to as *the quality without a name* or *the inner beauty of a building*. How certain objectives can be realized in architecture is, thus, found only through practical experience, which is then captured by patterns. For example, there are patterns describing how lighting in a room should be realized so that people feel comfortable and positive. Architects capture their knowledge gathered from existing buildings and feedback they received from users in patterns describing well-perceived building design. In this scope, each pattern describes one architectural solution for an architectural problem. It does so in an abstract format that allows the implementation in various ways. Architectural patterns, thus, capture the essential properties required for the successful design of a certain building area or function while leaving large degrees of freedom to architects.

Multiple patterns are connected and interrelated resulting in a *pattern language*. This concept of links between patterns is used to point to related patterns. For example, an architect reviewing patterns describing different roof types can be pointed to patterns describing different solutions for windows in these roofs and may be advised that some window solutions, thus, the patterns describing them cannot be combined with a certain roof pattern. For example, a *flat rooftop* cannot be combined with windows that have to be mounted vertically. Also, a pattern language uses these links to guide an architect through the design of buildings, streets, cities, etc., by describing the order in which patterns have to be considered. For example, the size of the ground on which a building is created may limit the general architecture patterns that should be selected first. After this, the number of floors, the aforementioned roofing style, etc., can be considered.

23.1.1.1 Patterns in Information Technology (IT) Solutions

In a similar way, the pattern-based approach has been used in IT to capture best practices in how applications and systems of applications should be designed. Examples are patterns for fault-tolerant software, general-application architectures, object-oriented programming, enterprise applications, or message-based application integration. Again, these patterns are abstract and independent of the programming language or runtime infrastructure used to form *timeless knowledge* that can be applied in various IT environments. In the domain of IT solutions, the desirable properties are portability, manageability, flexibility to make changes, and so on. The properties of IT solutions become apparent over time while an application is productively used, evolves to meet new requirements, has to cope with failures, or has to be updated to newer versions. During this lifecycle of an application, designers can reflect on the IT solution to determine whether it was well designed to meet such challenges.

23.1.2 Domain-Specific Decision Patterns

In the following, we discuss as illustrations, decision patterns for two domains or functional areas, namely, finance and customer relationship management (CRM). While the former is a formalized area to a large degree because of the statutory and regulatory requirements, the latter is defined and fine-tuned, across an extended period of operational experience, by the specific requirements of the business, offerings, and geographic region(s) in which the company operates.

23.1.2.1 Financial Decision Patterns

Financial management focuses on both the acquisition of financial resources on as favorable terms as possible and the utilization of the assets that those financial resources have been used to purchase, as well as looking at the interaction between these two activities. Financial planning and control is an essential part of the overall financial management process. Establishment of precisely what the financial constraints are and how the proposed operating plans will impact them are a central part of the finance function. This is generally undertaken by the development of suitable aggregate decision patterns like financial plans that outline the financial outcomes that are necessary for the organization to meet its commitments. Financial control can then be seen as the process by which such plans are monitored and necessary corrective action proposed when significant deviations are detected.

Financial plans are constituted of three decision patterns:

1. *Cash flow planning*: This is required to ensure that cash is available to meet the payments the organization is obliged to meet. Failure to manage cash flows will result in technical insolvency (the inability to meet payments when they are legally required to be made). Ratios are a set of powerful tools to report these matters. For focusing on cash flows and liquidity, a range of ratios based on working capital are appropriate; each of these ratios addresses a different aspect of the cash collection and payment cycle.

   The five key ratios that are commonly calculated are
   - Current ratio, equal to current assets divided by current liabilities
   - Quick ratio (or acid test), equal to quick assets (current assets less inventories) divided by current liabilities
   - Inventory turnover period, equal to inventories divided by cost of sales, with the result being expressed in terms of days or months
   - Debtors to sales ratio, with the result again being expressed as an average collection period
   - Creditors to purchases ratio, again expressed as the average payment period

   There are conventional values for each of these ratios (for example, the current ratio often has a standard value of 2.0 mentioned, although this has fallen substantially in recent years because of improvements in the techniques of working capital management, and the quick ratio a value of 1.0), but in fact these values vary widely across firms and industries. More generally helpful is a comparison with industry norms and an examination of the changes in the values of these ratios over time that will assist in the assessment of whether any financial difficulties may be arising.

2. *Profitability*: This is the need to acquire resources (usually from revenues acquired by selling goods and services) at a greater rate than using them (usually represented by the costs of making payments to suppliers, employees, and others). Although, over the life

of an enterprise, total net cash flow and total profit are essentially equal, in the short term, they can be very different. In fact, one of the major causes of failure for new small business enterprises is not that they are unprofitable but that the growth of profitable activity has outstripped the cash necessary to resource it. The major difference between profit and cash flow is in the acquisition of capital assets (i.e., equipment that are bought and paid for immediately, but that have likely benefits stretching over a considerable future period) and timing differences between payments and receipts (requiring the provision of working capital).

For focusing on longer-term profitability with short-term cash flows, profit to sales ratios can be calculated (although different ratios can be calculated depending whether profit is measured before or after interest payments and taxation). Value added (sales revenues less the cost of bought-in supplies) ratios can also be used to give insight into operational efficiencies.

3. *Assets*: Assets entail the acquisition and, therefore, the provision of finance for their purchase. In accounting terms, the focus of attention is on the balance sheet, rather than the profit and loss (P/L) account or the cash flow statement.

For focusing on the raising of capital as well as its uses, a further set of ratios based on financial structure can be employed. For example, the ratio of debt to equity capital (gearing or leverage) is an indication of the risk associated with a company's equity earnings (because debt interest is deducted from profit before obtaining profit distributable to shareholders). It is often stated that fixed assets should be funded from capital raised on a long-term basis, while working capital should fund only short-term needs.

It is necessary to be aware that some very successful companies flout this rule to a considerable extent. For example, most supermarket chains fund their stores (fixed assets) out of working capital because they sell their inventories for cash several times before they have to pay for them—typical inventory turnover is three weeks, whereas it is not uncommon for credit to be granted for three months by their suppliers.

There is, therefore, no definitive set of financial ratios that can be said to measure the performance of a business entity. Rather, a set of measures can be devised to assess different aspects of financial performance from different perspectives. Although some of these measures can be calculated externally, being derived from annual financial reports, and can be used to assess the same aspect of financial performance across different companies, care needs to be taken to ensure that the same accounting principles have been used to produce the accounting numbers in each case. It is not uncommon for *creative accounting* to occur so that acceptable results can be reported. This draws attention especially to the interface between management accounting (which is intended to be useful in internal decision making and control) and financial accounting (which is a major mechanism by which external stakeholders, especially shareholders, may hold managers accountable for their oversight).

Financial scandals, such as Enron and WorldCom have highlighted that a considerable amount of such manipulation is possible palpably within generally acceptable accounting principles (GAAPs). There is clear evidence that financial numbers alone are insufficient to reveal the overall financial condition of an enterprise. Part of the cause has been the *rules-based* approach of US financial reporting, in contrast to the *principles-based* approach adopted in United Kingdom. One result of the reforms that have followed these scandals has been a greater emphasis on operating information. In addition, legislation such as the Sarbanes–Oxley Act (SOX) in the United States has required a much greater disclosure of the potential risks surrounding an enterprise, reflected internally by a much greater emphasis on risk management and the maintenance of risk registers.

The finance function serves a boundary role; it is an intermediary between the internal operations of an organization and the key external stakeholders who provide the necessary financial resources to keep the organization viable. Decision patterns like financial ratios allow internal financial managers to keep track of a company's financial performance (perhaps in comparison with that of its major competitors), and to adjust the activities of the company, both operating and financial, so as to stay within acceptable bounds. A virtuous circle can be constructed whereby net cash inflows are sufficient to pay adequate returns to financiers and also contribute towards new investment; given sound profitability, the financiers will usually be willing to make additional investment to finance growth and expansion beyond that possible with purely internal finance. Conversely, a vicious cycle can develop when inadequate cash flows preclude adequate new investment, causing a decline in profitability, and so the company becomes unable to sustain itself.

23.1.3 CRM Decision Patterns

This section describes an overview of the statistical models–based decision patterns used in CRM applications as the guiding concept for profitable customer management. The primary objectives of these systems are to acquire profitable customers, retain profitable customers, prevent profitable customers from migrating to competition, and winning back *lost* profitable customers. These four objectives collectively lead to increasing the profitability of an organization.

CRM strategies spanning the full customer lifecycle are constituted of four decision patterns or models:

1. Customer acquisition: This involves decisions on identifying the right customers to acquire, forecasting the number of new customers, the response of promotional campaigns, and so on. The objectives of customer acquisition modeling includes identifying the right customers to acquire, predicting whether customers will respond to company promotion campaigns, forecasting the number of new customers, and examining the short- and long-run effects of marketing and other business variables on customer acquisition.

This is a conscious move from mass marketing of products to one that is focused on the end consumer. This is a direct result of increases in data collection and storage capabilities that have uncovered layer upon layer of customer differentiation. Differentiating and segmenting with regards to demographic, psychographic, or purchasing power-related characteristics became more affordable and possible, and eventually became necessary in order to keep up with competing firms. Although segment-level acquisition did not take this theory to the extent that one-to-one customer acquisition has, it reinforced a growing trend of subsets or groups of customers within a larger target market. Being able to collect, store, and analyze customer data in more practical, affordable, and detailed ways has made all of this possible. As firms have become more capable and committed with data analyses, offerings have become more specific, thus increasing the amount of choice for customers. This has in turn spurred customers to expect more choice and customization in their purchases. This continuous firm–customer interaction has consistently shaped segment-level marketing practices in the process to better understand customers.

The decision patterns would incorporate

- Differences between customers acquired through promotions and those acquired through regular means
- Effect of marketing activities and shipping and transportation costs on acquisition
- Impact of the depth of price promotions

- Differences in the impact of marketing-induced and word-of-mouth customer
- Acquisition on customer equity

2. Customer retention: This involves decisions on who will buy, what the customers will buy, when they will buy, and how much they will buy, and so on. During a customer's tenure with the firm, the firm would be interested in retaining the customer for a longer period of time. This calls for investigating the role of trust and commitment with the firm, metrics for customer satisfaction, and the role of loyalty and reward programs, among others. The objective of customer retention modeling includes examining the factors influencing customer retention, predicting customers' propensity to stay with the company or terminate the relationship, and predicting the duration of the customer–company relationship. Customer retention strategies are used both in contractual (where customers are bound by contracts, such as cell (mobile) phone subscription or magazine subscription) and noncontractual settings (where customers are not bound by contracts, such as grocery purchases or apparel purchases).

Who to retain can often be a difficult question to answer. This is because the cost of retaining some customers can exceed their future profitability and thus make them unprofitable customers. When to engage in the process of customer retention is also an important component. As a result, firms must monitor their acquired customers appropriately to ensure that their loyalty is sustained for a long period of time. Finally, identifying how much to spend on a customer is arguably the most important piece of the customer retention puzzle. It is very easy for firms to overcommunicate with a customer and spend more on his/her retention than the customer will ultimately give back to the firm in value.

The decision patterns would incorporate

- Explaining customer retention or defection
- Predicting the continued use of the service relationship through the customer's expected future use and overall satisfaction with the service
- Renewal of contracts using dynamic modeling
- Modeling the probability of a member lapsing at a specific time using survival analysis
- Use of loyalty and reward programs for retention
- Assessing the impact of a reward program and other elements of the marketing mix

3. Customer attrition or churn: This involves decisions on whether the customer will churn or not, and, if so, what will be the probability of the customer churning, and when. The objective of customer attrition modeling includes churn with time-varying covariates, mediation effects of customer status and partial defection on customer churn, churn using two cost-sensitive classifiers, dynamic churn using time-varying covariates, factors inducing service switching, antecedents of switching behavior, and impact of price reductions on switching behavior.

Engaging in active monitoring of acquired and retained customers is the most crucial step in being able to determine which customers are likely to churn. Determining who is likely to churn is an essential step. This is possible by monitoring customer purchase behavior, attitudinal response, and other metrics that help identify customers who feel underappreciated or underserved. Customers who are likely to churn do demonstrate *symptoms* of their dissatisfaction, such as fewer purchases, lower response to marketing communications, longer time between purchases, and so on. The collection of customer data is therefore crucial in being able to identify and capture such *symptoms* and that would help in analyzing the retention behavior and the choice of communication medium. Understanding who to save among those customers who are identified as being in the churn phase is again a question of cost vs. future profitability.

The decision patterns would incorporate

- When are the customers likely to defect
- Can we predict the time of churn for each customer
- When should we intervene and save the customers from churning
- How much do we spend on churn prevention with respect to a particular customer

4. Customer win-back: This involves decisions on reacquiring the customer after the customer has terminated the relationship with the firm. The objective of customer win-back modeling includes customer lifetime value, optimal pricing strategies for recapture of lost customers, and the perceived value of a win-back offer.

Identifying the right customers to win back depends on factors such as the interests of the customers to reconsider their choice of quitting, the product categories that would interest the customers, and the stage of customer life cycle and so on. If understanding what to offer customers in winning them back is an important step in the win-back process, measuring the cost of win-back is as important as determining who to win back and what to offer them. The cost of win-back, much like the cost of retention or churn, must be juxtaposed with the customer's future profitability and value to the firm.

23.2 Context-Aware Applications

Context is understood as "the location and identities of nearby people and objects, and changes to those objects." Initially, the term *context* was equivalent to the location and identity of users and objects. Very soon, though, the term expanded to include a more refined view of the environment assuming either three major components—computing, user, and physical environment—or four major dimensions, system, infrastructure, domain, and physical context. The interaction between the user and application was added by Dey (2001)*; according to them, a context is "any information that can be used to characterize the situation of an entity." An entity should be treated as anything relevant to the interaction between a user and an application, such as a person, a place, or an object, including the user and the application themselves and, by extension, the environment the user and applications are embedded in. Thus, a system is context-aware if it uses context to provide relevant information and/or services to the user, where relevancy depends on the user's task.

An ontology is a formal, explicit specification of a shared conceptualization, that is, an abstract model of some phenomenon in the world that identifies the relevant concepts of that phenomenon (*explicit* means that the type of concepts used and the constraints on their use are explicitly defined, and *formal* refers to the fact that the ontology should be machine readable). Given that ontologies are a promising instrument to specify concepts and their interrelations, they can provide a uniform way for specifying a context model's core concepts as well as an arbitrary amount of subconcepts and facts, altogether enabling contextual knowledge sharing and reuse in a ubicomp system. Ontologies are developed to provide a machine-processable semantics of information sources that can be communicated between different agents (software and humans). It is a necessity to decouple the process of context acquisition and interpretation from its actual use, by introducing a consistent, reliable, and secure context framework that can facilitate the development of context-aware applications.

Context-aware features include using context to

1. Present information and services to a user
2. Automatically execute a service for a user
3. Tag information to support later retrieval

In supporting these features, context-aware applications can utilize numerous different kinds of information sources. Often, this information comes from sensors, whether they are software sensors detecting information about the networked, or virtual, world or hardware sensors detecting information about the physical world. Sensor data can be used to recognize the usage

---

* Dey, A.K. Understanding and using context. *Personal and Ubiquitous Computing Journal*, (1), 4–7, 2001.

situation, for instance, from illumination, temperature, noise level, and device movements. A context-aware application can make adaptive decisions based on the context of interaction in order to modulate the information presented to the user or to carry out semantic transformation on the data, like converting text to speech for an audio device. The promise and purpose of context awareness are to allow computing systems to take action autonomously and enable systems to sense the situation and act appropriately. For example, in context-aware mobile applications, location is the most commonly used variable in context recognition as it is relatively easy to detect. User activity is much more difficult to identify than location, but some aspects of this activity can be detected by placing sensors in the environment.

Location is the most commonly used piece of context information, and several different location detection techniques have been utilized in context-awareness research. Global positioning system (GPS) is a commonly used technology when outdoors, utilized, for example, in car navigation systems. Network cellular ID can be used to determine location with mobile phones. Measuring the relative signal strengths of Bluetooth and WLAN hotspots and using the hotspots as beacons are the frequently used techniques for outdoor and indoor positioning. Other methods used indoors include ultrasonic or infrared-based location detection. Other commonly used forms of context are time of day, day of week, identity of the user, proximity to other devices and people, and actions of the user.

Context-aware device behavior may not rely purely on the physical environment. While sensors have been used to directly provide this physical context information, sensor data often need to be interpreted to aid in the understanding of the user's goals. Information about a user's goals, preferences, and social context can be used for determining context-aware device behavior as well. Knowledge about a user's goals helps prioritize the device actions and select the most relevant information sources. A user's personal preferences can offer useful information for profiling or personalizing services or refining information retrieval. The user may also have preferences about the quality of service issues such as cost efficiency, data connection speed, and reliability, which relate closely to mobile connectivity issues dealing with handovers and alternative data transfer mediums.

Finally, social context forms an important type of context as mobile devices are commonly used to support communication between two people and used in the presence of other people.

Challenges of context-aware systems include the following:

- A main issue regarding context-aware computing is the fear that control may be taken away from the user. Apart from control issues, privacy and security issues arise. The main parameters of context are user location and activity, which users consider as part of their privacy. Users are especially reluctant to exploit context-aware systems, when they know that private information may be disclosed to others.

- There is a gap between how people understand context and what systems consider as context. The environment in which people live and work is very complex; the ability to recognize the context and determine the appropriate action requires considerable intelligence.
- A context-aware system cannot decide with certainty which actions the user may want to be executed; as the human context is inaccessible to sensors, we cannot model it with certainty.
- A context-aware system cannot be developed to be so robust that it will rarely fail, as ambiguous and uncertain scenarios will always occur and even for simple operations exceptions may exist.
- A context-aware system can add more and more rules to support the decision-making process; unfortunately, this may lead to large and complex systems that are difficult to understand and use.
- A context-aware application is based on context information that may be imperfect. The ambiguity over the context soundness arises due to the speed at which the context information changes and the accuracy and reliability of the producers of the context, like sensors.

It is a challenge for context-aware systems to handle context, which may be nonaccurate or ambiguous, in an appropriate manner—more information is not necessarily more helpful; context information is useful only when it can be usefully interpreted.

23.3 Context-Aware Mobile Applications

A mobile application is context aware if it uses context to provide relevant information to users or to enable services for them; relevancy depends on a user's current task (and activity) and profile (and preferences). Apart from knowing who the users are and where they are, we need to identify what they are doing, when they are doing it, and which object they focus on. The system can define user activity by taking into account various sensed parameters like location, time, and the object that they use. In outdoor applications, and depending on the mobile devices that are used, satellite-supported technologies, like GPS, or network-supported cell information, like GSM, IMTS, and WLAN, is applied. Indoor applications use RFID, IrDA, and Bluetooth technologies in order to estimate the users' position in space. While time is another significant parameter of context that can play an important role in order to extract information on user activity, the objects that are used in mobile applications are the most crucial context sources.

In mobile applications, the user can use mobile devices, like mobile phones and PDAs and objects that are enhanced with computing and communication abilities. Sensors attached to artifacts provide applications with information about what the user is utilizing. In order to present the user with the requested information in the best possible form, the system has to know the physical properties of the artifact that will be used (e.g., artifact screen's display characteristics), the types of interaction interfaces that an artifact provides to the user need to be modeled (e.g., whether artifact can be handled by both speech and touch techniques), and the system must know how it is designed. Thus, the system has to know the number of each artifact's sensors and their position in order to gradate context information with a level of certainty. Based on information on the artifact's physical properties and capabilities, the system can extract information on the services that they can provide to the user.

In the context-aware mobile applications, artifacts are considered as context providers. They allow users to access context in a high-level abstracted form, and they inform other application's artifacts so that context can be used according to the application needs. Users are able to establish associations between the artifacts based on the context that they provide; keep in mind that the services enabled by artifacts are provided as context. Thus, users can indicate their preferences, needs, and desires to the system by determining the behavior of the application via the artifacts they create. The set of sensors attached to an artifact measure various parameters such as location, time, temperature, proximity, and motion—the raw data given by its sensors determine the low-level context of the artifact. The aggregation of such low-level context information from various homogenous and nonhomogenous sensors results into a high-level context information.

23.3.1 Ontology-Based Context Model

This ontology is divided into two layers: a common one that contains the description of the basic concepts of context-aware applications and their interrelations representing the common language among artifacts, and a private one that represents an artifact's own description as well as the new *knowledge or experience* acquired from its use. The common ontology defines the basic concepts of a context-aware application; such an application consists of a number of artifacts and their associations. The concept of artifact is described by its physical properties and its communication and computational capabilities; the fact that an artifact has a number of sensors and actuators attached is also defined in our ontology. Through the sensors, an artifact can perceive a set of parameters based on which the state of the artifact is defined; an artifact may also need these parameters in order to sense its interactions with other artifacts as well as with the user. The ontology also defines the interfaces via which artifacts may be accessed in order to enable the selection of the appropriate one. The common ontology represents an abstract form of the concepts represented, especially of the context parameters, as more detailed descriptions are stored into each artifact's private ontology. For instance, the private ontology of an artifact that represents a car contains a full description of the different components in a car as well as their types and their relations.

The basic goal of the proposed ontology-based context model is to support a context management process, based on a set of rules that determine the way in which a decision is made and are applied to existing knowledge represented by this ontology. The rules that can be applied during such a process belong to the following categories: rules for an artifact's state assessment that define the artifact's state based on its low- and high-level contexts, rules for local decisions that exploit an artifact's knowledge only in order to decide the artifact's reaction (like the request or the provision of a service), and finally rules for global decisions that take into account various artifacts' states and their possible reactions in order to preserve a global state defined by the user.

23.3.2 Context Support for User Interaction

The ontology-based context model that we propose empowers users to compose their own personal mobile applications. In order to compose their applications, they first have to select the artifacts that will participate and establish their associations. They set their own preferences by associating artifacts, denoting the sources of context that artifacts can exploit, and defining the interpretation of this context through rules in order to enable various services. As the context acquisition process is decoupled from the context management process, users are able to create their own mobile applications avoiding the problems emerging from the adaptation and customization of applications like disorientation and system failures.

The goal of context in computing environments is to improve interaction between users and applications. This can be achieved by exploiting context, which works like implicit commands and enables applications to react to users or surroundings without the users' explicit commands. Context can also be used to interpret explicit acts, making interaction much more efficient. Thus, context-aware computing completely redefines the basic notions of interface and interaction. In this section, we present how our ontology-based context model enables the use of context in order to assist human–computer interaction in mobile applications and to achieve the selection of the appropriate interaction technique. Mobile systems have to provide multimodal interfaces so that users can select the most suitable technique based on their context. The ontology-based context model that we presented in the previous section captures the various interfaces provided by the application's artifacts in order to support and enable such selections. Similarly, the context can determine the most appropriate interface when a service is enabled. Ubiquitous and mobile interfaces must be proactive in anticipating needs, while at the same time working as a spatial and contextual filter for information so that the user is not inundated with requests for attention.

Context can also assist designers to develop mobile applications and manage various interfaces and interaction techniques that would enable more satisfactory and faster closure of transactions. Easiness is an important requirement for mobile applications; by using context according to our approach, designers are abstracted from the difficult task of context acquisition and have merely defined how context is exploited from various artifacts by defining simple rules. Our approach presents an infrastructure capable of handling, substituting, and combining complex interfaces when necessary. The rules applied to the application's context and the reasoning process support the application's adaptation. The presented ontology-based context model is easily extended; new devices, new interfaces, and novel interaction techniques can be exploited into a mobile application by simply incorporating their descriptions in the ontology.

23.4 Location-Based Service (LBS) Applications

Location-based services are services that are sensitive to and take advantage of the location of the service user. Any service that makes use of the location of the user can be called an Location-based service. The location of a person can be determined using a GPS receiver or other technologies, now available in many mobile phone platforms. This position determination technology (PDT) is generally carried by the person, from which the location must be provided to the Location-based service provider. Today, the Location-based services are generally hosted in the network, which may pose performance and scalability issues.

The uptake of mobile phones with PDT capabilities continues to grow, and most mobile phone users have a phone that can be traced with good accuracy and a lower cost. This new technology has given the Location-based service market a greater push. LBSs can be divided into four categories:

1. Business to business
2. Business to consumer
3. Consumer to business
4. Consumer to consumer

The business-to-business services include fleet tracking and courier tracking. Business-to-consumer services include pushed ads based on the location, where a user will receive ads most relevant to the location. Consumer to business services include location-based search, where a user is searching for the nearest restaurant, petrol pump, and so forth. A consumer-to-consumer service is the friend finder service where the user will be alerted if his friend is within a few meters (see Table 23.1).

TABLE 23.1

Location-Based Services (LBSs) Classification

B2B	B2C	C2C	C2B
Fleet and freight, tracking, etc.	Discounts, ads, special events, etc.	Find a friend, primary schools, etc.	Find a gas station, community events, etc.
Trigger Services	**Tracking and Monitoring**	**Location-Based Information**	**Assistance Services**
E-commerce, payment information, advertising, etc.	Fleet management, telematics, asset tracking, etc.	Traffic and navigation, entertainment, mapping, etc.	Personal/vehicle emergency, roadside assistance, alarm management, etc.
Push Services		**Pull Services**	
Travel directions, taxi hailing, m-commerce, etc.		Zone alerts, traffic alerts, etc.	

23.4.1 LBS System Components

LBS is an intersection of the three technologies, namely, new information and communication technologies (NICTSs) such as the mobile telecommunication system and handheld devices, the Internet, and Geographic information systems (GIS) with spatial databases.

LBS gives the possibility of a two-way communication and interaction. Therefore, the user tells the service provider his actual context, like the kind of information he needs, his preferences, and his position. This helps the provider of such location services to deliver information tailored to the user needs. If the user wants to use an Location-based service, different infrastructure elements are necessary.

Here are the basic components in LBS:

1. *Mobile devices*: A tool for the user to request the needed information. The results can be given by speech, using pictures, text, and so on. Possible devices are PDAs, mobile phones, laptops, and so on, but the device can also be a navigation unit of car or a toll box for road pricing in a truck.

2. *Communication network*: The second component is the mobile network, which transfers the user data and service request from the mobile terminal to the service provider and then the requested information back to the user.

3. *Positioning determination technology (PDT) component*: For the processing of a service, the user position usually has to be determined. The user position can be obtained either by using the mobile communication network or by using the global positioning system (GPS). Further possibilities to determine the position are WLAN stations, active badges, or radio beacons. The latter positioning methods can

be especially used for indoor navigation, like in a museum. If the position is not determined automatically, it can be also specified manually by the user.

4. *Service and application provider*: The service provider offers a number of different services to the user and is responsible for the service request processing. Such services offer the calculation of the position, finding a route, searching yellow pages with respect to position, or searching specific information on objects of the user interest (e.g., a bird in wild life park).

5. *Data and content provider*: Service providers will usually not store and maintain all the information that can be requested by users. Therefore, geographic base data and location information data will usually be requested from the maintaining authority (e.g., mapping agencies) or business and industry partners (e.g., yellow pages, traffic companies).

Based on the information delivery method, we identify three types of LBS: pull, push, and tracking services. In the case of a pull service, the user issues a request in order to be automatically positioned and to access the LBS he or she wants. A use-case scenario demonstrating a pull service is the following. A tourist roams in a foreign city and wants to receive information about the nearest restaurants to his or her current location. Using a mobile device, the tourist issues an appropriate request (e.g., via SMS [short message service] or WAP [wireless application protocol]), and the network locates his or her current position and responds with a list of restaurants located near it. On the contrary, in the case of a push service, the request is issued by the service provider and not the user. A representative example of push services is location-based advertising, which informs users about products of their interest located at nearby stores. In this service, users submit their shopping preference profiles to the service provider and allow the provider to locate and contact them with advertisements, discounts, and/or e-coupons for products of interest at nearby stores. So, in this case, the service provider is the one who pushes information to the user. Finally, in a tracking service, the basic idea is that someone (user or service) issues a request to locate other mobile stations (users, vehicles, fleets, etc.).

From a technological point of view, LBSs are split into two major categories depending on the positioning approach they use to locate mobile stations. There is the handset-based approach and the network-based approach. The former approach requires the mobile device to actively participate in the determination of its position, while the latter relies solely on the positioning capabilities of elements belonging to the mobile network. For both of these approaches, several positioning techniques have been developed or are under development. What distinguish them from one another are the accuracy they provide and the cost of their implementation. The most popular network-based positioning techniques are cell-global-identity (CGI) methods, timing advance (TA), uplink time of arrival (TOA), and angle of arrival (AOA), while the most popular handset-based positioning techniques are observed time difference of arrival (OTDOA), enhanced observed time difference (E-OTD), and assisted Global Positioning System (A-GPS). The accuracy provided by some of these techniques in different coverage areas of the mobile network is presented in Tables 23.2 and 23.3.

23.4.2 LBS System Challenges

Despite the appealing idea of using user location information to provide highly personalized and intelligent services, there are certain challenges that should be addressed in order for LBS to succeed.

We can divide these challenges into three categories:

1. Technological challenge for LBS is the capability to create easy-to-use and satisfying services. There is much talk concerning what would be the most suitable user interface and type of service (pull or push) in terms of user satisfaction. For example, in the case of push-based services, a user is not required to manually issue queries in order to get the information he or she seeks. The system automatically informs him or her based on the current location and a list of preferences listed in the user's profile. The problem is that in this way, user intent cannot be perfectly captured and the user may be frequently disturbed by out-of-context information. So, despite the easiness of usage (no or minimal interface), user satisfaction is not assured. On the other hand, in pull-based LBS, in which clients have to poll the server for updates, the users may experience difficulties in using these services because cell phones, PDAs, and wearable computers are less suitable for browsing and query-based information retrieval due to their limited input device capabilities. All these restrictions along with the unpredictability in mobile environments (disconnections, frequent context differentiations, etc.) have to be taken very carefully into account when designing LBSs.

   Some of the implied requirements are the following:

   a. A less intensive use of the mobile network and a minimal volume of transmitted data

   b. The possibility of off-line operation

   c. Simple and user-friendly interfaces and limited and well-specified amounts of presented information content

   Therefore, it becomes apparent that LBS will not succeed in attracting users without implementing sophisticated techniques based on

TABLE 23.2

Location Enablement Technologies

Technology	Description	Advantages	Disadvantages
Network based			
Cell of origin (COO)	Information generated about the cell occupied by a user	RF technology Inexpensive—uses existing network No handset modification Fast implementation No consumer behavior change	Low resolution
Angle of Arrival (AOA)	Measures angle of signal from mobile device to cell towers Minimum of two cell sites required	FF technology No handset modification No consumer behavior change	Expensive network modifications required Resistance toward more antennas in neighborhoods Line-of-sight constraint Medium resolution (not less than 150 m)
Time Distance of Arrival (TDOA)	Triangulates at least three stations to measure and compare arrival time of signal from a user	RF technology No handset modification No consumer behavior change	Line-of-sight constraint Expensive Medium resolution Appropriate for CDMA
Enhanced cell ID (E-CID)	Software-based solution that compares list of cell sites available to user and checks for overlaps	RF technology Line of sight not required Moderate cost to upgrade	Works only with GSM Some modification required in handset and network
Handset based			
Global Positioning System (GPS)	Radio navigation system comprising low-orbit satellites. Triangulation by measuring the time to communicate with three satellites	FF technology Very accurate, 1–5 m, 95% precision Not dependent on network	Line-of-sight issues Significant handset Handset modification May require consumer behavior change modifications
Hybrid technology			
Enhanced observed time difference (E-OTD)	Similar to TDOA, but handset calculates the location	RF technology Accuracy of 50–125 m Some behavior change	Suited for GSM only Network and handset modification Cell coverage necessary
Assisted global positioning system (A-GPS)	Processing done by network while using the satellites	RF technology Moderate modification to handset Line-of-sight constraint reduced	Significant changes to network

TABLE 23.3

Accuracy and TIFF for Several Location Techniques

Technique	Network Impact	Terminal Impact	Accuracy	TTFF
Cell ID	None	None	250 m–20 km	<1 s
Signal strength	None	None	100 m–10 km	<1 s
TOA/TDOA	Medium	Low/medium	40–150 m	<1 s
AOA/DOA	High	None	50–150 m	<1 s
Fingerprint	High	None	50–150 m	Seconds
GPS	Low	Very high	3–50 m	Seconds
Hybrid systems	Depends on the techniques hybridized	Depends on the techniques hybridized	3–100 m	Seconds
Ultrawide baud	Dedicated infrastructure	Very high	10–50 cm	<1 s

carefully designed interfaces and/or detailed knowledge of customer profiles, needs, and preferences. So, given existing technical limitations such as device capabilities and access speeds combined with human limitations such as reduced consideration sets and the need for speed and convenience, in order for LBSs to succeed, they will need to deliver relevant, targeted, and timely information to consumers at the time and place of their choice.

Also, from a database perspective, LBSs raise critical challenges such as spatial and temporal query processing because the continuous movement of users or objects leads to the need for fast and frequent or continuous updates to the databases. Some of the most important database research challenges brought to the surface by LBS are the following:

i. *Support for nonstandard-dimension hierarchies*: In LBS, the geographical area may be divided into multidimensional regions following the pattern of network coverage. Until now, geographical area representation models used by data warehouses were in the form of completely balanced trees (strict hierarchies), which cannot capture irregularities like those that frequently occur in mobile networks (e.g., the same region covered by more than one base station).

ii. *Support for imprecision and varying precision*: Varying precision means that the location of the same user may be pinpointed with different accuracies depending on the positioning technology used while he or she is roaming from network to network. Imprecision means that the location data for the trace of a specific user may be incomplete (e.g., a user may have gone out of the network coverage or may have switched off the device for some time). So, varying precision and imprecision should be carefully handled by employing intelligent query-processing techniques, especially for queries on complete user traces.

iii. *Support for movement constraints and transportation networks*: Most of the time, users move on certain routes as defined by transportation networks (e.g., railways, roads), and their movement is blocked depending on the morphology of the land (e.g., mountains). The incorporation of such constraints into query resolution may offer increased positioning accuracy to LBS despite the potentially low-accuracy positioning technology used.

iv. Support for spatial data mining on vehicle movement.

v. Support for continuous location change in query-processing techniques.

2. *Ethical challenges*: From an ethical point of view, a critical challenge is to protect user privacy. LBS can potentially intrude on customer privacy. The adoption of LBS is highly dependent on the successful confrontation of digital frauds, attempts of intrusion in customer databases with sensitive data and profiles, and the threat of unauthorized or uncontrolled resale of location information. It has also been shown that a privacy-intruding service (e.g., an always-on tracking service), despite its usability, is not desirable by users since it does not allow them to switch it off whenever they want. So, when designing an LBS and in order for the service to be adopted, the provider should take into account very seriously the user's concerns on privacy.

3. *Business challenges*: Capitalizing on the promise of LBS requires developing sustainable and viable business models for offering such services. Unfortunately, until today, there has been little effort on developing a framework with which to identify the most appropriate business models for the large variety of LBSs. The major obstacle for this arises from the fact that there is a multitude of players participating in the provision of such services forming a complex value network.

The main categories under which these players are grouped are the following:

- Application developers and content providers
- Service providers and network providers
- Hardware manufacturers

The roles of all these different actors or players are many times conflicting if not competitive, and fairness in revenue sharing is viewed differently by each actor. In this context, it is difficult to determine which activities should be performed by which actor (e.g., should the network operator develop its own services or outsource them to more focused application providers) or to identify which actor should be the dominant one in the business model (i.e., the operator providing access to its customer base, the content or service provider offering the actual service, or the location technology vendor offering the enabling positioning equipment).

23.5 Summary

Considering the tremendous growth in mobile-based services and applications, context-aware applications are envisaged to emerge as an area with high-growth business potential in the future. None of these services and applications will be viable without being enabled by cloud computing and big data. This chapter introduces the fundamentals of context-aware systems with specific context of the mobile-based applications that is presently witnessing the highest growth in the consumer market space. Mobile services and applications have the potential to be the *killer apps* for the coming boom in the m-business. The later part of the chapter discusses aspects of location based services (LBSs) and attendant challenges.

Appendix: Future of Moore's Law

On April 19, 1965, Gordon Moore, the cofounder of Intel Corporation, published an article in *Electronics Magazine* entitled "Cramming More Components onto Integrated Circuits" in which he identified and conjectured a trend that computing power would double every 2 years (this was termed as *Moore's law* in 1970 by the CalTech professor and VLSI pioneer, Calvin Mead). This law has been able to predict reliably both the reduction in costs and the improvements in computing capability of microchips, and those predictions have held true (Figure A.1).

In 1965, the amount of transistors that fitted on an integrated circuit could be counted in tens. In 1971, Intel introduced the 4004 microprocessor with 2300 transistors. In 1978, when Intel introduced the 8086 microprocessor, the IBM PC was effectively born (the first IBM PC used the 8088 chip)—this chip had 29,000 transistors. In 2006, Intel's Itanium 2 processor carried 1.7 billion transistors. In the next 2 years, we will have chips with over 10 billion transistors. What does this mean? Transistors are now so small that millions of them could fit on the head of a pin. While all this was happening, the cost of these transistors was also exponentially falling, as per Moore's prediction (Figure A.2).

In real terms, this means that a mainframe computer of the 1970s that cost over $1 million had less computing power than your iPhone has today. The next generation of smartphone we will be using in the next 2–3 years will have 1 GHz processor chips. That is roughly one million times faster than the Apollo Guidance Computer. Theoretically, Moore's law will run out of steam somewhere in the not too distant future. There are a number of possible reasons for this. Firstly, the ability of a microprocessor silicon-etched *track* or circuit to carry an electrical charge has a theoretical limit.

At some point when these circuits get physically too small and can no longer carry a charge or the electrical charge *bleeds*, then we will have a design limitation problem. Secondly, as successive generations of chip technology are developed, manufacturing costs increase. In fact recently, Gordon Moore said that each new generation of chips requires a doubling in cost of the manufacturing facility as tolerances become tighter. At some point, it will theoretically become too costly to develop the manufacturing plants that produce these chips. The usable limit for semiconductor process technology will be reached when chip process geometries shrink to be smaller than 20 nanometers (nm) to 18 nm nodes. At those nodes, the industry will start getting to the point where semiconductor manufacturing tools are too expensive to depreciate with volume production; that is, their costs will be so

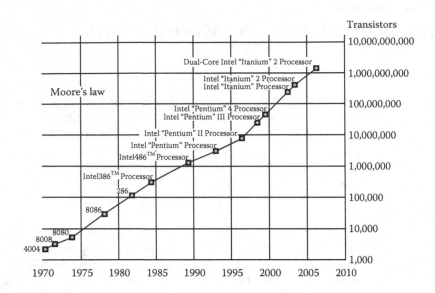

FIGURE A.1
Increase of the number of transistors on an Intel chip.

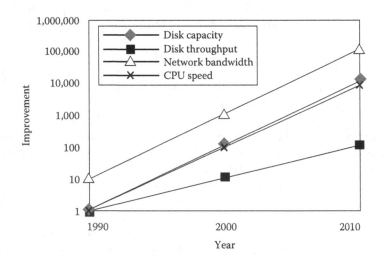

FIGURE A.2
Hardware trends in the 90s and the current decade.

high that the value of their lifetime productivity can never justify it. Lastly, the power requirements of chips are also increasing. More power being equivalent to more heat equivalent to bigger batteries implies that at some point, it becomes increasingly difficult to power these chips while putting them on smaller platforms.

A.1 Cloudware and Moore's Law

Deploying cloudware entails transferring processing to the cloud. What are the economic issues of moving a task from one computer to another? A computation task has four characteristic dimensions:

- Computation—transforming information to produce new information
- Database Access—access to reference information needed by the computation
- Database Storage—long-term storage of information (needed for later access)
- Networking—delivering questions and answers

The ratios among these quantities and their relative costs are pivotal: it is fine to send a GB over the network if it saves years of computation, but it is not economical to send a kilobyte question if the answer could be computed locally in a second. Consequently, on-demand computing is only economical for very CPU-intensive (100,000 instructions per byte or a CPU day-per gigabyte of network traffic) applications; for most other applications, pre-provisioned computing is likely to be more economical.

But considering the trends, one can conjecture that Moore's law will continue to remain valid not for the processing chips *per se*, but for the cost of cloudware: *cloudware i.e Internet-accessible computing power would double every 2 years.*

References

Akerkar, R. (Ed.), *Big Data Computing* (CRC Press, 2014).

Alexander, C., *The Timeless Way of Building* (Oxford University Press, 1979).

Alexander, C. et al., *A Pattern Language* (Oxford University Press, 1977).

Bean, J., *SOA and Web Services Interface Design: Principles, Techniques and Standards* (San Francisco, CA: Morgan Kaufmann Press, 2010).

Berman, J.J., *Principles of Big Data: Preparing, Sharing and Analyzing Complex Information* (Morgan Kaufman, 2013).

Bernstein, P. and E. Newcomer, *Principles of Transaction Processing* (Burlington, MA: Morgan Kaufmann, 2nd edn., 2009).

Bessis, N. and C. Dobre, *Big Data and Internet of Things: A Roadmap for Smart Environments* (Springer, 2014).

Bunker, G. and D. Thomson, *Delivering Utility Computing: Business-Driven IT Optimization* (Wiley, 2006).

Burgess, M., *In Search of Certainty: The Science of Our Information Infrastructure* (CreateSpace, 2013).

Buyya, R., J. Broberg, and A. Goscinski (Eds.), *Cloud Computing: Principles and Paradigms* (Hoboken, NJ: Wiley, 2011).

Chappell, D., *Enterprise Service Bus* (Sebastopol, CA: O'Reilly, 2004).

Chatarjee, S. and J. Webber, *Developing Enterprise Web Services: An Architect's Guide* (Upper Saddle River, NJ: Prentice Hall, 2003).

Chin,. A. and D. Zhang (Eds.), *Mobile Social Networking: An Innovative Approach* (Springer, 2014).

Coulouris, G., J. Dollimore, T. Kindberg, and G. Blair, *Distributed Systems: Concept and Design* (Boston, MA: Addison-Wesley, 2011).

Coyne, R., *Logic Models of Design* (University College of London, 1988).

Daconta, M.C., L.J. Obrst, and K.T. Smith, *The Semantic Web: A Guide to the Future of XML, Web Services, and Knowledge Management* (Wiley, 2003).

Dey, A.K., *Understanding and Using Context. Personal and Ubiquitous Computing Journal*, 1, 4–7, 2001.

Domingue, J., D. Fensel, and J.A. Hendler (Eds.), *Handbook of Semantic Web Technologies* (Springer, 2011).

Drogovtsev, S.N. and J.F.F. Mendes, *Evolution of Networks—From Biological Nets to the Internet and WWW* (Oxford University Press, 2003).

Duggan, D., *Enterprise Software Architecture and Design: Entities, Services, and Resources* (Hoboken, NJ: Wiley, 2012).

Finkelstein, C., *Enterprise Architecture for Integration: Rapid Delivery Methods and Technologies* (Artech House, 2006).

Frischmann, B.M., *Infrastructure: The Social Value of Shared Resources* (Oxford University Press, 2012).

Furht, B. and A. Escalante (Eds.), *Handbook of Data Intensive Computing* (Springer, 2011).

Gorton, I. and D.K. Gracio (Eds.), *Data Intensive Computing: Architectures, Algorithms and Applications* (Cambridge University Press, 2013).

Graham, I., *Requirements Modeling and Specification for Service Oriented Architecture* (Chichester, U.K.: Wiley, 2008).

Harney, J., *Application Service Provider (ASPs): A Manager's Guide* (Boston, MA: Addison-Wesley, 2003).

Hentrich, C. and U. Zdun, *Process-Driven SOA: Patterns for Aligning Business and IT* (Boca Raton, FL: Auerbach Publications, 2011).

Hugos, M. and D. Hulitzky, *Business in the Cloud: What Every Business Needs to Know About Cloud Computing* (Wiley, 2011).

Hwang, K. et al., *Distributed and Cloud Computing: From Parallel Processing to the Internet of Things* (Morgan-Kaufmann, 2011).

Jackson, K., *OpenStack Cloud Computing Cookbook* (Birmingham, U.K.: PACKT Publishing, 2012).

Juric, M., R. Nagappan, R. Leander, and S. Jeelani Basha, *Professional J2EE EAI* (Birmingham, U.K.: Wrox Press, 2011).

Juric, M., P. Sarang, R. Loganathan, and F. Jennings, *SOA Approach to Integration: XML, Web Services, ESB, and BPEL in Real-World SOA Projects* (Birmingham, U.K.: PACKT Publishing, 2007).

Kale, V., *Implementing SAP CRM: The Guide for Business and Technology Managers* (London, U.K.: Auerbach Publication, 2014).

Kupper, A., *Location-Based Services: Fundamentals and Operation* (Wiley, 2005).

Kwok, Y.-K.R., *Peer-to-Peer Computing: Applications, Architecture, Protocols, and Challenges* (CRC Press, 2012).

Lewis, T.G., *Network Science: Theory and Practice* (Wiley, 2009).

Linthicum, D., *David Linthicum's Guide to Client Server and Intranet Development Guide* (New York: John Wiley, 1997).

Loke, S., *Context-Aware Pervasive Systems: Architectures for a New Breed of Applications* (Boca Raton, FL: Auerbach Publications, 2006).

Luckham, D., *Event Processing for Business: Organizing the Real Time Enterprise* (Hoboken, NJ: Wiley, 2012).

Mahmood, Z. and S. Saeed (Eds.), *Software Engineering Frameworks for the Cloud Computing Paradigm* (Springer, 2013).

Maier, M.W. and E. Rechtin, *The Art of Systems Architecting*, 3rd ed. (CRC Press, 2009).

Manes, A.T., *Web Services: A Manager's Guide* (Boston, MA: Addison-Wesley, 2003).

Marinescu, D., *Cloud Computing: Theory and Practice* (Boston, MA: Morgan Kaufmann, 2013).

Marino, J. and M. Rowley, *Understanding SCA (Service Component Architecture)* (Addison-Wesley, 2010).

McGovern, J., O. Sims, A. Jain, and M. Little, *Enterprise Service Oriented Architectures: Concepts, Challenges, Recommendations* (Dordrecht, the Netherlands: Springer, 2006).

Minoli, D., *A Networking Approach to Grid Computing* (Hoboken, NJ: Wiley, 2005).

Newcomer, E. and G. Lomow, *Understanding SOA with Web Services* (Upper Saddle River, NJ: Addison-Wesley Professional, 2004).

Pacheco, P., *An Introduction to Parallel Programming* (Amsterdam, the Netherlands: Morgan Kaufmann, 2011).

Palfrey, J. and U. Gasser, *Interop: The Promise and Perils of Highly Interconnected Systems* (Basic Books, 2012).

Pant, K. and M. Juric, *Business Process Driven SOA Using BPMN and BPEL: From Business Process Modeling to Orchestration and Service Oriented Architecture* (Birmingham, U.K.: PACKT Publishing, 2008).

Papazoglou, M.P. and P.M.A. Ribbers, *E-Business: Organizational and Technical Foundations* (Wiley, 2006).

Puder, A., A. Romer, and F. Pilhofer, *Distributed System Architecture: A Middleware Approach* (Amsterdam, the Netherlands: Morgan Kaufmann, 2005).

Renso, C., S. Spaccapietra, and E. Zimanyi (Eds.), *Mobility Data: Modeling, Management, and Understanding* (Cambridge University Press, 2013).

Ruh, W., F. Maginnis, and W. Brown, *Enterprise Application Integration: A Wiley Technical Brief* (New York: Wiley, 2001).

Sadasivam, G.S. and R. Shankarmani, *Middleware & Enterprise Integration Technologies* (New Delhi, India: Wiley, 2010).

Sage, A.P. and W.B. Rouse, *Handbook of Systems Engineering and Management* (Wiley, 2009).

Sarbazi-Azad, H. and A.Y. Zomaya, *Large Scale Network-Centric Distributed Systems* (Wiley, 2013).

Schmidt, D., M. Stal, H. Rohnert, and F. Buschmann, *Pattern-Oriented Software Architecture Volume 2: Patterns for Concurrent and Networked Objects* (Wiley, 2000).

Scoble, R. and S. Israel, *Age of Context: Mobile, Sensors, Data and the Future of Privacy* (Patrick Brewester, 2014).

Sherif, M.H. (Ed.), *Handbook of Enterprise Integration* (Boca Raton, FL: Auerbach Publications, 2010).

Smatani, G., *B2B Integration: A Practical Guide to Collaborative E-Commerce* (London, U.K.: Imperial College Press, 2002).

Tapscott, D., D. Ticoll, and A. Lowy, *Digital Capital: Harnessing the Power of Business Webs* (Boston, MA: Harvard Business School Press, 2000).

West, B.J. and P. Grigolini, *Complex Webs: Anticipating the Improbable* (Cambridge University Press, 2011).

Yang, L., A. Waluyo, J. Ma, L. Tan, and B. Srinivasan, *Mobile Intelligence* (Hoboken, NJ: Wiley, 2010).

Zhou, H., *The Internet of Things in the Cloud: A Middleware Perspective* (CRC Press, 2013).

Index

Printed in the United States
by Baker & Taylor Publisher Services